AF294687

Hans Gumhalter

# Power Supply in Telecommunications

Springer

*Berlin
Heidelberg
New York
Barcelona
Budapest
Hong Kong
London
Mailand
Paris
Tokyo*

# Hans Gumhalter

# Power Supply in Telecommunications

Third, Completely Revised Edition
with 263 Figures and 45 Tables

 Springer

Hans Gumhalter
Siemens AG
Hofmannstraße 51
D-81359 München
Germany

ISBN-13:978-3-642-78405-7      e-ISBN-13:978-3-642-78403-3
DOI: 10.1007/978-3-642-78403-3

2nd edition: "Power Supply Systems in Communications Engineering"
© Siemens Aktiengesellschaft and John Wiley and Sons 1986

Library of Congress Cataloging – in – Publication Data
Gumhalter, Hans.
    [Stromversorgungssysteme der Kommunikationstechnik.  English]
    Power supply in telecommunications / Hans Gumhalter. – 3rd.
completely rev. ed.
        p.  cm.
    Includes bibliographical references and index.
    ISBN-13:978-3-642-78405-7
    1. Telecommunication systems – Power supply.   I. Title.
TK5102.5.G68513   1995
621.382'32––dc20





Production: PRODUserv Springer Produktions-Gesellschaft, Berlin
Typesetting: Macmillan India Ltd., Bangalore
SPIN 10037229     61/3020 – 543210 – Printed on acid-free paper.

This book is dedicated to the memory
of my dear late father

Felix Gumhalter,
born 6 January 1905
and deceased 30 December 1993,
Retired Post and Telegraph Superintendent,
Vienna, Austria.

I owe him everything in my life.
His works and his humanity will be
remembered by all who knew him.

Hans Gumhalter

# Preface

An important part of any communications system is its power supply system. The smooth running of all communications depends directly on the quality of the power supply and thus on the operational reliability of the ever more complex equipment and devices used for the purpose.

It is the intention of this book to explain present-day technology in the supply of power for telecommunications comprehensively and in detail with thyristor and transistor-controlled converters, as developed by manufacturers and applied by users, and explains the circuit techniques and relationships with the aid of extensive generally conceived illustrations. The book also deals with the subjects of automatic control, grounding (earthing) and protection techniques. Information is given on the planning to battery and earthing installations.

The individual chapters are constructed and written so that the book will be useful for those readers seriously wishing to become involved in the subject as well as for those having to deal directly with the equipment such as in planning, installation, commissioning and servicing.

The selection of material and method of presentation are based on knowledge and experience gained in the context of training measures for power supply systems.

Thanks are due especially to Dr. Dietrich Berndt for his energetically assistance with Chap. 6 and following companies for generous assistance and permission to use documentation, descriptions, operating instructions, circuit diagrams and photos: Gustav Klein GmbH & Co KG, Accumulatorenfabrik Sonnenschein GmbH, Stamford A.C. Generators Newage International Company, AD. Strüver KG (GmbH & Co), Siemens AG, Varta AG, Hagen Batterie AG.

Munich, July 1995                                                      Hans Gumhalter

# Contents

X    Contents

Reliability and quality are requirements which are particularly important for power supplies, as they take up a key position within the whole assembly. The functioning of the whole system depends completely on them.

In the event of temporary loss of power or complete mains failure, measures need to be taken to maintain the operational dependability of the communications systems. This is in general ensured by the use of lead batteries (accumulators) as energy stores. If the network current is not available for a longer period of time the feeding can be performed by standby power supply systems.

Telecommunication power supply systems are classified according to:

- type of power supply system;
- type of automatic control for the individual devices, units and equipment.

Important aspects of a telecommunications supply include:

- observance of the requested tolerance range for the supply voltage, namely:
  between no load and rated load,
  with surges in load,
  with mains voltage fluctuations,
  with changes in mains frequency;
- maintenance of the purity of the direct voltage in accordance with the specifications, i.e. the limit values for superimposed alternating voltage must not be exceeded;
- a power supply free from interruption as far as possible;
- availability of sufficient monitoring, protective, limiting and signalling devices;
- simplicity of extension;
- ecology, ease of recycling;
- economy;
- small dimensions and light weight;
- ease of installation and maintenance;
- design according to relevant specifications and international regulations.

## 1.2 Basic Types of Power Converter

*Four basic types of power converter* are used for telecommunications power supply systems in the control or conversion of electrical power (Fig. 1.1).

*With rectifiers* (Fig. 1.1a) an alternating current can be converted into a direct current. Energy flows from the alternating current system to the direct current one.

*D.C./D.C. converters* (Fig. 1.1b) enable direct current of a given voltage and polarity to be converted into direct current of another constant or changeable voltage and/or polarity. In today's power supply systems d.c/d.c converters are often used to obtain component voltages from an 'energy transport voltage' of, for example, 48 V or 60 V.

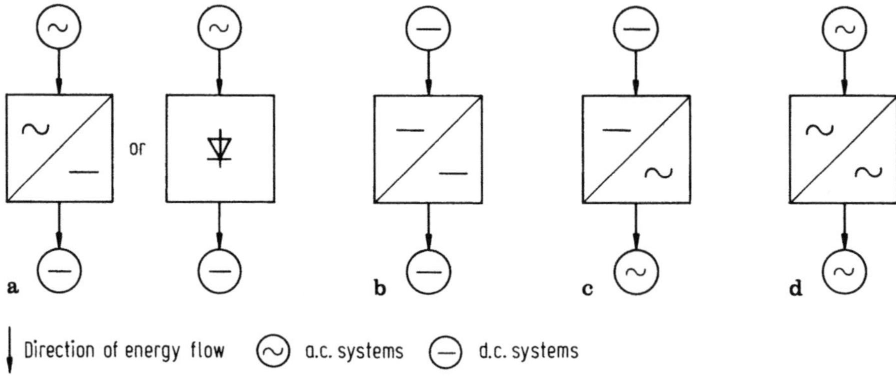



**Fig. 1.1a–d.** Basic types of power converter. **a** Rectifiers; **b** d.c./d.c. converters; **c** inverters; **d** a.c./a.c. converters

*Inverters (d.c./a.c. converters)* (Fig. 1.1c) are used for converting direct current into alternating current.

*A.C./A.C converters* (Fig. 1.1d) enable alternating current of a given voltage, frequency and number of phases to be converted into alternating current of another voltage and/or frequency and/or number of phases. Converters can be further divided into externally commutated converters (with natural commutation) and self-commutating converters (with forced commutation).

*Externally commutated converters* require an external alternating voltage source to provide the commutating voltage for the duration of the commutating period, if necessary through a transformer. In supply-commutated converters this voltage source is the supply mains, and in load-commutated converters it is the load. The class of externally commutated converters includes, for example, rectifiers and a.c./a.c. converters.

*Self-commutating converters* do not need an external alternating voltage source for commutation. In these converters, the effect is produced by a voltage in the commutating circuit; this is either provided by an energy storage device (normally a capacitor) associated with the converter, or produced by an increase in the resistance of the converter valve to be turned off through its control facility. Self-commutating converters include, for example, d.c./d.c. converters and inverters.


## 1.3 Elements of a Power Converter

The typical elements of a power converter with the energy conversion flow are shown in Fig. 1.2.

From the *energy source* the nonregulated (noncontrolled) energy flow reaches the *power section*. There the energy is converted by means of power transistors or thyristors. The controlled energy flow is then passed on to the *load* (consumer,

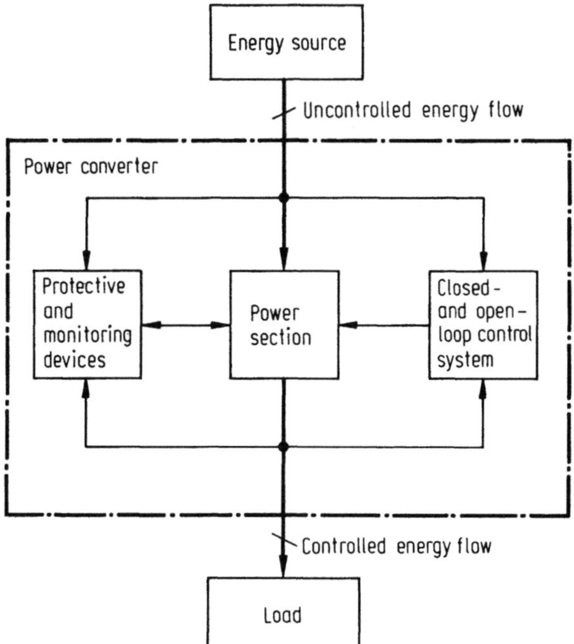


**Fig. 1.2.** Elements of a power converter

communication system). The *control system* performs the comparator function and modifies the trigger pulses to the power section. The *protective* and *monitoring devices* are provided for internal protection of the device as well as of the load, e.g.:

- mains monitoring,
- mains overvoltage protectors,
- undervoltage monitoring,
- overvoltage monitoring,
- overvoltage cutoff devices,
- current limiting devices,
- overcurrent and short circuit cutoff devices,
- ripple contents monitoring,
- fuses
- automatic circuit breakers and
- frequency monitoring.

## 1.4 Energy Storage

Figure 1.3 shows the symbol for an energy storage device. The *battery* (accumulator), as a reliable emergency power source for d.c. loads, takes up electrical energy when charged and stores it as chemical energy. On discharging, the chemical energy is reconverted into electrical energy.

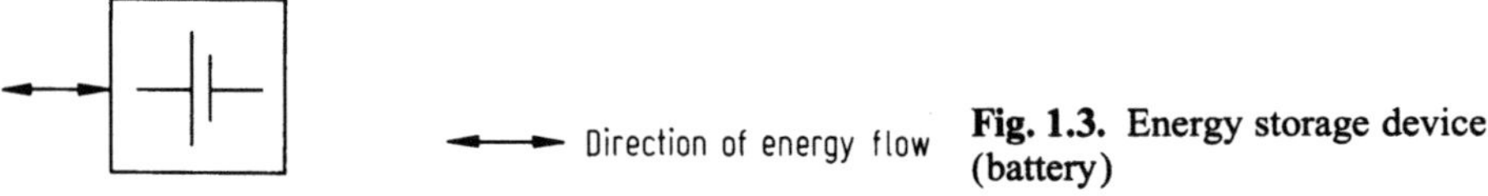



**Fig. 1.3.** Energy storage device (battery)

There are *lead-acid batteries* and *nickel-cadmium* (Ni–Cd) *batteries*; they are made in different designs and with different properties.

The fields of application are:

- stationary (standby batteries),
- motive power (traction batteries) and
- starting.

It is the stationary lead battery in particular that is important for telecommunications power supply. It provides a standby energy for a certain period (bridging time). Starter batteries are also applied as standby power supply systems.

Because of their different electrical characteristics, the individual designs of lead batteries can be categorized, in relation to different bridging times, as follows:

- batteries for short-term loading ( <1 h) and
- batteries for long-term loading (capacitive loading, >1 h).

For short-term loading, for example, according to DIN[2] the designs GroE[3] and OGi-Block[4] are available, while the designs OPzS[5] and OGi-Block (or OGiE) are used for long-term loading.

Apart from the conventional lead-acid batteries with liquid electrolytes, 'maintenance-free, valve-regulated (sealed)' lead-acid batteries are also now available; they differ from traditional types in their 'fixed' electrolytes (gel technology or absorbent-glass-mat technology). Instead of vent plugs, the cells are provided with safety valves which open under excessive pressure. The batteries are available in the pasted-plate (OGiV) and tubular-plate (OPzV) forms of construction.

In addition, non-rechargeable backup batteries (generally lithium) are used for safeguarding memory contents.


## 1.5 Primary Power Sources

*Diesel generators* (Fig. 1.4a) provide a permanent (a.c.) standby. It is customary to distinguish, according to design, between generators (e.g. single generators) which function as an emergency during system startup, and installations with two or more generators alternating to provide a permanent power supply when no public power source is available (island networks).

---

[2] DIN: German Industrial Standards (Deutsche Industrie Normen).
[3] GroE: Planté plate in narrow construction.
[4] OGi: stationary grid-plate (pasted-plate).
[5] OPzS: stationary cells with tubular (ironclad) positives plates.

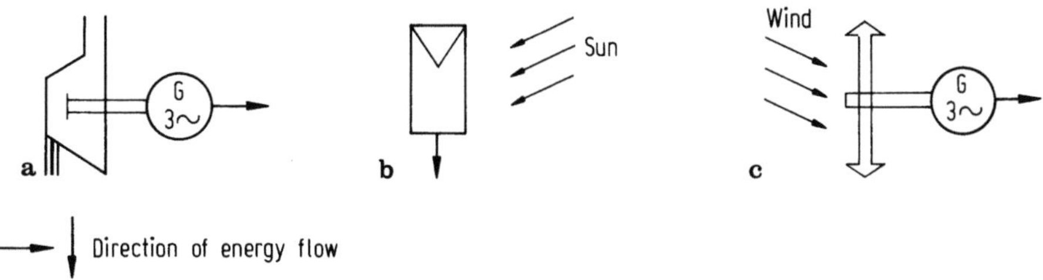



**Fig. 1.4a–c.** Examples of primary power sources. **a** Diesel generator; **b** solar generator; **c** wind-driven generator

With a *solar-powered generator* (Fig. 1.4b) solar energy can be converted directly into (d.c.) electrical power. The solar generator consists of several solar modules, each of which contains a large number of photovoltaic cells.

A difference is made between thick-film arrays (i.e. out of crystalline silicium) and thin-film arrays (i.e. out of amorphous silicium). Their light-sensitive surfaces are protected against mechanical damage by a special type of highly transparent glass with a surface of low reflectivity. At the reverse side of the module, a plastic-coated aluminium foil provides a moisture barrier, which simultaneously serves to dissipate heat. The optimal solar generator for a given application can be constructed by series and parallel combinations of solar modules.

A *wind-driven generator* (Fig. 1.4c) consists of a rotor or a turbine and a generator. It produces alternating current. There are several varieties of wind-driven generators. The most favourable locations for these installations are mountain ridges, open plains and coastlines or funnel-shaped valleys.

The *small steam turbine* (not shown) has a closed steam circulation; it can be operated on liquefied gas, natural gas, diesel fuel or kerosene. The turbines can be used either individually with a battery, or in parallel mode with or without battery.

*Thermoelectric generators* (not shown) consist of numerous thermocouples which convert heat gradients into electrical energy; they thus contain no moving parts. The higher the temperature difference between hot and cold ends, the greater will be the efficiency. The output power of thermoelectric generators depends greatly on the ambient temperature. Several devices can be connected together in parallel to form larger units and can be operated with or without a battery.

## 1.6 Power Supply Systems

Fig. 1.5 shows the outline of a typical telecommunications power supply system, while Fig. 1.6 shows an overall view of such a system.

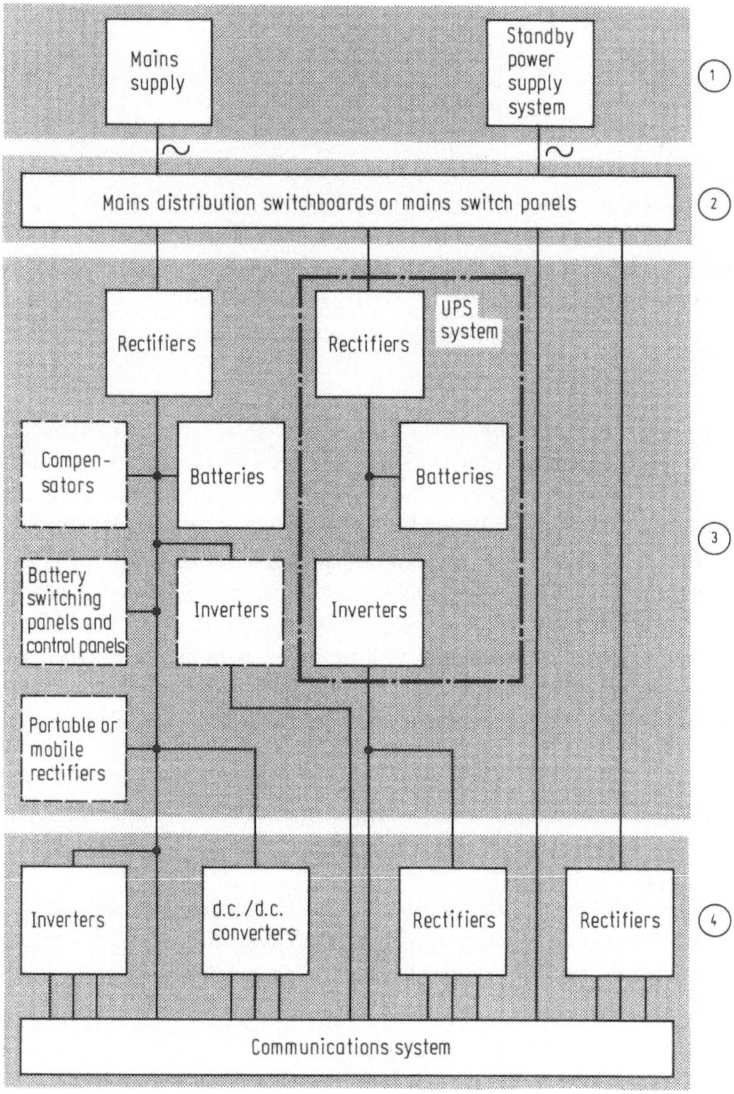

**Fig. 1.5.** Telecommunications power supply systems

The system can be divided into four levels:

(1) Mains supply and standby power supply system.[6]
(2) Mains distribution switchboards or mains switch panels.
(3) A central power supply.
(4) A decentralized power supply.

---

[6] Standby systems with internal combustion (generally diesel) engines.

**Fig. 1.6.** Telecommunications power supply system (photo by courtesy of Siemens AG)

## 1.6.1 Mains Supply and Standby Power Supply Systems (1)

The necessary electrical power is usually taken from the public mains. As there is the possibility of power failure with this mains supply, a standby fixed or mobile power supply is used, where necessary. Both, the mains supply and the standby power supply system, provide single-phase or three-phase alternating voltage.

Trouble other than power failure can also occur with the mains supply, e.g.

- failure of individual phases,
- excessive voltage or frequency fluctuations and
- excessive harmonics.

Further arrangements, apart from standby power supply systems, which bridge disturbances in the mains supply, may be found in (3), central power supply.

## 1.6.2 Mains Distribution Switchboards or Mains Switch Panels (2)

Mains distribution switchboards and mains switch panels are used for the distribution, switching and measurement of the mains voltage and for providing fuse protection. While power supply equipment of lower output can be linked with the mains supply via a mains distribution switchboard, rectifiers of greater output power must be connected to a mains switch panel.


### 1.6.3 Central Power Supply (3)

A basic distinction is made between a central and a decentralized power supply of a communications system. This distinction becomes increasingly difficult to maintain, however, in the case of modern power supplies (e.g. compact systems) employing hybrid converter modules.

*Rectifiers.* The rectifiers in a central power supply system feed the communications system, inverters, batteries and d.c./d.c. converters. These rectifiers are with thyristor power section (thyristor-controlled rectifiers) and convert the alternating current of the mains supply or standby power supply systems into direct current. Rectifiers with a transistor power section are also available (known as transistor-controlled rectifiers or switching-mode power supply, SMR). The rectifiers together with the batteries and any standby power supply systems, if present, must be capable of providing an uninterruptible supply of direct current for the communications system (fail-safe d.c. power supply system).

The most important modes of operation for rectifiers are:

- *rectifier mode* (without battery). In the event of a power failure, the communications system does not receive any supply voltage.
- *standby parallel mode* (with battery). Rectifiers, battery and communications system are connected in parallel.
- *changeover mode* (with battery). In the event of power failure the supply is switched from mains to battery operation without interruption.

The normal ratings for the d.c. output voltage provided by the rectifiers are 48 V and 60 V. Thyristors or switching transistors (bipolar or unipolar e.g. power MOS field effect transistors), as the final control element, together with their associated control equipment, ensure that changes in the power consumption of the communications system or fluctuations in the mains voltage or frequency result in the d.c. output from the rectifiers varying only within a permissible tolerance.

The essential components of thyristor or transistor-controlled rectifiers are:

- mains transformer,
- thyristor set or transistor bridge circuit,
- filter,
- closed-loop and open-loop control,
- protective and monitoring devices and
- signalling system.

Monitoring devices on the direct voltage side protect the equipment against excessive over- or undervoltages which would interfere with its proper working.

There are electronic mains monitoring devices to protect rectifiers operating on three-phase mains. These monitoring devices detect the failure of one or more phases or mains under- or overvoltage, and switch off the relevant rectifier. The device automatically switches the rectifier back into circuit when normal conditions have been restored.

In small or medium power installations, rectifiers with power switching transistors [see (4), decentralized power supplies, switching power supplies and components, rectifier modules], batteries and distributing devices are built into racks and arranged in the immediate vicinity of the communication system as compact power supply. Requirements are similar to those for thyristor-controlled rectifier units.

The main advantages of having the power supply equipment in the same room or having the power integrated into the telecommunications system are:

- reduction of the costs of transporting energy,
- reduction of the costs of materials like distribution cables,
- reduction of the costs of installation,
- reduction of the total maintenance cost and
- reduction of the training duration of the staff.

New transistor-controlled rectifiers are specially designed to draw sinusoidal current from the mains with a power factor and displacement factor that is very close to 1.0.

*Batteries.* If the rectifier fails to provide a supply, a continued supply to the communications system can be ensured with the aid of batteries (accumulators). Sometimes a standby power supply system, mentioned in (1), can also be used.

*Portable and mobile rectifiers.* Portable and mobile power supply equipment with the technical characteristics of stationary rectifiers can replace the fixed system.

*Inverters.* Inverters (d.c./a.c. converters) are used when a communications system requires an alternating current supply. In this way the demand on the central power supply for an a.c. voltage free from interruptions can be met using batteries and rectifiers. The a.c. voltage is needed to feed peripheral devices in OMC (operation and maintenance centre) such as computers, pageprinters and memories.

*Compensators.* In the event of a power failure a compensator, fed by the battery, supplies a gradually rising booster voltage in series with the battery voltage. Thus, despite a falling battery voltage, the supply voltage for the communication system remains within the permissible tolerance, even with a power failure.

There are also compensators for additional or opposing voltages. These devices thus act as step-up or step-down converters, allowing the supply of consumers with a narrow tolerance from one power supply (whose tolerance range is 40–75 V). These compensators can raise or lower the supply voltage by a maximum of 7 V.

*Battery switching panels and control panels.* A distinction is made between battery switching panels with control and battery switching panels (without control).

Battery switching panels with control contain not only the power circuits but also the complete control, monitoring and signalling systems for the power supply. They can be connected to a maximum of two batteries. No *central control panel* is necessary with these power supply systems.

Battery switching panels contain only the power circuits. Each battery must have its own battery switching panel. A central control panel can also be necessary for each system. Arranged within this panel are the complete control, monitoring and signalling systems for the power supply. Many systems have no central control panel since the devices mentioned are contained in the rectifier units.

Small to medium-sized power supply systems can also be constructed without battery switching panels, control panels and compensators. Some systems are even supplied without a battery. The ultimate design depends on what communications system the power supply is being planned for, what type series of rectifiers has been decided upon and what reliability requirements are to be observed.

### 1.6.4 Decentralized Power Supply (4)

Normally modern communications systems require a number of low supply voltages (component voltages of, for example 5 and 12 V) with narrow tolerances and differing polarities. These are obtained from the central supply voltage (in this case energy transport voltage) of, for example, 48 V.

It would be uneconomic to provide the various component supply voltages centrally. With small supply voltages the current would increase. The voltage drop on the lines would therefore also increase, leading to a current distribution system with disproportionately large cross-sections. It would also be necessary for each component supply voltage to have its own supply line.

A decentralized power supply system enables the power supply to likewise expand when a communications system is extended. In addition, the effect of disturbance is more confined.

The output voltage is controlled by power transistors in switching mode at a clock frequency of 20 kHz or higher (usually 60 kHz).

When the power supply is decentralized different systems exist for providing component voltages:

*Systems for d.c. input voltage:*

– Provision of direct voltages with d.c/d.c. converter and
– Provision of alternating voltages with inverter.

*Systems for a.c. input voltage:*

– Provision of direct voltages with rectifier.

*Power supply units with d.c. input voltage:*

– d.c./d.c. converter, Direct current converters are also called: d.c. chopper converters or regulators respectively and d.c. voltage transformers.

A distinction is made between three types of converters:

– single-ended fly back (blocking) converters,
– single-ended forward (flow) converters and
– push-pull forward (flow) converters.

D.C./D.C. converters can be further divided into step-down and step-up converters. The term 'step-down' is applied to converters whose output voltages are lower than their input voltages – the most frequent to be found in practice. In step-up converters – used mostly in solar power supply systems – the relationship is reversed. There are also converter's which can do both step up *and* step down. In the following, the step-down converter will be discussed in more detail.

From the rectifier unit of the central power supply, the d.c. voltage arrives at the d.c./d.c. converter of the decentralized power supply; this voltage is 'chopped' into a square wave a.c. voltage by a rapid switching transistor. The alternating voltage passes through a transformer and is then rectified and filtered. The d.c. output voltage is stabilized by a regulating circuit, which regulates the duty cycle of the transistor switch (pulse width control).

The tolerance limits for supply voltages to the d.c./d.c. converter are very wide. Direct current converters generally process d.c. input voltages from 40 to 75 V. The d.c. output voltages, as already explained, are held constant at the desired component voltage levels within a narrow range of tolerance.

– *Inverters.* In certain cases decentralized power supply facilities also include inverters which provide alternating voltage from the energy transport direct voltage from the central power supply. The alternating voltage is needed to feed peripheral devices such as computers, page printers and memories.

*Power supplies with a.c. input voltage*

– *Rectifiers.* Rectifiers with transistor power section (transistor-controlled rectifiers) are linked either with the inverters of the central power supply or directly with the mains supply. Here they are 'cycled' elements of the switching mode power supply. There is alternating voltage at the input of the switching mode power supplies which is rectified with a mains input rectifier circuit. There a d.c./d.c converter circuit follows.

All power converters must comply with radio interference suppression regulations. The terms 'limit-value class' and 'radio interference (suppression) level' belong to the wider area of 'electromagnetic compatibility' (EMC). EMC signifies that an item of electrical equipment functions properly in its electromagnetic environment (interference immunity) and does not interfere magnetically with other equipment (interference emission).

In this connection, earthing is an important aspect of protection measures against overvoltage and interference. Equipment is also provided with overvoltage- protection elements such as, for example, surge arresters, metaloxide varistors, suppressor diodes and zener diodes. At this stage 'electrostatically sensitive devices' (ESD) should be mentioned, referring to components and assemblies which are liable to be damaged by electrostatic effects. These components (e.g. MOS) and assemblies must be adequately protected against damage and destruction due to electrostatic discharges.

## 1.7 UPS Systems

In the case of uninterruptible power supply (UPS) systems (no-break a.c. power supply systems) the loads are supplied continuously, i.e. without interruption. Regardless of their form, such systems can be designed as 'single-block systems', 'single-block systems with passive redundancy' (of the mains) or 'multi-block parallel systems with passive and active redundancy' (Fig. 1.7).

In a static UPS system, each UPS block consists of a rectifier unit, a battery and a inverter (see Figs. 1.5 and 1.8).

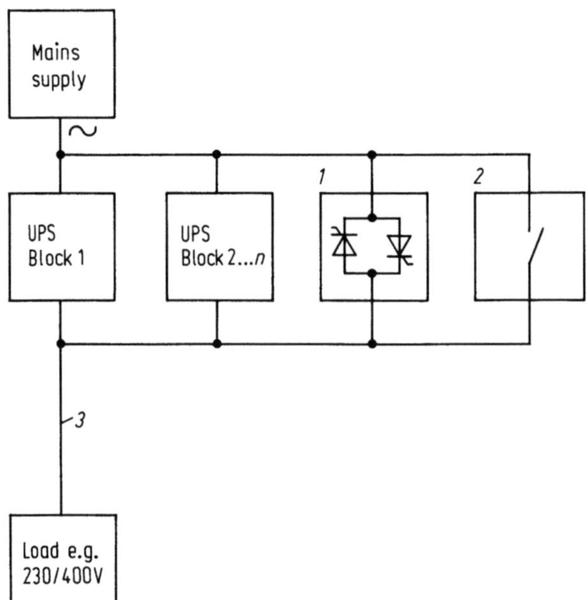


**Fig. 1.7.** (Static) Uninterruptible power supply system UPS

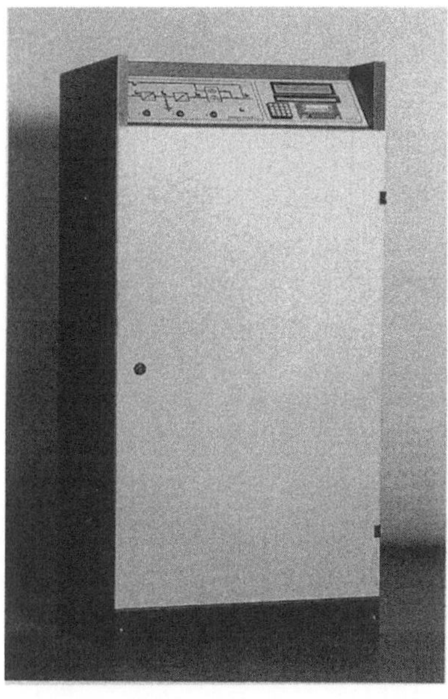

**Fig. 1.8.** Static uninterruptible power supply system, for single phase, series 110, 20 kVA (photo by courtesy of Gustav Klein GmbH & Co. KG)

In normal operation the mains supplies the uninterruptible power supply blocks, which in turn supply the alternating voltage to the 'safe' (fuse) bar and thus to the load. For maintenance work the whole UPS system can be made dead by means of the manual bypass.

The number of uninterruptible power supply blocks connected in parallel consists of at least one block more than is necessary for the load.

When there are a number of blocks in an uninterruptible power supply system, then, in the event of trouble with one of them, only the faulty block is separated from the safe bar. In the example given there remains a further block, which continues supplying the load.

If the output of the remaining block is not sufficient to supply the load or if the second block is out of order, the revert-to-mains unit switches to the mains without interruption.

## 1.8 Mains-independent Hybrid Power Supply Systems

Hybrid systems, consisting for example of wind-powered generator, solar generator, diesel generator and battery, offer a combination of extremely and high reliability low maintenance (Figs. 1.9 and 1.10).

When the battery is discharged, the diesel generator starts automatically, supplies the consumer and charges the battery. The battery capacity is chosen in such a way that the diesel generator seldom needs to take over the function of power supply. This in turn permits relatively long maintenance intervals. Existing diesel supplies can be complemented by adding wind-driven or solar generators,

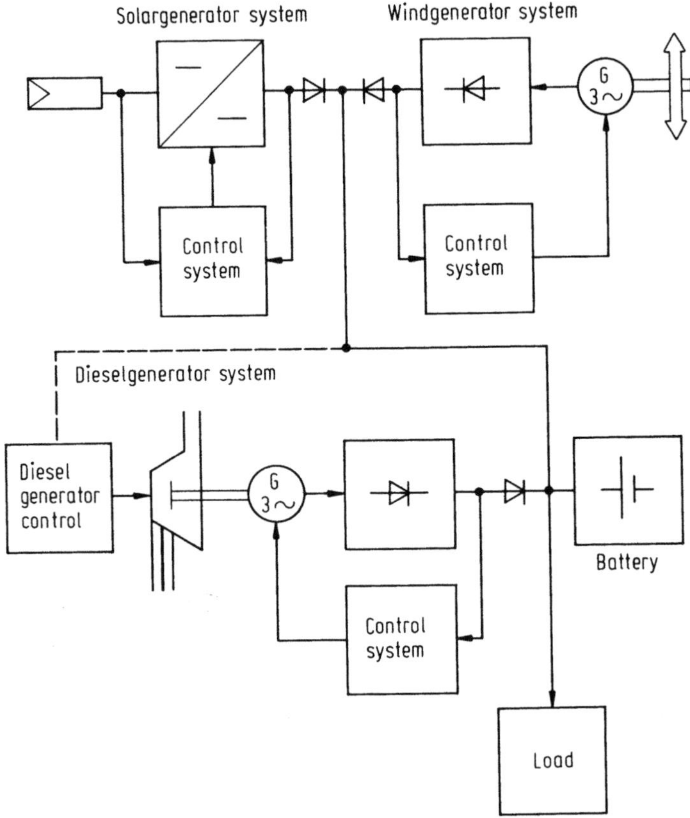



**Fig. 1.9.** Mains-independent hybrid power supply system

allowing a further savings in fuel consumption and maintenance. Devices used for communications, control, and rectification are housed in shelters with the diesel generators and batteries.

The following combinations are especially suitable in practical applications: solar generator, diesel generator and battery, and solar generator, wind-driven generator and battery. These systems are used as power supplies for transmission systems.

## 1.9 Historical Review

This introduction to the supply of power for telecommunications systems will be rounded off with a review of the prominent steps which led to its development. It can facilitate the understanding of new techniques and technologies.

**Fig. 1.10.** Wind-generator installation and solar generator of a mains-independent hybrid power supply system (photo by courtesy of Siemens AG)

*Galvanic* or *voltaic cells* were the first sources of electrical energy.

- The development of a power supply for a telecommunications systems began with the invention of the telephone in 1861 by Philipp Reis and the improvements made by Alexander G. Bell and David E. Hughes (1876).
- The first public telephone network in Germany with a hand-operated exchange was put into operation in Berlin as early as 1881. This was a local battery system (LB system) in which each subscriber extension had its own power supply in the form of a primary cell (dry battery).
- For reasons of economy and operating reliability a switch was made from the LB system to the central or common battery system (CB system) on introduction of the first automatic systems (around 1900). With this system the current sources were located centrally with the switching system and, instead of primary cells, secondary cells (lead accumulators or storage batteries) came into use (invented in 1859 by the Frenchman G. Planté and based on the work of Sinsteden who had introduced lead into secondary batteries in 1854).
- Around the same time as the introduction of automatic systems a decision was made for today's public alternating current supply.
- The mode of operation with the CB system was initially with two batteries, also called the charge-discharge mode (battery mode),
- Later parallel mode (floating mode) replaced operation with two batteries. For reliability two batteries are normally connected in parallel. At first a rotat-

ing converter still fed the battery and communications system. The converter, battery and communications system were always connected in parallel,

– Invention of the mercury arc rectifier (1902) by P. Cooper-Hewitt replaced the rotating converter, which was well known to be inefficient and required much maintenance. As with today's thyristor or transistor- controlled rectifiers, it was already possible with the mercury arc rectifier to keep the d.c. output voltage constant by modifying the firing point (trigger pulse).

– Arrival of the polycrystalline semiconductor around 1930 permitted construction of the first 'dry rectifier' (copper oxide rectifier) for the power range up to 60 V/3 A. The mercury arc rectifier was used for higher powers,

– Around 1934 the selenium rectifier (selenium diode) was developed – like the copper oxide rectifier, also based on polycrystalline semiconductor material. The selenium rectifier with its higher specific load capacity replaced the copper oxide rectifier. It is in fact more sturdy than the mercury arc rectifier, but is not suitable for maintaining a constant output voltage. For this reason the mercury arc rectifier was preferred to the selenium rectifier until around 1949 when there were suitable control elements for selenium rectifiers – transductors, also called magnetic amplifiers. There thus came into being the first magnetically controlled rectifiers. The importance of the mercury arc rectifier declined rapidly from that time on.

– The magnetically controlled rectifiers already had an automatic control circuit with magnetic regulating inductors for control. The automatic control circuit contains units working solely on a magnetic basis, e.g. magnetic controllers, transductors, etc.

– With mastery of the behaviour of single crystals there appeared in 1948 the first single-crystal semiconductor devices – germanium diode and germanium transistor (invented by J. Bardeen, W.H. Brattain and W. Shockley, USA). The technology now existed for the development of transistor controllers instead of magnetic controllers in magnetically controlled rectifiers.

– Magnetically controlled rectifiers with transistor controllers appeared around 1960 providing better control behaviour.

– For smaller and medium-sized private automatic branch exchange (PABX) systems there were also uncontrolled and phase-controlled rectifiers in addition to the magnetically controlled ones. It is well known with uncontrolled rectifiers that the d.c. output voltage drops as the load increases and is also dependent on frequency and mains voltage fluctuations.

– Uncontrolled rectifiers are suitable for current strengths of about 0.7 to 3 A.

– Phase-controlled rectifiers were 'self-controlling'. Like magnetically controlled rectifiers, they kept the d.c. output voltage constant despite mains voltage fluctuations and changing load conditions throughout the communications system. Changes in frequency, however, do influence the d.c. output voltage. Phase-controlled rectifiers were used for current strengths from 1.5 to 25 A.

– From around 1955 there have existed single-crystal semiconductor devices based on silicon, namely the silicon diode and the silicon transistor. The silicon diode has a higher reverse voltage, greater load capacity and a steeper characteristic than the selenium diode. That is why with magnetically controlled rectifiers the rectifier stack was later designed with silicon diodes,

- In 1956 in the United States, Moll, Tanenbaum *et al.* developed a variant of the transistor, which can be considered a silicon-controlled rectifier. In Germany, by analogy, the hot-cathode gas-filled tube (thyratron) was first called the silicon-hot-cathode gas-filled tube, but later since the early 1960s it has been called a *thyristor*. It has brought about a turning point in rectifier techniques and is today one of the predominant components in power electronics,

- When in 1965 medium-sized PABX systems (telephone system 400 E) came out using ESK crosspoint switching technique, thyristor-controlled rectifiers could then already be used for the power supply (3, 5, 10 and 16 A).

- In 1973 a 48 V/40 A thyristor-controlled rectifier came into the market for the KS 3000 E communications system (large-sized PABX, 1974 already saw thyristor-controlled rectifiers for up to 1000 A being built.

- Since 1976 integrated circuits have been available for rectifiers.

- The first switching-mode power supplies came out in 1977 after power transistors were developed. These are now also being used in new communications systems.

- In the last few years there have also appeared MOS-field-effect power transistors and special forms of fast-switching thyristor, such as turn-off thyristors (GTO). Digital integrated circuits are being used on an ever-increasing scale in closed-loop and open-loop control sections. Information processing is carried out increasingly by means of microprocessors. All these latter components lead to smaller and lighter equipment, with better efficiency and less noise generation.

# 2 Requirements of Telecommunications Systems on the Power Supply

## 2.1 D.C. Power Supplies

### 2.1.1 Level of the Direct Voltages

Table 2.1 contains the data on power supplies which are important in connection with the features of communications systems. Line 1 of Table 2.1 shows the *rated voltages* for communications systems. A distinction is made between 24, 48 and 60 V systems. Modern systems are usually 48 or 60 V, the *positive pole* normally being *earthed*. The choice of voltages also depends on the safety regulations of the country in question. In Germany, direct voltages over 120 V and alternating voltages over 50 V are classed as 'dangerous contact voltages'.

At the output of any telecommunications power supply system is the *'central supply voltage'* for the communication system (load) and for the battery (Fig. 2.1; see also Table 2.1, line 2).

Examples of *lead-acid battery voltages* are:

- 2.0 V/cell (nominal voltage),
- 2.23 V/cell (trickle (float) charging voltage),
- 2.33 V/cell (charging voltage) and
- up to 2.7 V/cell (initial charging voltage).

The central supply voltage and the *operating voltage* differ by the voltage drop $\Delta U$ over the distribution lines.

The voltages for conventional communications systems are normally fed directly. Modern systems, on the other hand, additionally require *decentralized supply voltages*, also termed component or electronic voltages (e.g. 5 V, 12 V). These secondary voltages are obtained from the central supply voltage via d.c./d.c. converters. The central supply voltage is then called energy transport voltage.

### 2.1.2 Tolerance for Direct Voltages

For each communications system the *tolerance* is shown for the operating voltage given in Table 2.1, line 3. The *bottom, continuously permissible voltage* is determined by the necessary reliability against a wrong connection when putting through the call and by maintenance of the connected call. The *top, continuously permissible voltage* depends on the heat generated by the components.

The *top critical voltage* must not be exceeded even for a short time, as components may be destroyed (e.g. semiconductor elements, Table 2.1, line 4).

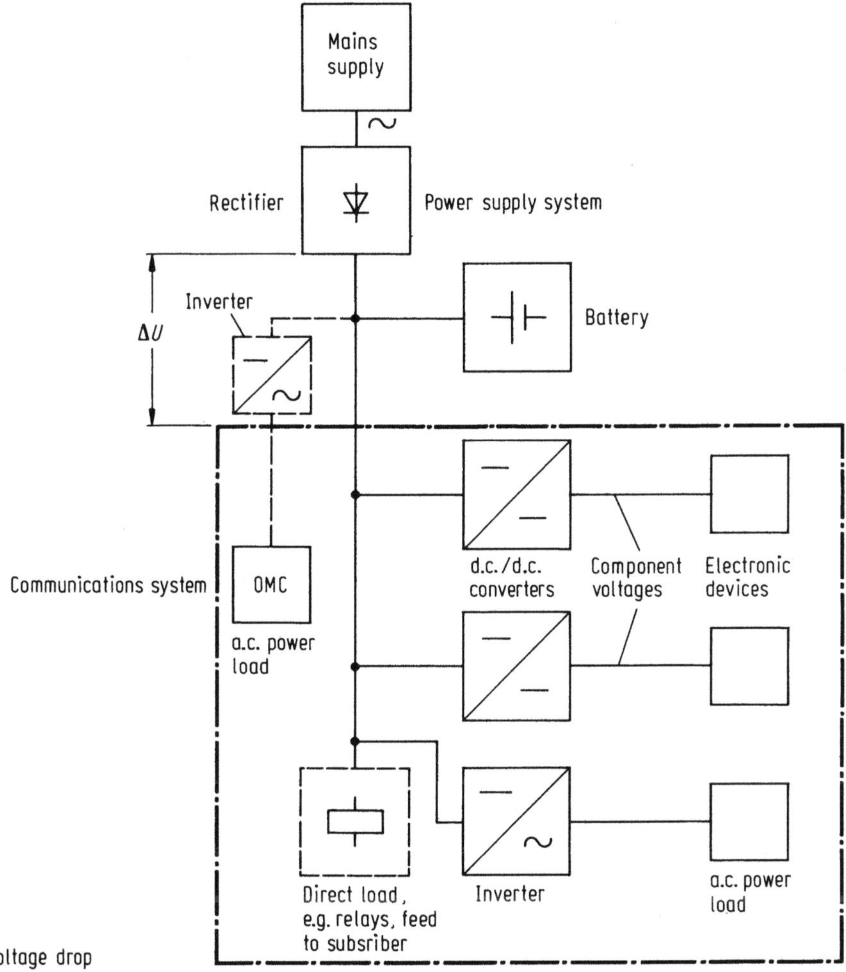



**Fig. 2.1.** Telecommunications power supply system with battery and load

A *bottom critical voltage* of $\leq 40$ V ($t \leq 1$ ms) is specified, for example, for 48-V and 60-V EWSD systems.

In the case of modern communications systems the observance of a d.c. output voltage tolerance of $\leq \pm 0.5\%$ or $\leq \pm 1\%$ (static tolerance range), possible nowadays with rectifiers, is necessary only with respect to a battery connected in parallel. A tolerance range of $\pm 4\%$ must be allowed in case of sudden changes in load or mains supply voltage because of the dynamic behaviour of rectifiers.

**Table 2.1.** Data concerning the characteristics of power supply systems relevant to communications systems

| | Communications systems (examples) | | | | | |
| --- | --- | --- | --- | --- | --- | --- |
| | EWSA | EWSA | EWSD/ KN System | EWSD[a] | EMS | Transmission systems |
| 1 Rated voltage in V | 48 | 60 | 48 | 60 | 48 | 48    60 |
| 2 Central supply voltage in 'normal' mode in Volts | Direct loads 51<br>D.C./D.C. converters 58 | Direct loads 62<br>D.C./D.C. converters 67 | 56 (53.5) | 67 | 53.5 | 53.5    67 |
| 3 Continuously permissible tolerance for operating voltage measured at the system's subassemblies or components in Volts | Direct loads 44 to 53 | 57 to 64<br>D.C./D.C. converter 40 to 75 | 44 to 58 | 50 to 71 | 42 to 58 | 42 to 75 |
| 4 Top critical voltage measured at the system's subassemblies of components ($t = 0$ s) in Volts | $\leq$ 60 | $\leq$ 75 | $\leq$ 80 | $\leq$ 80 | $\leq$ 70 | $\leq$ 100 |
| 5 Frequency-weighted interference voltage in accordance with A-filter-curve (CCITT) in mV | Direct loads $\leq$ 2<br>D.C./D.C. converters: no conditions | | | | $\leq$ 0.5 | $\leq$ 2 |
| 6 Superimposed alternating voltage (without frequency weighting) in mV | $\leq$ 195 | | | | $\leq$ 150 | $\leq$ 195 |

| 7 | Degree of radio interference | Limit value class A according to VDE 0878 respectively<br>Limit class B according to VDE 0871 | | | | | | |
|---|---|---|---|---|---|---|---|---|
| 8 | Maximum voltage drop $\Delta U$ overall in Volts (design basis) | 2.7 | Without Cmp<br>2.4<br>———<br>With Cmp<br>3.4 | 2.7[b] | 3.4 | 2 | 2.7 | 3.4 |
| 9 | Lead-acid battery, number of cells | 26 | 30 | 25 (24) | 30 | 24 | 24 | 30 |

Cmp compensator (for additional voltage),

EWSA analog electronic switching system,

EWSD digital electronic switching system,

EMS electronic modular system (PABX),

KN communications network (PABX),

[a] Predominantly for systems of the German Postal Administration (Telekom Network)

[b] or 1.8 V (for explanation see Fig. 2.3).

When the *battery* is *discharging* the operating voltage depends on the battery's characteristic curve. If a voltage with a narrower tolerance is required, a compensator[1] must be added.

Direct voltages are monitored by voltage monitoring systems built either into the rectifiers or into the control devices of the battery switching panels or control panels.

An alarm is triggered if there is a drop below the set values for the bottom voltage limits. An alarm is also triggered and the rectifier switched off if the top voltage limit is exceeded, e.g. due to a fault in the control system.

### 2.1.3 Purity of Direct Voltages

The direct voltage taken at the output of a power supply system always has a superimposed alternating voltage which intrudes into the voice circuits of the communications system. In different ways hum interference is produced depending on the frequency of the alternating voltage and on the transmission characteristics of the telephone circuit. Apart from the requirement that the direct voltage must be as constant as possible, its purity is particularly important for the error-free transmission of information.

The superimposed alternating voltage is made up of a combination of frequencies. These individual frequencies cause varying degrees of interference. The purity of the direct voltage is defined by international provisions of the CCITT (Comité Consultatif International Télégraphique et Téléphonique/International Telegraph and Telephone Consultative Committee) so that a constant quality is ensured in the case of communications across a number of national or international networks. A weighting curve has been defined for the degree of interference of the individual frequencies. This is the CCITT 'A-filter curve'. The individual maximum permissible values for interference voltage (line 5) and the superimposed alternating voltage (line 6) for various communications systems can be found in Table 2.1.

A psophometer (circuit noise meter) is used to measure the superimposed alternating voltage. When measuring, a distinction is normally made between:

– the interference voltage, $U_{int}$, the alternating voltage at the power supply output weighted with A-filter;
– the superimposed voltage, $U_{sa}$, the unweighted alternating voltage at the power supply output;
– the noise voltage, $U_n$, the alternating voltage weighted with A-filter on the voice wires of a telephone connection;
– the extraneous voltage, $U_e$, the unweighted alternating voltage on the voice wires of a telephone connection.

The extraneous voltage $U_e$ is proportional to the superimposed alternating voltage $U_{sa}$ as the noise voltage $U_n$ is proportional to the interference voltage $U_{int}$.

---

[1] In general only for 60 V systems of the German Postal Administration Telekom Network.

Within the framework of a detailed consideration of the superimposed alternating voltage on the direct current side there now follows an explanation of the terms alternating voltage component, ordinal number, alternating voltage content, etc.

An alternating voltage is superimposed on the direct voltage $U_d$ (arithmetic mean). It consists of sinusoidal components of different frequencies $vf$, where

> $v$    is the ordinal number and
>
> $f$    is the frequency.

The ordinal numbers $v$ of these components are integral multiples of the pulse number $p$:

$$v = kp \quad \text{with } k = 1, 2, 3, \ldots$$

where

> $k$    is the distortion factor

The ideal alternating voltage component of the ordinal number $v$ has, at full modulation (control angle $\alpha = 0°$), the r.m.s value:

$$U_{vi} = \frac{\sqrt{2}}{v^2 - 1} U_{di}$$

where

> $U_{di}$    is the ideal direct voltage.

With partial modulation the values for $U_{vi\alpha}$ (ideal alternating voltage component of the ordinal number $v$ as a function of the control angle $\alpha$) increase as the control angle increases.

With thyristor-controlled rectifiers it is possible, e.g. in the case of capacitive loading, for the current to have gaps as a function of the degree of modulation. In contrast with the ideally stabilized direct current, the behaviour of current and voltage differs in 'gap mode' (control angle $\alpha$ very large).

The alternating current content $w$ is at a minimum value with full modulation and reaches its maximum value with least modulation. It follows from this that the alternating voltage content $w$ of the direct voltage $U_d$ is primarily dependent on the control angle $\alpha$ and increases as the control angle becomes larger.

The ideal alternating voltage content (ideal ripple) $w_{i\alpha}$ is the ratio of the r.m.s. value of the superimposed ideal alternating voltage to the ideal direct voltage:

$$w_{i\alpha} = \frac{U_{si\alpha}}{U_{di}} = \frac{\sqrt{\Sigma U_{vi\alpha}^2}}{U_{di}}$$

**Table 2.2.** Ideal superimposed alternating voltage on the direct current side with full modulation (control angle $\alpha = 0°$)

| $v$ | $vf$ | Pulse number $p$ | | |
|---|---|---|---|---|
| | | 2 | 6 | 12 |
| | | $\frac{U_{vi}}{U_{di}}$ [%] | | |
| 2 | 100 | 47.14 | – | – |
| 4 | 200 | 9.43 | – | – |
| 6 | 300 | 4.04 | 4.04 | – |
| 8 | 400 | 2.24 | – | – |
| 10 | 500 | 1.43 | – | – |
| 12 | 600 | 0.99 | 0.99 | 0.99 |
| 14 | 700 | 0.73 | – | – |
| 16 | 800 | 0.55 | – | – |
| 18 | 900 | 0.44 | 0.44 | – |
| 20 | 1000 | 0.35 | – | – |
| 22 | 1100 | 0.29 | – | – |
| 24 | 1200 | 0.25 | 0.25 | 0.25 |

$U_d$   Direct voltage (100 %).
$v = 2$ to 24 Ordinal number of superimposed alternating voltage component.

where

$U_{si\alpha}$   is the superimposed ideal alternating voltage when modulated with the control angle $\alpha$.

For the ideal alternating voltage content at full modulation:

$$w_i = \frac{U_{si}}{U_{di}} = \frac{\sqrt{\Sigma U_{vi}^2}}{U_{di}}$$

the values for pulse numbers 2, 6 and 12, taking into account all even ordinal numbers $v = 2$ to 24 with their assigned frequencies, are to be taken from Table 2.2.

It is seen that as the ordinal number $v$ rises, the quotient $U_{vi}/U_{di}$ becomes smaller.

The pulse number of the thyristor circuit determines the ordinal number of the alternating voltage components occurring. $U_{vi}/U_{di}$ is dependent only on the ordinal number respectively frequency, but not on the pulse number $p$ of the rectifier.

### 2.1.4 Filter Sections

*Filter sections* (low-pass filters) are necessary with rectifiers for maintaining the required purity of the d.c. output voltage. At a low frequency the attenuation is very slight (pass band). As the frequency becomes higher, the attenuation increases accordingly (stop band).

With surges in load the storage effect of a filter section, in particular with large capacitances, permits within certain limits the attenuation of fluctuations in the d.c. output voltage. The effect of filtering can be visualized with the aid of the simplified circuit shown in Fig. 2.2. There is a (stepped-down) alternating voltage at the transformer output. Parallel with the output of the rectifier circuit is the charging capacitor, $C_L$, which acts as an energy store. The amplitude of the discharge current and the size of the capacitor influence the level of the direct voltage. If there is a small load at the output, the charging capacitor is scarcely discharged and the ripple will be slight, whereas with a large load at the output an (equally large) capacitor gives off a large part of the stored energy per cycle. Thus the ripple increases.

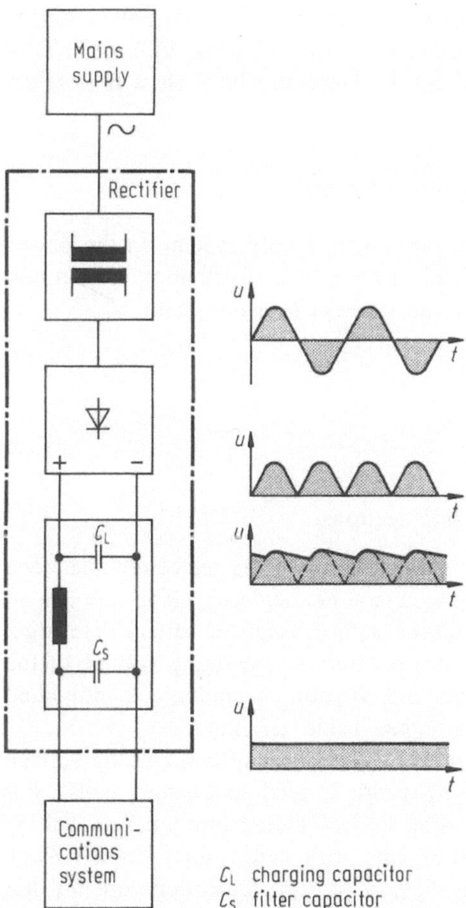


$C_L$  charging capacitor
$C_S$  filter capacitor

**Fig. 2.2.** Filtering the d.c. output voltage of a rectifier

### 2.1.5 Degree of Radio Interference and Limit Classes

Interference voltages can be transmitted along lines as through the air by the propagation of radio waves. This interference can be heard in the case of radio reception and seen in the form of stripes, etc., in the case of television reception. For this reason, all electrical equipment, including power converters, must be equipped for radio *interference suppression.*

Radio interference is reduced as much as necessary by suppression using capacitors and chokes. The various degrees of radio interference are distinguished according to the level of the remaining radio interference voltage. The frequency of the interference is especially important for this procedure.

Power converters for telecommunications systems are normally suppressed in accordance with the norm VDE 0878 to limit class A (for residential areas). Power converters had previously been designed to conform to suppression degree N (normal, VDE 0875).

The *limit class* designates the assignment of high-frequency equipment and systems to different limit values. In the case of telecommunications power supply equipment causing continuous interference of more than 10 kHz, VDE 0871/DIN 57871 permits three limit classes, A, B, and C. Normally limit class B is taken as the basis when designing equipment.


### 2.1.6 Power Distribution System and Voltage Drops

The effects of interference voltages from the power supply system on the power distribution system, feedback from the load to the power distribution system and the mutual interference of the loads must all be kept to a minimum.
A power distribution system normally comprises:

– power supply lines,
– earth lines,
– protective lines,
– cutout devices (fuses) and
– diodes and capacitors for decoupling line sections.

The voltage drop occurring on the supply lines between the telecommunications power supply system and the equipment cannot be neglected. This applies in particular in the case of mains power failures with subsequent battery discharge.

Taking into consideration the minimum permissible operating voltage of the system, the voltage drops were specified and distributed among the individual segments of the power distribution system (see Table 2.1, line 8).

As an example, Fig. 2.3 shows the power distribution system and the voltage drops for the 48-V switching system EWSD (with 25 lead-acid battery cells). For 48-V systems in a digital environment using DLUs (digital line unit) as well as for all 60-V systems, the main distribution line with battery line has a voltage drop of 1.9 V and a total voltage drop $\Delta U_G$ of 2.7 V. Exceptions are the use of LTGAs (line/trunk group analog) and DLUs in 48-V systems, whose range

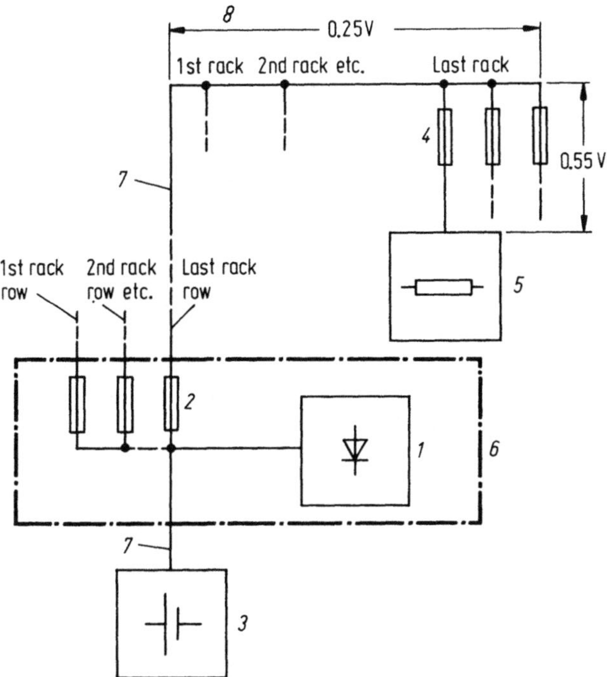

**Fig. 2.3.** Power distribution system and voltage drops in a 48 V communications system, e.g. EWSD, with 25-cell lead battery; total voltage drop $\Delta U_G$ max.: 2.7 V (1.8 V).
*1* Rectifier, *2* Load fuse, *3* Battery (25 cells), *4* Rack fuses, *5* Load, *6* Telecommunications power supply system, *7* Main distributor cable with battery line: 1.9 V (1.0 V), Voltage drop within the racks: 0.8 V

conditions require a reduced voltage drop. Here a voltage drop over the main distribution line with battery line of 1 V and a total voltage drop of 1.8 V are appropriate.

### 2.1.7 Availability of the Power Supply

Availability $V$ is understood to be the *reliability* of the telecommunications power supply.

Availability with present-day systems is very high. It can be calculated as follows:

$$V = \frac{(MTBF)}{(MTBF + MTTR)}$$

where
$MTBF$   is the mean time between failures (in years),
$MTTR$   is the mean time to repair (in minutes).

The unavailability, $A$, is, correspondingly:

$$A = \frac{(MTTR)}{(MTBF + MTTR)}$$

The public mains usually serves as the primary source of energy for a telecommunications power supply. Although the reliability of the mains in Germany can be regarded as relatively high, reliable operation of communications systems is not possible without the provision of alternative power sources (battery and/or standby power supply system). In case of a power failure the alternative power supply source takes on the task of feeding the communications system. The capacity and the number of cells chosen for a battery depends on the time the battery is required to supply energy, the level of the discharge current and the permissible minimum voltage of the communications system. Thus it is possible to have an alternative power supply available for a limited time.

If a supply is to be guaranteed to be available without any limit on time, an alternative to the mains will be necessary. In this case the capacity of the battery can be made smaller, as this is only necessary for switching to the mains equivalent mode without interruption. As a backup supply a battery can normally be used for 6 to 8 hours. If a fixed standby power supply system is provided, the energy reserve required of the battery can be reduced to 4 hours.

Very small local exchanges in the German Postal Administration Telekom Network are provided with a battery reserve of up to 72 hours. These systems have no standby power supply systems.

Systems such as digital electronic switching systems (EWSD) require air conditioning units, whose operation is maintained during longer periods of power failure by emergency power generators. These generators also supply rectifiers and

**Table 2.3.** Requirements of a.c. Power Supplies for switching systems EWSD (Source: Siemens AG)

| | |
|---|---|
| Rated voltage (in V) | 230 single phase |
| Operating voltage (in V) | 220 to 240 V (measured on system boards) |
| Voltage limit | When supply is from the public a.c. system, overvoltage protection against transients has to be provided |
| Frequency (in Hz) | $50 \pm 1$ % |
| EMI suppression class | B as per VDE 0878 |
| Distortion factor | $< 5$ % |
| Load changes | Inrush currents up to $30 \times I_{rated}$ lasting several milliseconds may occur when power units are switched on |
| Distribution system | Each load group is supplied via a separate feeder |
| Voltage drop (in V) | Feeder: 4 |
| Earthing system | Units with non-fused earthing conductor (safety class I) are connected to the EWSD distributed earth electrodes |

other important loads. Therefore, an interim energy reserve time of the battery is planned for one or two hours.

## 2.2 A.C. Power Supplies

An example of the requirements for a.c. power supplies for the switching system EWSD is given in Table 2.3.

# 3 Operating Modes of a Direct Current Power Supply System

Figure 3.1 gives an overview of the operating modes of d.c. power supply systems currently in application. Table 2.1 applies to the operating voltages of communications systems.

## 3.1 Rectifier Mode

In the rectifier mode, also called the direct feed mode, there is *no* battery. The communications system is supplied with direct voltage directly from the mains via the rectifier (Fig. 3.2). The supply is interrupted for the duration of any power failure or in the event of a breakdown of the rectifier. The rectifier automatically switches on again on return of the mains. This mode is used with small to medium-sized communications systems when occasional interruptions in operation can be accepted.

## 3.2 Battery (Charge-Discharge) Mode

Because of its relatively low efficiency and the especially large strain on the battery, this mode of operation is used in today's telecommunications power supply systems only when the mains supply fails and consequently the continued presence of a.c. power must be insured by a mains-independent power supply system. In a typical case, two emergency power generators of relatively short operating time charge the battery through rectifiers. The generators actually run for a few hours only, whereas the load is continuously supplied from the battery. The maintenance intervals for the generators can thus be lengthened.

The battery mode (not illustrated in Fig. 3.1) can also be used in solar-generating or wind-driven generating systems.

## 3.3 Standby Parallel Mode

If the communications system is required to provide continuous unrestricted service during a power failure, or in the case of other troubles, a reserve of energy (preferably in the form of a lead battery) should be kept ready. In the parallel mode the rectifier, battery and communications system are constantly connected in parallel (Fig. 3.3). If the rectifier fails, the battery takes over the further supply

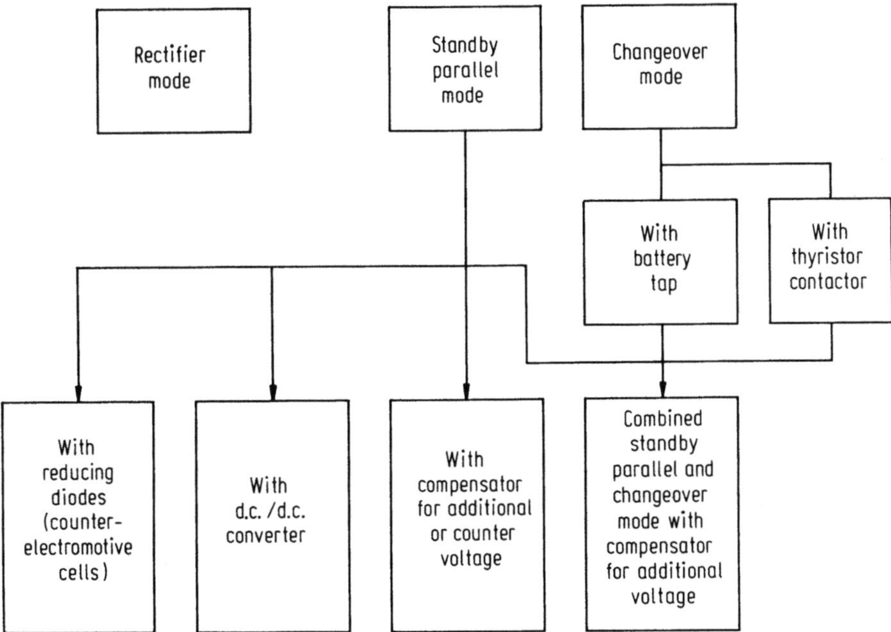



**Fig. 3.1.** Operating modes of d.c. power supply systems

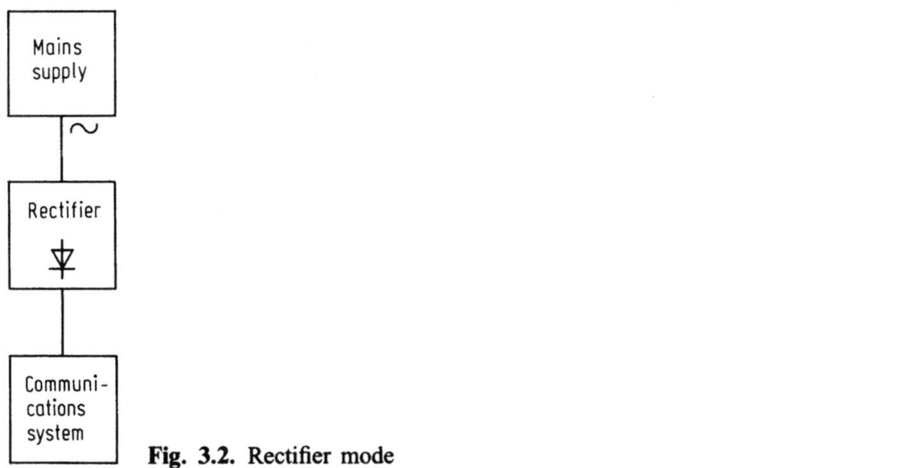



**Fig. 3.2.** Rectifier mode

of the communications system until the rectifier, e.g. on return of the mains, starts operating again. The rectifier then supplies the communications system again and also charges the battery.

In the parallel mode a distinction is made between the floating mode (not illustrated in Fig. 3.1) and the standby parallel mode. In the *floating mode* the

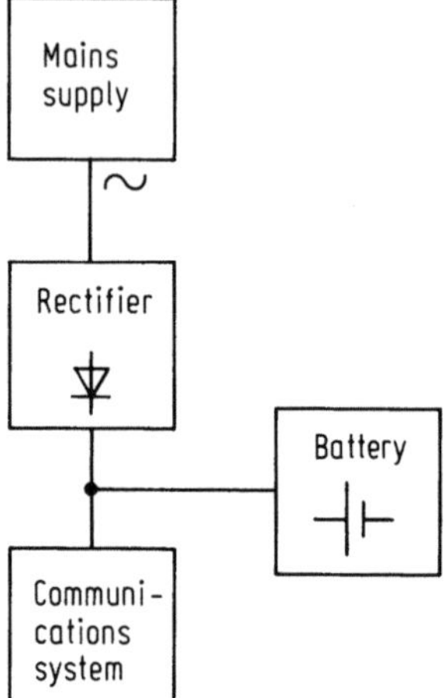


**Fig. 3.3.** Basic representation of the standby parallel mode

rectifier can handle the communications system's normal energy requirements, though it cannot deal with the peak current. In this case the battery provides the current over and above the rectifier's rated current ($I_{load} > I_{rated}$). If the energy requirement (outside busy hours) diminishes again, the battery takes up a charging or trickle charging current from the rectifier; in other words, the battery is either used as a supplementary power source or is being charged. In the floating mode it sometimes happens that the full battery capacity is not available. This means a shorter reserve time for bridging a power failure, unless a correspondingly larger battery capacity is chosen with regard to a certain reserve time. The life of the battery is shortened in the floating mode. That is why today the second variant of the parallel mode, viz. standby parallel mode, is generally chosen.

In the *standby parallel mode* the rectifier always covers the communications system's whole energy requirement. The battery is also supplied with a 'trickle (float) charge' by the rectifier. It is therefore available with its full capacity in case of a power failure or breakdown of the rectifier, provided the interval of time between the previous duty and renewed discharging has been sufficient to charge the battery.

In practice, for additional security, one more rectifier is provided ($n + 1$ redundancy) than is actually necessary to cover the power requirements of the load (Fig. 3.4) (Fail-safe d.c. power supply). This rectifier serves as reserve and battery-charging device (trickle (float) charging, recharging). If in exceptional cases the communications system requires a higher current than that rated for the rectifiers, this current is supplied by the battery.

To avoid the dependence on battery capacity and line length as regards filtering all rectifiers are smoothed and stabilized for the interference voltage values required by the communications system, regardless of the mode in which they are used (see Table 2.1).

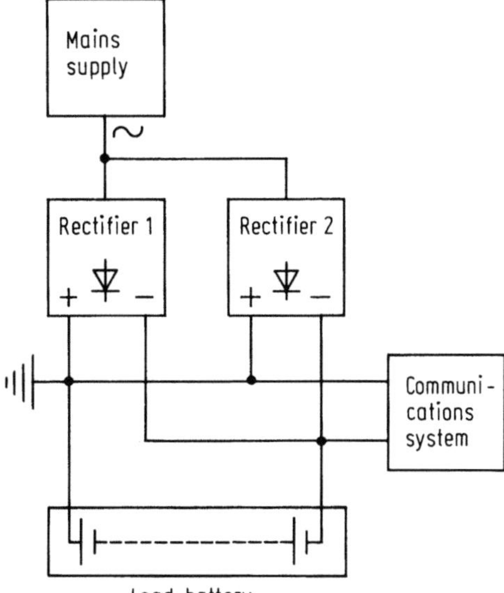


**Fig. 3.4.** Standby parallel mode

In present-day communications systems d.c./d.c converters and inverters whose permissible input voltage range is so great that they can be directly connected in parallel with the battery are increasingly used. Thus, the standby parallel mode can be used without further refinements for systems such as EWSD, since a wider operating voltage tolerance range is permissible here.

The advantages of the standby parallel mode are:

- a longer battery life due to continuous trickle charging;
- the full capacity of the battery and the calculated reserve time are available in the event of power failure or system outages as, in normal operation, it is always fully charged;
- uninterrupted supply of the communications systems with no additional switching devices;
- load surges are to a certain extent compensated for by the battery; thus the battery relieves the communications system of load surges, since it is continually connected in parallel.

*Description of mode of operation.* During normal operation (Fig. 3.5, operating state 1) – with mains operative – the rectifier supplies the communications system. The trickle (float) charge voltage of 2.23 V/cell is applied to the battery. The customary number of lead battery cells for 48 V and 60 V systems is listed in Table 3.1 (see also Table 2.1).

Thus the rectifier devices for example, a 25-cell lead battery, deliver a voltage of about 56 V (tolerance e.g. ±0.5%) and for a lead battery with, e.g., 30-cell a voltage of about 67 V (tolerance e.g. ±0.5%) to the communications system and the battery connected in parallel.

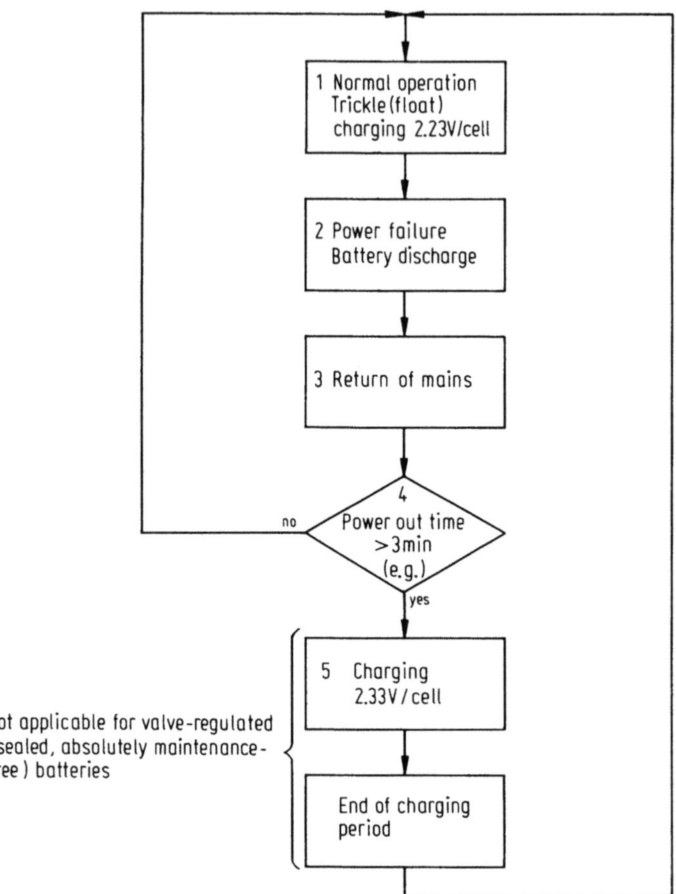


**Fig. 3.5.** Flow chart for standby parallel mode

**Table 3.1.** Number of lead battery cells

|       | 48-V systems | 60-V systems |
|-------|--------------|--------------|
| Cells | 24           | 30           |
|       | 25           | 31           |

On power failure (Fig. 3.5, operating state 2) the battery is discharged. The voltage to the communications system corresponds to the battery voltage (less voltage drop). After a short time the rated battery voltage of 2 V per cell is reached.

On return of the mains (Fig. 3.5, operating state 3) it is checked how long the mains power was off. If the duration of the failure was less than the preset time (of e.g. 3 min), then all rectifiers switch back to normal operation (operating state 1).

If the mains failure lasted longer than the present time ('of e.g. 3 min), then all rectifiers switch back on with higher voltage than normal (Fig. 3.5, operating state 5, charging with 2.33 V/cell).

The required charging voltage can be calculated as follows:

2.33 V multiply with number of cells.

After charging is completed (adjustable, according to system, e.g. up to 24 h) the rectifiers are switched back to normal operation (operating state 1).

There are also systems (e.g. rectifier modules) in which the duration of charging at 2.33 V/cell is made to depend on the duration of the mains failure:

| Main fault duration in minutes: | Duration of charging at 2.33 V/cell in hours (setting options): |
|---|---|
| > 2 | 2 |
| > 4 | 4 |
| > 6 | 6 |
| >10 | 10 |
| >14 | 14 |
| >18 | 18 |
| >24 | 24 |

*Remark*: If the mains fault was smaller than <2 min ······ < 24 minutes, after power return there is no charging 2.33 V/cell but 2.23 V/cell again. There is an automatic switch back from 2.33 V/cell to 2.23 V/cell after the adjusted charging time – of e.g. 2 h is over.

If valve-regulated (absolutely maintenance-free, sealed) batteries are used, there is no switchover to charging at 2.33 V/cell (blocking of the charging characteristic). In addition to the variants of the standby parallel mode shown in Fig. 3.1, this operating mode can also be used in conjunction with solar, wind and hybrid power supply systems.

### 3.3.1 Standby Parallel Mode with Reducing Diodes (Counter electromotive Cells)

If a certain number of battery cells is provided for a communications system according to its minimum permissible operating voltage, which is reached when the battery finishes discharging and below which it must not fall (system-conditioned final voltage $U_{s\,min}$), the resultant supply voltage can become too high for the communications system, if it is connected in parallel. This is especially true of conventional systems with their narrow tolerance ranges.

*Description of mode of operation.* In normal operation the rectifier supplies the communications system. The voltage for a trickle charge of 2.23 V/cell is applied to the battery (Fig. 3.6). As the resultant voltage (2.23 V × number of cells) is too high for the communications system, it is reduced to the desired value by reducing diodes (counter electromotive cells). For this, the voltage drop (in the forward direction) of silicon diodes is used. The bridging contact K13 is

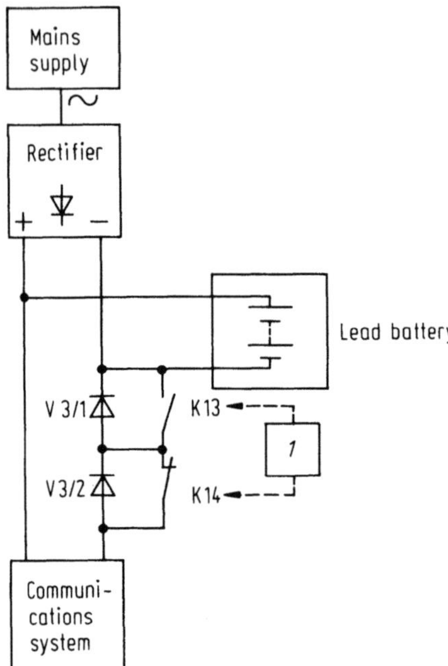


Fig. 3.6. Standby parallel mode with reducing diodes (normal operation). V 3/1 and V 3/2, each a number of silicon diodes connected in series and, additionally, in parallel. They are housed either in the rectifier unit or in the battery switching panels with control. K13 and K14 bridging contractors; *I* Reducing diode control

thereby opened with the group of reducing diodes V 3/1 thus becoming effective. The group of reducing diodes V 3/2 is bridged by the bridging contactor K14 and is thus ineffective. K13 and K14, and thus V 3/1 and V 3/2 as well, are activated voltage-dependently by the reducing diode control.

In the event of a power failure the battery takes over the provision of a power supply to the communications system without interruption. When the battery voltage falls to a certain value the group of reducing diodes V 3/1 are voltage-dependently bridged by the bridging contactor K13 so that the whole battery voltage is now available for supplying the communications system.

On return of the mains the battery is supplied with 2.33 V/cell for rapid recharging (hour range). For this purpose a further group of reducing diodes (V 3/2) is looped in additionally to the V 3/1 group. Both bridging contactors K13 and K14 are opened and therefore both groups of reducing diodes V 3/1 and V 3/2 are effective. In this way the supply voltage for the communications system is kept within the permissible limits, even when in this operating state (e.g. for 48-V systems to 51 V ±2% or 60-V systems 62 V ±2%).

At the end of charging there is a switch back to normal operation with trickle charging.

### 3.3.2 Standby Parallel Mode with d.c./d.c. Converter

Voltages can be stepped up or down by using centralized or decentralized d.c./d.c. converters (Fig. 3.7). This mode of operation can also be used if it is necessary to supply 60-V loads from a 48-V system.

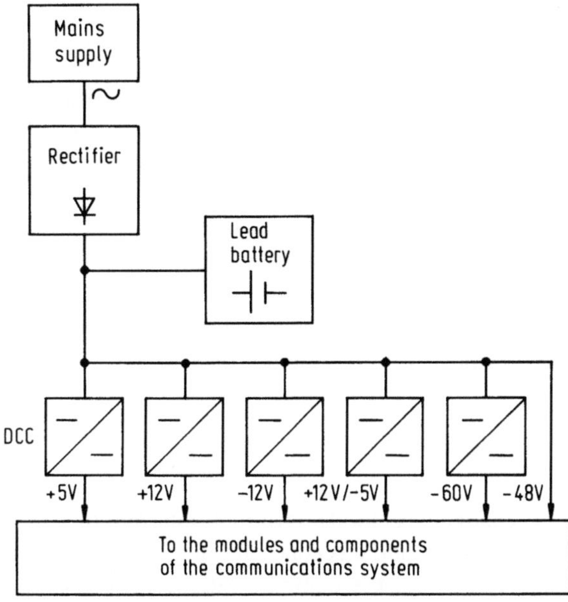



**Fig. 3.7.** Standby parallel mode with d.c./d.c. converters

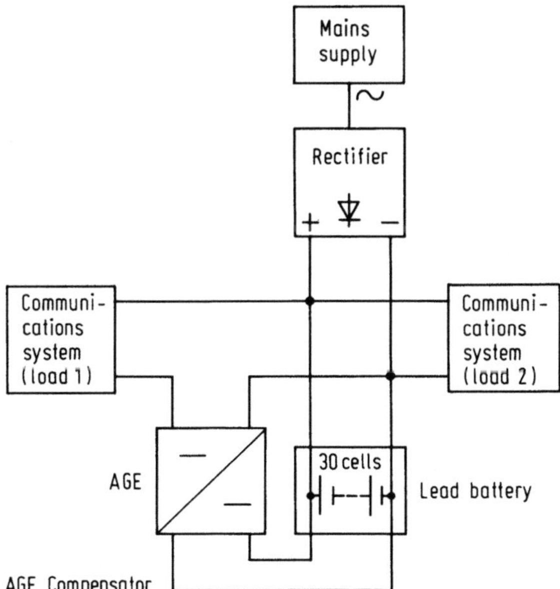



**Fig. 3.8.** Standby parallel mode with compensator for additional and counter voltages (normal operation)

### 3.3.3 Standby Parallel Mode with Compensators for Additional and Counter Voltage

If conventional 62-V direct consumer groups (load 1) requiring low tolerance operating voltages are present alongside the more modern communications systems, compensators for additional or counter voltages 120 A (or 240 A) ±7 V can be inserted (Fig. 3.8). In such a case the compensator is *continually* in operation; it must compensate for voltage differences between the communications system and the battery during charging, trickle charging and discharging of the battery.

The compensator supplies '62-V consumers', load 1 (which require a narrow tolerance range) from a '67-V power supply system' (tolerance range 40–75 V). It can produce a maximum additional voltage of 7 V or a maximum voltage drop of 7 V. This device thus acts as a voltage stepup or stepdown. During charging or trickle charging, the operating voltage, which is too high for the 62-V load, is reduced to 62 V by a two-stage group of reducing diodes. In battery discharging mode several d.c./d.c. converters (single-ended forward converters) connected in parallel augment the falling battery voltage by a maximum of 7 V.

## 3.4 Changeover Mode

In the normal operating state of changeover mode (without interruption) one rectifier supplies the communications system and a second one supplies the battery (Fig. 3.9). The battery is only connected with the communications system in the event of a power failure.

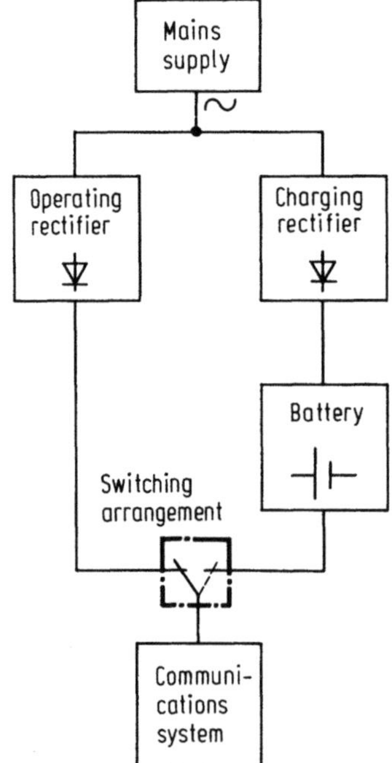


**Fig. 3.9.** Basic representation of the changeover mode

### 3.4.1 Changeover Mode with Battery Tap

The changeover mode with battery tap (Fig. 3.10) is a variant of the changeover mode without interruption and is mainly used in larger power supply sources up to 10 000 A. The importance of this operating mode has declined in recent years since it is predominantly used for supplying conventional communications systems.

*Description of mode of operation.* During normal operation (Figs. 3.10 and 3.11, operating state 1), with the mains available, the system is supplied by rectifier 1 (operating rectifier), while at the same time the trickle charging voltage is applied to the battery by rectifier 2 (charging rectifier). Only so much current flows across the battery that its inherent discharging is covered and the battery thus remains fully charged.

On power failure transfer to battery occurs with the aid of a battery tap (operating state 2). If the mains voltage fails or the supply energy from the rectifier is absent due to some other trouble, the communications system is supplied via tapping diode V7 from a certain number of the battery's cells unit the battery discharge contactor K11 closes.

With the battery tap (Table 3.2) it is possible to transfer to the battery operation *without interruption.*

It is characteristic that the tapping diode V7 is *only* conducting during the time the battery discharge contactor is changing over from normal operation to the battery operation. Under all other operating conditions it is polarized in the reverse direction.

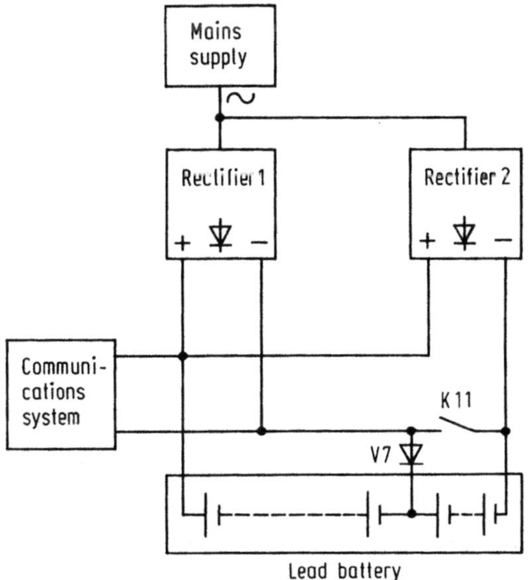


**Fig. 3.10.** Changeover mode with battery tap (normal operation). V7 tapping diode, K11 battery discharge contactor, rectifier 1: operating rectifier, rectifier 2: charging rectifier

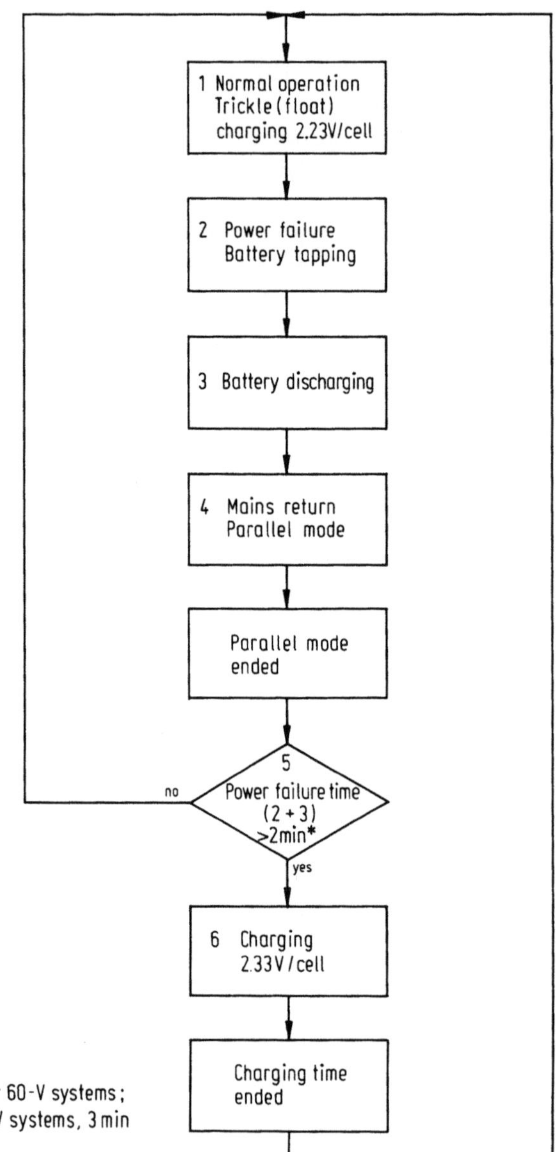


**Fig. 3.11.** Flow chart for changeover mode with battery tap

The position of the battery tap is selected so that even when charging, for example, 26 battery cells with 2.33 V/cell no higher voltage occurs than the d.c. output voltage of rectifier 1. There would otherwise be a passage of current from the battery to the communications system via the tapping diode. On the other hand, there must be a sufficient number of cells up to the tap to ensure that

**Table 3.2.** List of battery tap positions and cell numbers

| System | Battery tap at the |
| --- | --- |
| With 25-cell lead battery | 21st cell |
| With 26-cell lead battery | 22nd cell |
| With 30-cell lead battery | 26th cell |
| With 31-cell lead battery | 26th cell |

when the tap is in use the voltage does not fall below the voltage specified for the communications system.

With almost all systems there is, in the case of power failure, an *immediate* switch to battery discharging via the tapping diode. Current passes through the tapping diode only during an interval of some 100 ms.

An exception are 48 V systems with a 26-cell battery and tapping at the 22nd cell. In these systems the switch is voltage-dependent, i.e. only when the voltage has reached a certain minimum during discharge is the discharge contactor released, thereby changing over to battery, discharging all cells. This prevents the possibility of a too high battery voltage reaching the system.

After the battery discharge contactor K11 has closed, the communications system is supplied from the whole battery (all cells, operating state 3: *Battery discharging*). During a power failure the voltage of the communications system is the same as the battery voltage, i.e. the supply voltage falls in accordance with the battery discharge characteristic depending on the size of the load and duration of discharging. If necessary, a compensator can be used in addition (see section 3.5).

The rectifiers automatically switch on after the mains voltage returns (operating state 4: *Mains return, Parallel mode*). For a certain adjustable transition time (e.g. 30 min to 3 h) rectifiers 1 and 2 are connected in parallel with the battery and communications system, thereby ensuring that the rectifiers are already supplying energy before the battery is separated.

If the *duration of power failure* (2+3) has been *shorter than 2 min* (for 60-V systems) or *shorter than 3 min* (for 48-V systems) and the parallel mode time is at an end, (operating state 5 – decision on *power failure time*), there is a switch back to *normal operation* (see 1). The contact of the battery discharge contactor K11 breaks and the battery is thus separated again from the communications system.

If the *duration of power failure* (2+3) has been *longer than 2 min* (for 60-V systems) or *longer than 3 min* (for 48-V systems) and if the parallel mode time has elapsed, there is a switch to *battery charging* (2.33 V/cell, operating state 6 – *battery charging*).

Here too, once the parallel mode time has expired, the contact of the battery discharge contactor K11 breaks and the battery is thereby separated from the communications system. Rectifier 1, as in the parallel mode (see 4), continues supplying the communications system. Rectifier 2 takes on charging the battery with 2.33 V/cell. After an adjustable time (e.g. up to 24 h) rectifier 2 switches back to trickle charging (2.23 V/cell, see 1). Thus, normal operation is restored.

*Initial charging.*[1] The battery can be charged at up to 2.7 V/cell for the initial (forming, commissioning) charging of uncharged batteries or the subsequent special treatment of batteries after damage (special charging). In this case, e.g. in systems with a battery switching panel, the appropriate battery switch is set to initial charging by hand. One of the rectifiers is also switched from the load bar to the special charging bar.

The communications system must be separated in this special operating mode from the rectifier and battery, which is set up for charging. This is possible in 'manual operation', but not in 'automatic operation' of the rectifier unit.

### 3.4.2 Changeover Mode with Thyristor Contactor

In addition to the changeover mode described in Section 3.4.1, the changeover mode with thyristor contactor (connecting device, KET60) (see Fig 3.12) will be briefly explained in the following.

The changeover mode with thyristor contactor is a variant of the changeover mode without interruption.

When performing a changeover from analog communications systems to digital ones there are frequently conflicts caused by the different permissible voltage tolerance ranges. This means that, for example, normal consumers need a low tolerance range of 62 V, whereas the new systems may have high tolerances of power supply adaption.

In power supply systems for 67 V/70 V and 62 V consumers (see Sect. 3.5) combined standby parallel and changeover mode, the non-interrupted supply of the 62 V consumers is ensured by the 30-cell battery. If there is a voltage drop below 59 V of the 62 V-bar the thyristor contactor links the 67 V-bar to the 62 V-bar so that the battery also takes over the supply of the 62 V-consumer. This linking of the two bars is performed by the thyristor contactor almost without interruption.

At the same time the characteristics of the 67 V/70 V charging rectifiers is lowered to 62 V. The battery voltage drops within a short period of time to a value within the 62 V-tolerance range.

The thyristor is designed for 630 A/d.c. permanent nominal current. In order to avoid the voltage drop via the thyristor, the permanent current is taken over by the contactor switched in parallel to the thyristor. Thus the thyristor is reset. In order to ensure a fault-free separation of the two voltage bars there are several time stages before the resetting so that the load behaviour and the situation of the network and power supply system can be supervised and evaluated electronically.

---

[1] In new systems it is generally not used anymore.

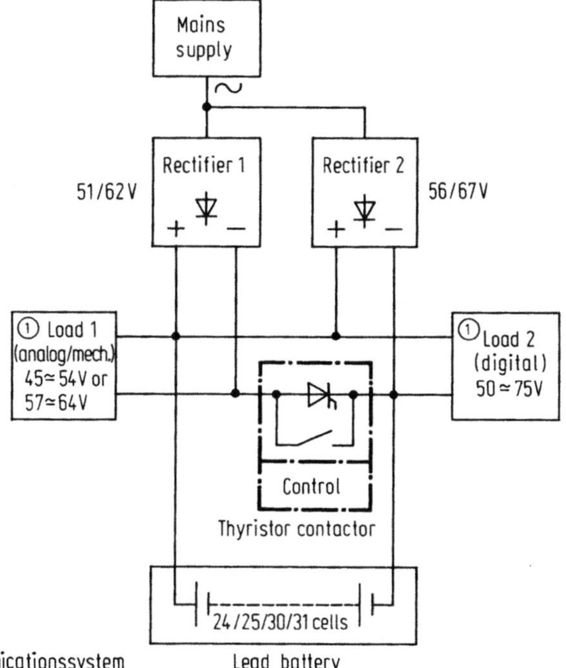


**Fig. 3.12.** Changeover mode with thyristor contactor (KET 60). rectifier 1: operation rectifier, rectifier 2: charging rectifier

## 3.5 Combined Standby Parallel and Changeover Mode with Compensator for Additional Voltage

The combined standby parallel and changeover mode (Figs. 3.13 and 3.14) represents a mixture of the standby parallel mode (Fig. 3.4) and the changeover mode (Fig. 3.10).

Compared with Fig. 3.10 there is also a load 2 in Fig. 3.13 in addition to load 1. Load 1 represents telecommunications equipment requiring a supply voltage with narrow tolerances. Therefore, the changeover mode is used here. Rectifier 1 normally supplies load 1 with constant direct voltage. A compensator is inserted in the supply line so that the supply voltage remains constant within the tolerance range. The compensator (supplied from the battery) supplies an additional voltage of up to 7 V in the case of power failure. A bridging contactor K2 is then opened. The more the battery voltage decreases in the course of its discharging, the more the compensator's additional voltage increases.

The voltage that is fed to or is given off by the battery is always applied to load 2. Load 2 refers to for example d.c./d.c. converters and inverters, which permit wide tolerances in the supply voltage (e.g. 40 to 75 V). For this reason the standby parallel mode is used here.

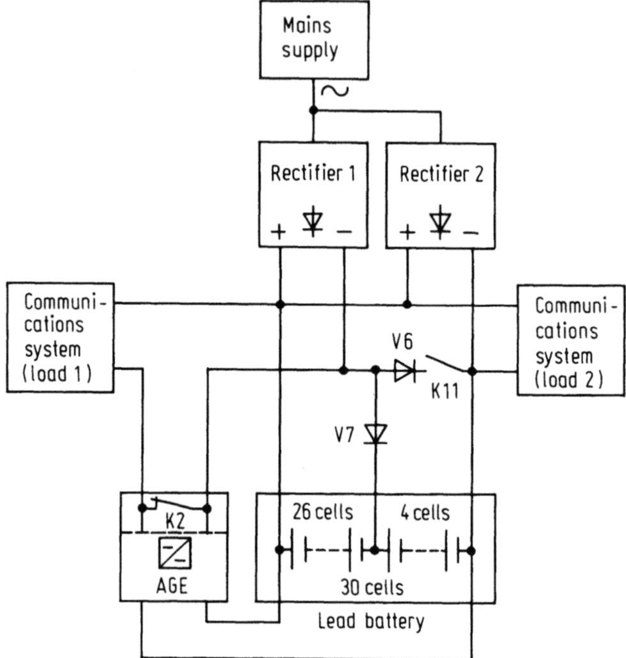



**Fig. 3.13.** Combined standby parallel and changeover mode (normal operation). V6 decoupling diode, V7 tapping diode, K2 compensator's bridging contactor, K11 battery discharge contactor, AGE compensator, rectifier 1: operating rectifier, rectifier 2: operating rectifier and charging rectifier

Compared with Fig. 3.10, Fig. 3.13 also shows the tapping diode V7 and, in addition, the decoupling diode V6. This diode V6 prevents rectifier 1 from participating in the charging of the battery in the operating state of parallel mode. The operation (Fig. 3.14) of the combined standby parallel and changeover mode is described similar to section 3.4.1: changeover mode with battery tap.

## 3.6 Assignment of Operating Modes to Communications Systems

Table 3.3 reviews the standby parallel mode mentioned in section 3.3 and also gives examples of individual series of rectifiers, battery switching panels and control panels, as well as communications systems.

**Fig. 3.14.** Flow chart for combined standby parallel and changeover mode

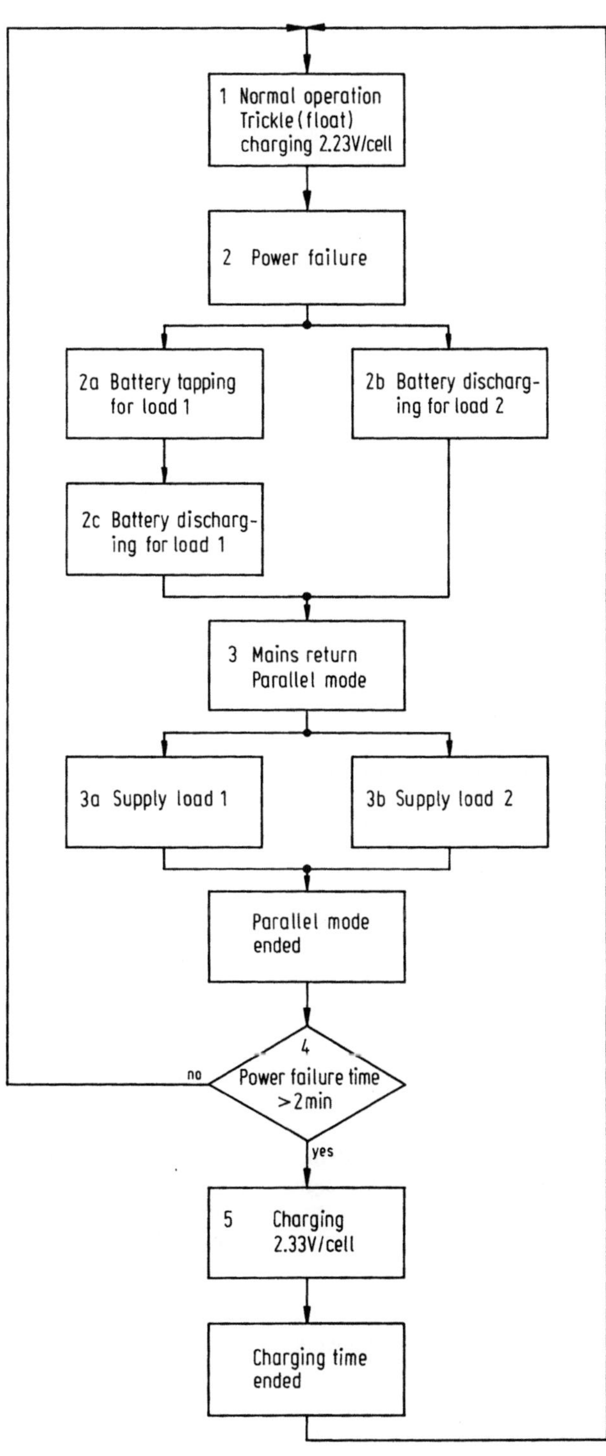
1 Normal operation
Trickle (float)
charging 2.23V/cell

2  Power failure

2a Battery tapping
for load 1

2b Battery discharg-
ing for load 2

2c Battery discharg-
ing for load 1

3  Mains return
Parallel mode

3a Supply load 1

3b Supply load 2

Parallel mode
ended

4
Power failure time
>2min
no
yes

5    Charging
2.33V/cell

Charging time
ended

**Table 3.3.** Assignment of d.c. supply modes to communications systems

| Mode | Standby parallel mode | | | | | |
|---|---|---|---|---|---|---|
| Equipment[a] series/ Equipment's rated current | GR 3/ 100 A | GR 11/ 100 A | GR 12/ 200 A<br>GR 20/ 30, 100 A<br>GR 40/120 A | GR 12/ 200, 500, 1000 A<br>BF/ 1500, 2000 A | GR 10/[b] 50, 100, 200 A<br>BS/200, 400, 600, 1000, 2000 A | GR 10/ 500, 1000 A<br>BF/2000, 3000 A<br>SF |
| Communications system | Power supply systems: rated current/ lead battery | | | | | |
| EMD   48 V | | 100 to 300 A 25 cells | | | | |
| 60 V | | 100 to 300 A 31 cells | | | | |
| ESK   48 V | | 100 to 300 A 25 cells | | | | |
| EWSA 48 V | | 100 to 300 A 25 cells | | | | |
| EWSD 48 V | | | 30 to 5000 A 24 or 25 cells | 800 to 10 000 A 24 or 25 cells | | |
| 60 V[c] | | | | | 200 to 2000 A 30 cells | 2000 to 10 000 A 30 cells |
| EMS   48 V 600/ 12 000 | 100 to 300 A 24 cells | | | | | |
| KN   48 V system | | | 30 to 600 A 24 or 25 cells | | | |

EMD noble metal uniselector motor switch-switching system, ESK cross point switching system, EWSA analog electronic switching system, EWSD digital electronic switching system, EMS electronic modular system (PABX), KN communications network (PABX), GR rectifier, BF battery switching panel, BS battery switching panel with control device, SF control panel.
[a]Source Siemens AG.
[b]GR 10/50, 100 A only for 60-V systems.
[c]Mainly for German Postal Administration Telekom Network.

## 3.7 Further Modes

There are still some d.c. supply modes apart from those discussed in sections 3.1
to 3.5. The most important of these will be presented in sections 3.7.1 to 3.7.4.

### 3.7.1 Changeover Mode with Voltage Gates

A variant of the changeover mode without interruption is the changeover mode
with voltage gates (Fig. 3.15), also called the counter voltage technique. Instead
of the tapping diode there are two groups of voltage gates, V1 and V2, here for
the uninterrupted changeover to battery discharging.

In normal operation with trickle charging rectifier 1 feeds the 60 V communications system with a voltage of 62 V. Rectifier 2 supplies the 30-cell battery
with a trickle charge voltage of 67 V. Voltage gate V2 (charging level 2) in
this operating condition is bridged by bridging contactor K2. Contact K1, on the
other hand, is open.

For this reason a voltage drop of some 5 V now occurs at voltage gates
V1 (charging level 1). Only a small current is flowing. In this way the battery
is prevented from feeding the communications system as long as rectifier 1 is
delivering current.

In the event of a power failure contactor K1 makes contact and connects
the battery to the communications system. The transition from mains to battery

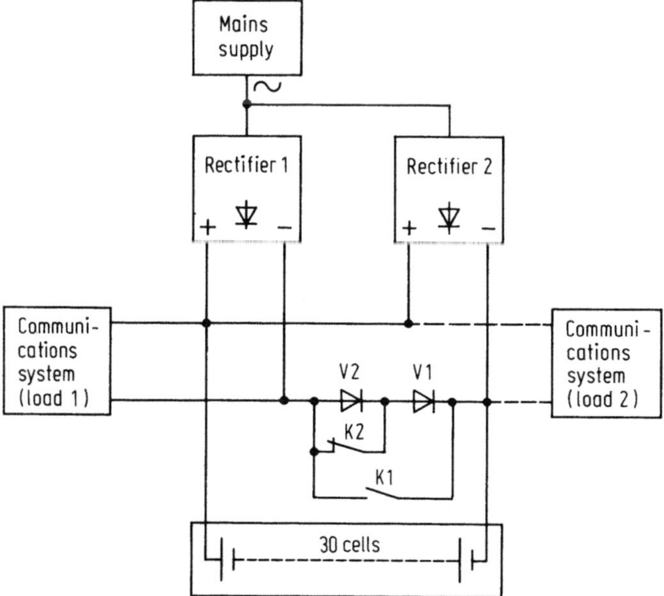


**Fig. 3.15.** Changeover mode with voltage gates (normal operation)

operation takes place without interruption, because the battery is feeding the communications system via voltage gate V1 during the switching time of contact K1.

In the operating condition charging rectifier 2 is changed over to a d.c. output voltage of 70 V. Now both voltage gates V2 and V1 are brought into action by the opened bridging contactors K2 and K1. A voltage drop of some 3 V also occurs at gate V2.

### 3.7.2 Parallel Mode with 'Floating' Charge Method

The parallel mode with 'floating' charge method is a variant of the parallel mode. In normal operation both rectifiers 1 and 2, both batteries 1 and 2 and the communications system are constantly in parallel (Fig. 3.16). A voltage of 2.05 to 2.1 V/cell is fed to the battery (some 63 V for 30 cells).

Sometimes no reducing diodes are required with this mode, as the operating voltage of 63 V lies within the permissible tolerance range of many communications systems. With this method 'equalizing charging' of, for example, the 2.4 V/cell must be done at regular intervals. For this rectifier 1 can be switched to the charging bar. This also applies to battery 2. Rectifier 1 is switched to a higher d.c. output voltage and battery 2 is thus charged. During this time rectifier 2 and battery 1 continue to work normally.

After the equalizing charging of battery 2 there is a switch back to normal operation.

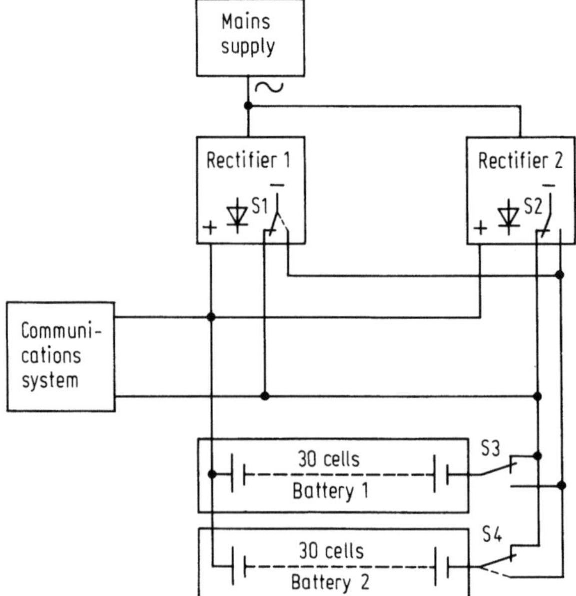


**Fig. 3.16.** Parallel mode with 'floating' charge method (normal operation)

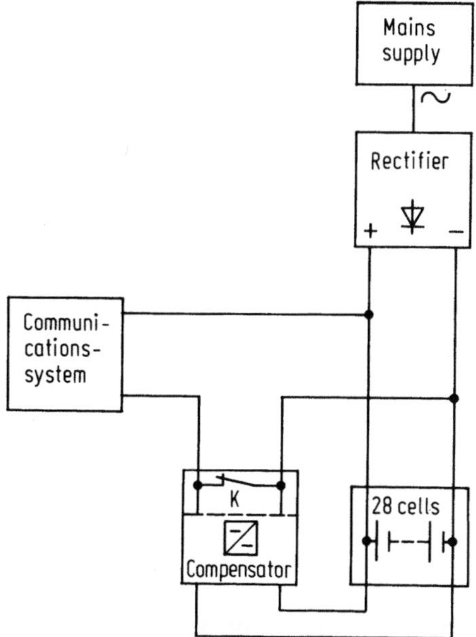



**Fig. 3.17.** Parallel mode with reduced number of battery cells and compensator (normal operation)

### 3.7.3 Parallel Mode with Reduced Number of Battery Cells and Compensator

In this mode a 28-cell battery is used, for example, instead of one with 30 cells. The resultant trickle charging voltage lies within the tolerance range of most communications systems (Fig.3.17).

In normal operation the rectifier supplies the communications system and the 28-cell battery with the trickle charging voltage. The compensator in this mode is bridged by contactor K.

In the event of a power failure the compensator is cut in by breaking contact K. The lower the battery voltage becomes, the more the (automatically controlled) additional voltage rises. Thus the voltage at the communications system remains constant even when the battery is discharging (e.g. with the 60 V system at about 61 V).

### 3.7.4 Parallel Mode with End Cells

The 'end-cell technique' (cubicle technique) is a variant of the parallel mode (Fig. 3.18). In normal operation the 'main rectifier' supplies the communications system and in parallel with it the 23 'main cells' of, for example, a 26-cell battery with trickle charging voltage via the made K1 contact (about 51.3 V). The 'end-cell rectifier' has only to supply the three end cells (in the case of trickle charging with some 6.7 V). In the event of a power failure contactor

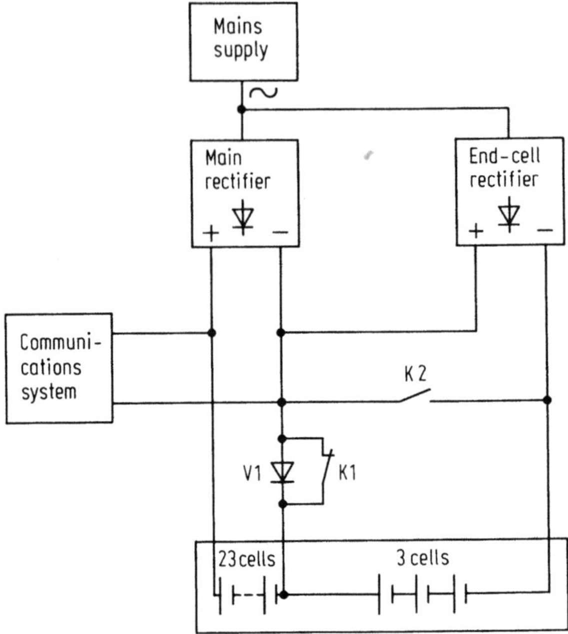


**Fig. 3.18.** Parallel mode with end cells (normal operation)

K1 opens and K2 closes, which thereby switches to battery discharging without interruption.

For charging the main battery cells the charging voltage must lie within the tolerance range of the communications system.

# 4 Operating Modes of an Alternating Current Power Supply System

The operating modes used for a.c. loads in communications systems are listed in Table 4.1.

## 4.1 Mains Mode

In the mains mode the loads are supplied directly from the public mains (Fig. 4.1). There is no standby power supply.

## 4.2 A.C. Changeover Mode

In the event of a power failure the a.c. changeover mode provides a switch to an a.c. source standing by or running jointly (Fig. 4.2). Return of the mains results in switching back and renewed supply from the mains.

A distinction is made between standby power supplies operating with an interruption $> 1$ s and those operating with an interruption $< 1$ s.

### 4.2.1 A.C. Changeover Mode with Interruption (> 1 s)

In general, about 10 to 15 seconds are required for changeover from mains to standby mode.

### 4.2.2 Standby Power Supply Systems

*Emergency power supply.* The standby power supply system (standby generating set) is an alternative system 'available without limit on time' with a generator and internal combustion engine (Figs. 4.3 and 4.4).

In normal operation the loads are supplied directly from the mains. In the event of a mains failure there is a changeover to the standby power supply system. For this it is necessary to start the internal combustion engine which is coupled with a generator. Once the rated speed has been reached the load is further supplied with alternating voltage via the generator. After return of the mains there is a switch back to mains operation.

Automatic control and monitoring comprises, for example:

- control of the mains and generator contactors,
- automatic starter,

**Table 4.1.** Operating modes of an a.c. power supply system

| Mains mode (without standby power supply) | |
| --- | --- |
| A.C. changeover mode with interruption $> 1$ s | Standby power supply systems<br>Standby power supply available without limit on time with generator and internal combustion engine<br>Starting mode<br>Mains-independent island power supply<br>Alternating continuous mode |
| A.C. changeover mode with interruption $< 1$ s | Rapid standby systems<br>Standby power supply available without limit on time with generator and internal combustion engine<br>Joint mode |
| | Static standby power supply systems<br>Standby power supply available with limit on time with inverter and battery[a]<br>Starting or joint mode |
| A.C. changeover mode without interruption | Immediate standby systems<br>Standby power supply available without limit on time with generator, electric motor, flywheel and internal combustion engine<br>Continuous mode |
| | Standby power supply systems with rotating converter<br>Standby power supply available with limit on time with generator, d.c. motor and battery[a]<br>Continuous mode |
| | Static standby power supply systems with inverter (with revert-to-mains unit)<br>Standby power supply available with limit on time with inverter connected to fail–safe d.c. power supplies with revert-to-mains unit[a]<br>Continuous mode |
| | Static standby power supply systems with inverter (without revert-to-mains unit)<br>Standby power supply available with limit on time with inverter connected to fail–safe d.c. power supplies without revert-to-mains unit[a]<br>Continuous mode |
| | Static uninterruptible power supply (UPS) systems<br>Standby power supply available with limit on time with rectifier, inverter and battery[a]<br>Continuous mode |

[a]If necessary, additional standby power supply system. This then ensures an alternative power supply without any limit on time.

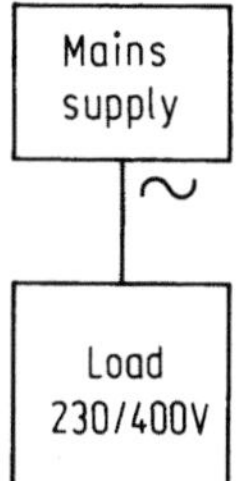


**Fig. 4.1.** Mains mode

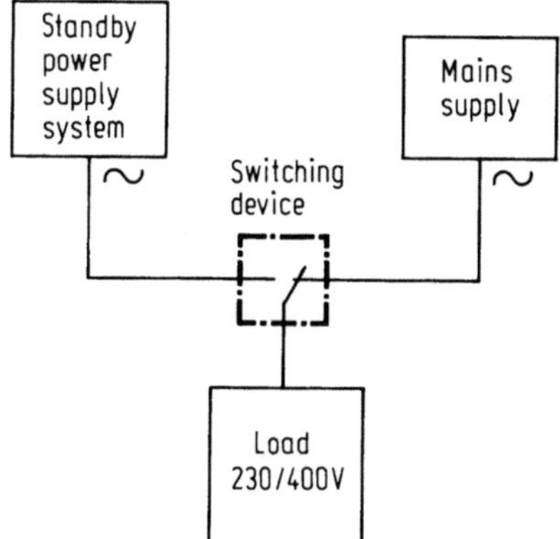


**Fig. 4.2.** A.C. changeover mode

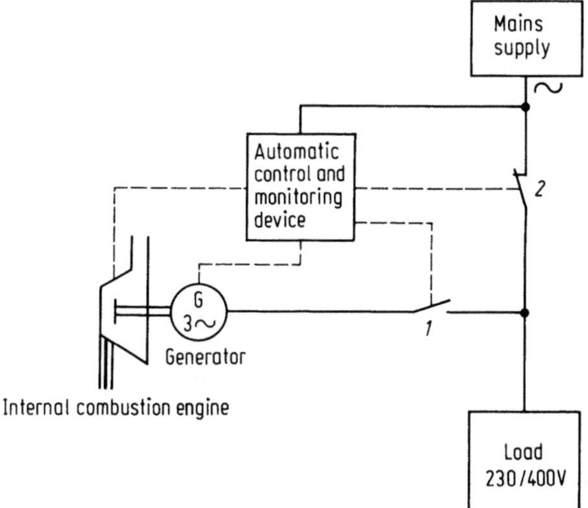


**Fig. 4.3.** Standby power supply system. 1 Generator contactor, 2 Mains contactor

**Fig. 4.4.** Standby power supply system, 1250 kVA (Photo by courtesy of Siemens AG)

- monitoring mains voltage,
- monitoring three–phase sequence (clockwise rotating field),
- monitoring generator voltage,
- monitoring generator frequency,
- monitoring generator start-up failure,
- monitoring lubricating oil pressure failure,
- monitoring overheating,
- monitoring overcurrent,
- monitoring lubricating oil level (too low?),
- monitoring starter battery for overvoltage,
- monitoring circuit breakers,
- monitoring fuel service tank level (too low?),
- monitoring fuel service tank level (too high?),
- monitoring reserve fuel tank level (too low?),
- remote signalling,
- indication (e.g. by LEDs) and
- operation (e.g. by key).

The control system can be set to any of the four operating modes: 'automatic control', 'manual control', 'test' and 'off'.

*Island power supply.* If no public mains is available, it is possible to have two (Fig. 4.5) or three generators working alternately (e.g. each generator for 24

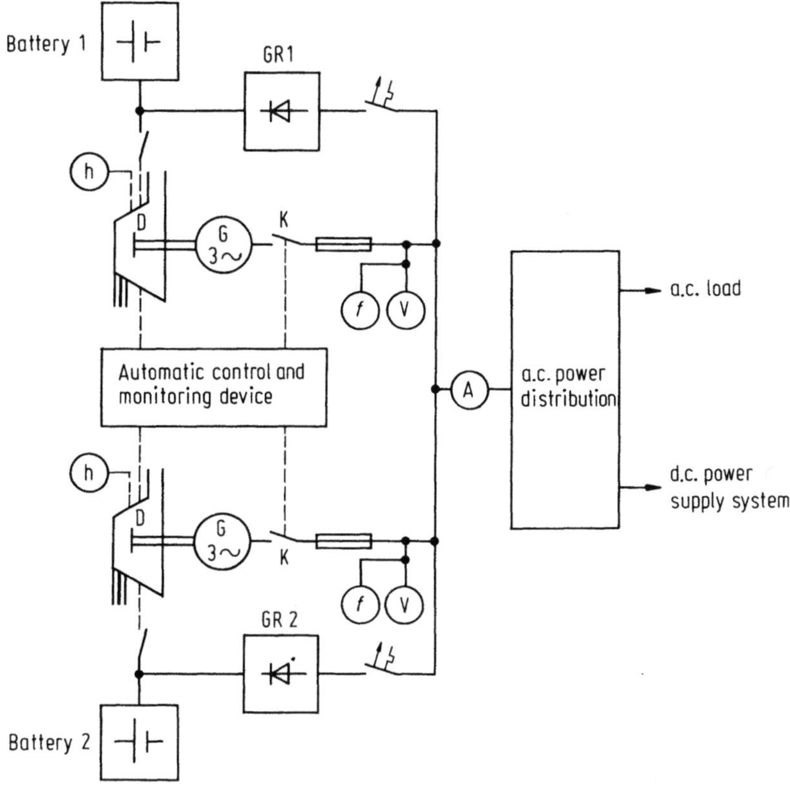


**Fig. 4.5.** Mains-independent island power supply system with diesel generators. Batt. 1,2 Starter battery, GR 1,2 charging rectifier for starter battery, h elapsed time meter, K generator contactor

hours). With the exception of the diesel generators, all components and modules are housed in a switching cabinet. If an additional public mains is provided in the future, the switching apparatus can be designed to accommodate it.

### 4.2.3 A.C. Changeover Mode with Interruption (< 1 s)

With the aid of additional facilities operation can be maintained with a rapid changeover to the alternative a.c. supply 'without impairment of the load'.

*Rapid standby system.* An alternative power supply system with a generator and internal combustion engine 'available without limit on time ' is known as a rapid standby system. It enables loads to be supplied which permit interruptions of max. 0.2 to 0.3 s in the event of a power failure (Fig. 4.6).

In normal operation the mains supplies the load and an asynchronous motor. The latter keeps the synchronous generator and flywheel at a speed just below

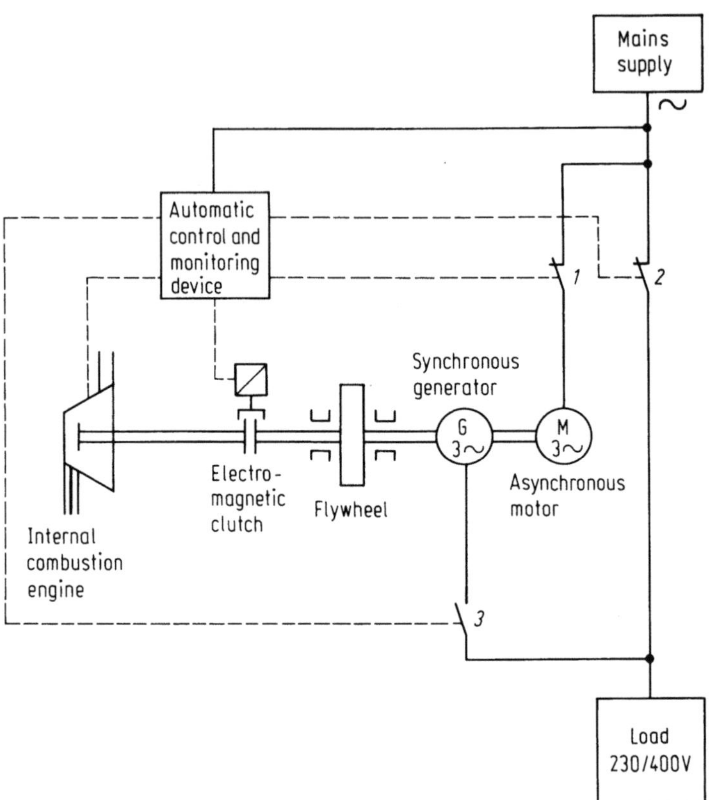


**Fig. 4.6.** Rapid standby system

the generator's rated speed. The flywheel can be separated from the internal combustion engine by means of an electromagnetic clutch.

In the event of a power failure a control system breaks contacts 1 and 2, makes contact 3 and starts the internal combustion engine. The flywheel is connected to the engine by means of the clutch. The energy stored in the flywheel is used to start the engine. In doing so the flywheel is also driving the synchronous generator. The engine now running at its rated speed together with the synchronous generator takes over and supplies the load. After return of the mains there is a switch back to normal operation.

*Static standby power supply system.* The static standby power supply system is considered as one with an inverter and battery that is 'available for a limited time'.

In normal operation the mains supplies the load directly and the battery is on trickle charging (Fig. 4.7).

In the event of a power failure the battery is switched to the inverter (*1*). This ensures a continued supply to the load (*2*). The supply is interrupted for a short time with any switching operation.

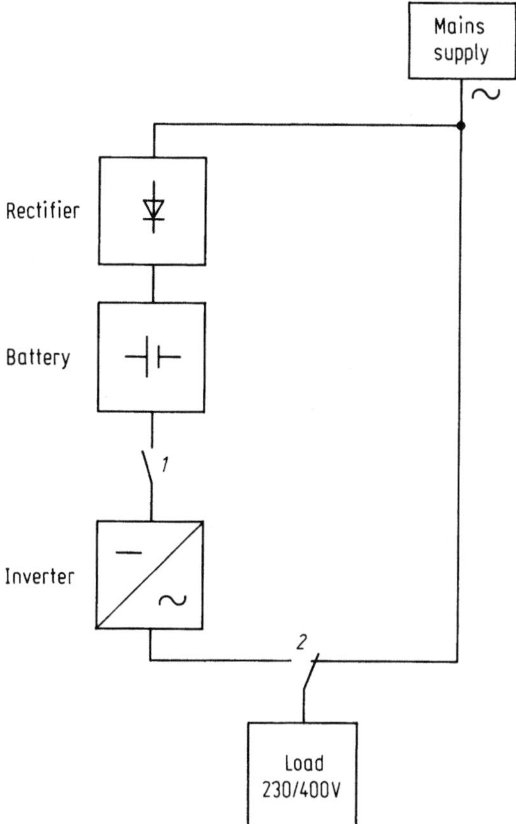


**Fig. 4.7.** Static standby power supply system

## 4.3 Uninterruptible a.c. Changeover Mode

If this operating mode is used, the load can be supplied without interruption (see section 1.7, Fig. 1.7).

One block of an uninterruptible power supply (UPS) can consist of:

– motor, generator, flywheel, internal combustion engine, or
– rectifier, battery, d.c. motor, generator, or
– rectifier, battery, d.c./a.c. inverter (static UPS systems, see Fig. 4.13).

For a *single block system without revert-to-mains unit* there is neither passive nor active redundancy.

If a *single block system* is equipped with a *revert-to-mains unit*, passive redundancy is present. In the event of trouble with the uninterruptible power supply block there is a switch back without interruption to the mains using the revert-to-mains unit.

In the *multi-block parallel system* with passive and active redundancy, the passive redundancy is augmented by an active one. There is active redundancy when the number of UPS blocks connected in parallel consists of at least one block more than is necessary for the load (see Sect. 1.7, Fig. 1.7).

### 4.3.1 Immediate Standby System

The immediate standby system is often called a flywheel diesel converter system. It represents an alternative power supply without interruption and without any limit on time with a generator, electric motor, flywheel and internal combustion engine.

In normal operation the mains supplies a synchronous or asynchronous motor via the closed contact. The motor drives the synchronous generator and also keeps the flywheel at the rated speed (Fig. 4.8). The synchronous generator supplies alternating voltage to the load. The electromagnetic clutch is out and the flywheel is thus separated from the engine.

In the event of a power failure an automatic control system breaks the contact and lets in the clutch. The flywheel continues driving the synchronous or asynchronous motor and at the same time gets the internal combustion engine started. There is now a link between the engine and generator. The alternative power supply is ensured and is not limited in time. After return of the mains there is a switch again (without interruption) to mains operation.

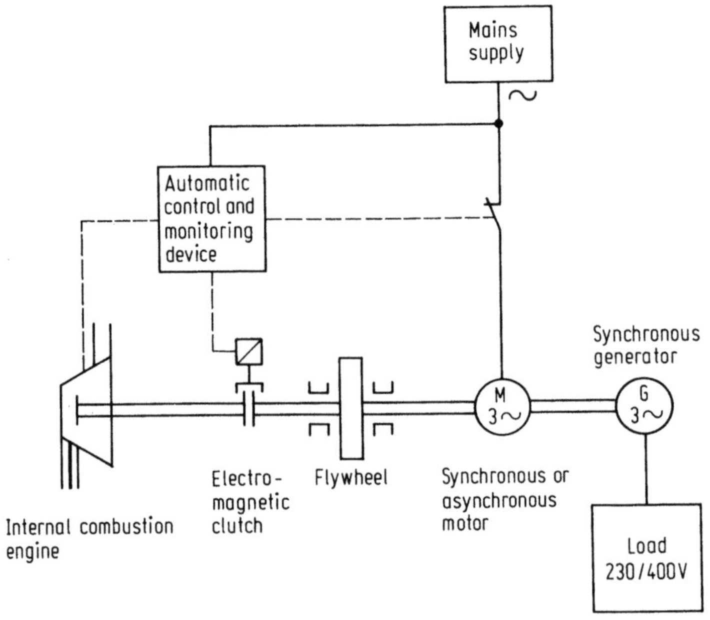


**Fig. 4.8.** Immediate standby system

### 4.3.2 Uninterruptible Standby Power Supply System with Rotating Converter

These systems are uninterruptible standby power supplies, available for a limited time with a generator, d.c. motor, rectifier and battery. Shown in Fig. 4.9 as an example is a single-block uninterruptible power supply with rotating converter in the battery standby parallel mode, with passive redundancy through a revert-to-mains unit. In normal operation the mains supplies a rectifier which drives a d.c. motor and at the same time feeds the battery. The d.c. motor is linked with a generator. The latter supplies the load via the cutoff unit.

In the event of a power failure the battery takes over the supply without interruption. The d.c. motor is then driven from power supplied by the battery instead of from the rectifier.

This is a standby power supply whose availability is limited in time by the capacity of the battery.

If trouble arises with the rectifier, battery, d.c. motor or generator, the contact of the cutoff unit is broken and there is a switch back without interruption

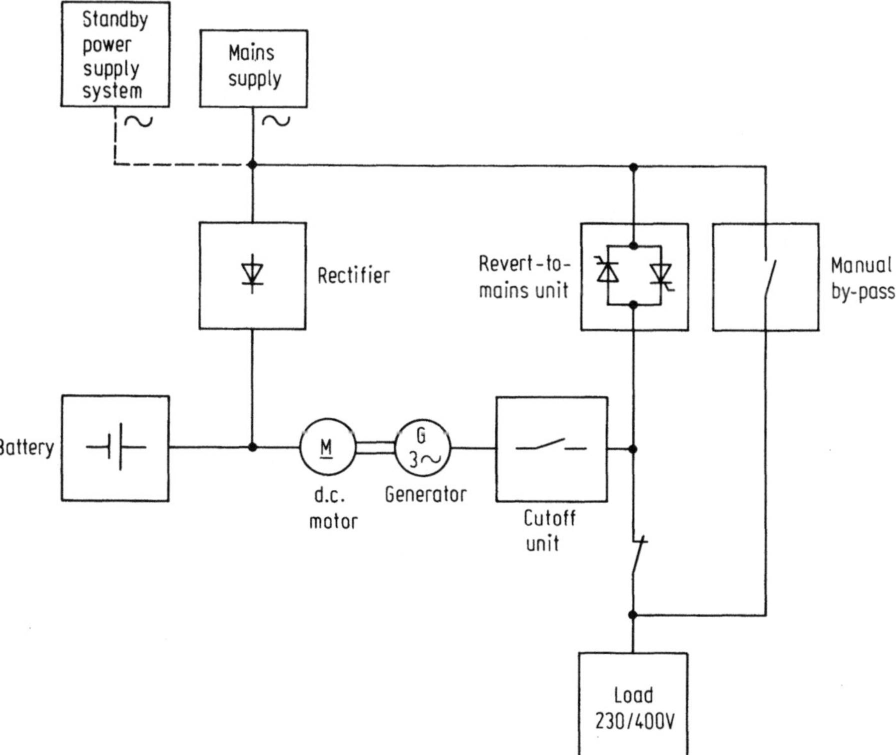


**Fig. 4.9.** Single-block uninterruptible power supply system with rotating converter in battery standby parallel mode with passive redundancy through revert-to-mains unit

to mains operation via the revert-to-mains unit. With this system there is also a manual by-pass as an additional safety measure. With this setup the whole uninterruptible power supply system can be isolated.

There can also be a standby power supply system in addition to the alternative power supply which is limited in time. In the event of a power failure there is a changeover to the standby power supply system. With this variant the battery capacity needs only to be designed for short bridging times. If there is also a standby power supply system, then there can of course be an alternative power supply without any limit on time.


### 4.3.3 Static Standby Power Supply Systems with Inverters Connected to Fail-safe d.c. Power Supplies and with Revert-to-mains Unit

*Modes of Operation.* The inverter can be operated in starting, joint or continuous mode.

- In *starting mode* the load is supplied from the mains. The inverters are available as standby, but must first start up when a power failure occurs, resulting in a brief transfer delay (cold standby).
- In *joint mode* the load is supplied from the mains. The inverters operate in idle mode, are synchronized with the mains, and act as standby (hot standby).
- In *continuous mode* the load is supplied from inverters running synchronous to the mains. The mains acts as a standby. There is no interruption during switchover (no break transfer).

As an illustration the frequently used continuous mode is as shown in Fig. 4.10. The advantages of the continuous mode of operation are:

- it is more advantageous for electrical components, since these are continually carrying current,
- there are two power supplies available, the inverter and as standby the mains as passive redundancy and
- there are no transfer delays, i.e. the operation corresponds to that of a UPS system.

In *normal operation* (Fig. 4.10) the mains supply is present and feeds the rectifier, which in turn feeds the inverter and battery. The electronic contactor T 206 is closed and T 207 is open. The inverter delivers a voltage of, for example 225 V/50 Hz to the load. There are also systems for three-phase a.c. power (e.g. 400 V).

In the event of *power failure* the battery within the fail-safe d.c. power supply will be discharged, i.e. the inverter will continue to be supplied without interruption. The load continues to be supplied from the inverter, switched to internal clock, via the thyristor module. Reversion to mains is blocked. The same operating mode would be present if the mains were available but the permissible tolerance for voltage or frequency were exceeded. In case of *inverter failure* the

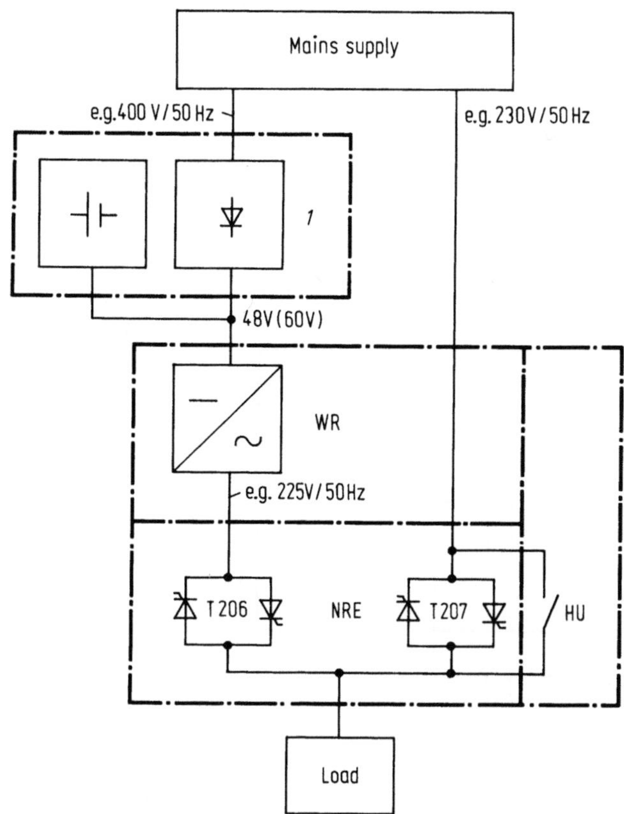


**Fig. 4.10.** System with inverter (e.g. 7.5 kVA) and with revert-to-mains unit (normal operation). HU Manual bypass, WR inverter, NRE revert-to-mains unit, T 206 thyristor module 'inverter', T 207 thyristor module 'mains' 1 Fail-safe d.c. power supply with rectifier and battery

thyristor module switches the load to the mains supply, provided this is available within the required tolerance. With the manual bypass the system can be switched to a voltage-free condition for maintenance.

### 4.3.4 Static Standby Power Supply Systems with Inverters Connected to Fail-safe d.c. Power Supplies and without Revert-to-mains Unit

Systems are supplied without revert-to-mains unit if the public mains is either not frequently available or does not generally satisfy the tolerance requirements. The same is true if the frequency required by the load deviates from that of the mains, or if, to save costs, the load to be supplied is not classified as particularly important to the operations. These systems may be supplied with or without a manual bypass. Fig. 4.11 shows a system without manual bypass.

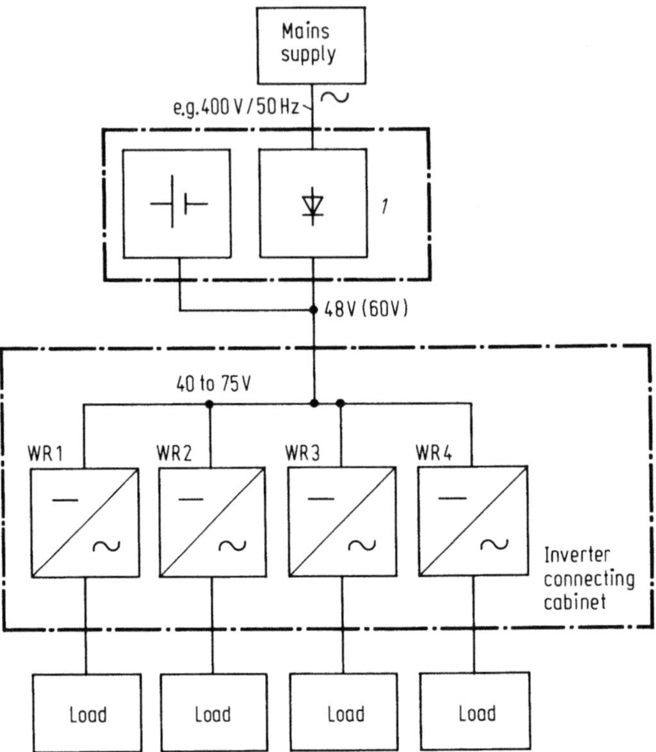



**Fig. 4.11.** Systems with inverters (without revert-to-mains unit). 1 Fail-safe d.c. supply with rectifier and battery, WR 1,2,3,4 inverter, e.g. 2.5 kVA, Load: 220 V/50 Hz (60 Hz)

In this system each inverter supplies an individual load group. Here there is no synchronizing. If an inverter fails, the corresponding load group can no longer be supplied. After this inverter is replaced by a spare unit, the load is again supplied.

### 4.3.5 Static Uninterruptible Standby Power Supply Systems

Sophisticated equipment and systems are now in use in factories and offices everywhere – backed-up by uninterruptible power supplies (UPS). They ensure reliability in all sectors of the economy and public institutions, wherever information and communication are at a premium and the slightest power interruption will have series consequences.

This system is considered as UPS, limited in time, with a rectifier section, inverter section and battery. Static UPS systems are built up with batteries and, if needed to further raise redundancy, with standby power supply systems. Such systems are found, for example, in ground stations for telecommunications satellites, in the data switching and data processing systems.

Many of these systems require an uninterrupted supply of power if data loss and system crashes are to be avoided. UPS systems ensure that voltage fluctuations, brief interruptions and failures of the mains (Fig. 4.12) do not affect the load, irrespective of whether the faults in question are irregularities in the millisecond range or power failures for minutes at a time. Only when the availability of all components is assured, can full benefit be obtained from the equipment and systems. Power supplies to critical loads can be reliably guaranteed only when a UPS system is an integral part of the whole set-up. And what applies to DP equipment and systems applies all the more to control systems and data transmission networks, automated production lines, communications systems, protection systems, etc.

When the mains supply is present, the load is supplied with constant voltage and frequency through the rectifiers and inverters. If the power fails, the battery immediately delivers power to the inverters. If an inverter fails, a revert-to-mains unit switches the load back to the mains without delay. Thus the UPS system provides an uninterruptible supply of power to the load.

The power ratings of UPS systems range from $< 0.5$ kVA up to 500 kVA. If eight of the largest single devices are connected in parallel, 4.000 kVA may be attained. The efficiency of the devices can be as high as 93 %. Because of a short settling time they also permit surges in load of up to 100 % of the rated power, while maintaining all tolerances.

The substantial reliability of such systems, can be further increased by raising the level of redundancy. It is well known that an improvement in active redundancy is achieved if the number of UPS blocks connected in parallel is increased by at least one block more than is normally for the load.

In normal operation the system works with all UPS blocks in the partial load range. It is only after the failure of a block that the remaining ones are fully loaded.

Passive redundancy is achieved with the aid of the revert-to-mains unit, which in the event of trouble directly switches the mains to the 'safe' (fuse) bar and thereby to the loads (cf. Fig. 1.7).

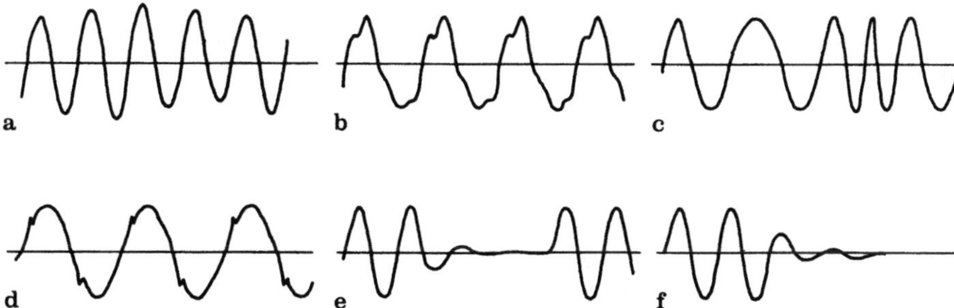

**Fig. 4.12a–f.** Examples of mains disruption: **a** voltage variations, **b** distorted voltage waveforms, **c** frequency variations, **d** superimposed noise voltages, **e** short interruptions, **f** mains failures. (Source: Siemens AG)

Thyristors or power transistors (phase-angle – or pulse-width control) are used as final control elements in the devices rectifiers and inverters. They are available in a variety of designs and series. UPS systems can also be employed as frequency converters.

*Applications.* The standby UPS systems provide an uninterruptible supply of single or three-phase power at, for example, 230 V or 400 V to essential loads. The frequencies involved are mostly 50 or 60 Hz. For example:

– communications systems,
– ground stations for telecommunications satellites,
– data switching systems,
– memories,
– hard copy terminals,

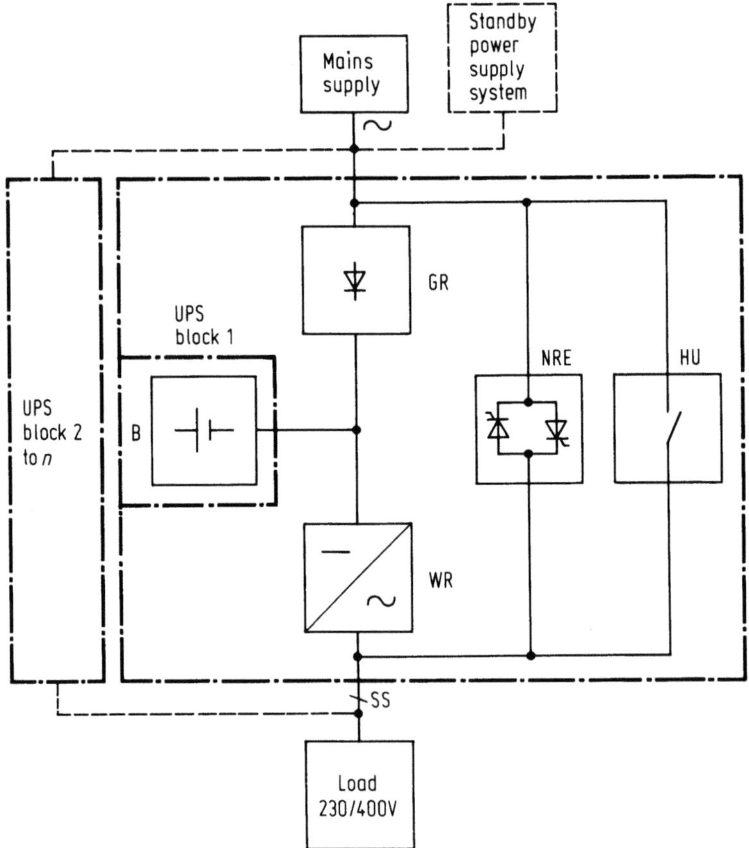


**Fig. 4.13.** Static uninterruptible power supply system in parallel mode. SS safe (fuse) bar, GR rectifier, B battery, WR inverter, NRE revert-to-mains unit, HU manual bypass

– printers,
– security systems and
– data processing systems.

*Operating Mode.* UPS systems are generally operated in the *continuous mode.* Figure 4.13 shows an example of a *two-block* or *multi-block* system.

A UPS system comprises the following components: rectifier (GR), inverter (WR), revert-to-mains unit (NRE), (also called static transfer switch or static bypass switch SBS) and battery (B), (Fig. 4.14).

In normal operation the rectifier section is supplied from the mains; within the uninterrupted power supply block this has to supply the inverter section and the battery. The rectifier converts the single- or three-phase power of the mains to d.c. Out of this the inverter produces a single- or three-phase a.c. voltage of constant amplitude and frequency. The inverter is connected with the safe (fuse) bar and thereby with the load. In the event of a power failure or failure of the rectifier, the battery, which is connected in parallel with the intermediate d.c. circuit, supplies

**Fig. 4.14.** Static uninterruptible power supply system 120 kVA, interior view, covers removed. (Photo by courtesy of Siemens AG)

the power to run the inverter, thus ensuring a completely uninterruptible supply of power to the load. The battery capacity is usually selected for bridging times of 10 to 60 min. After return of the mains, the rectifier automatically resumes operation, supplying the inverter while simultaneously recharging the battery.

In case of excessive overload or malfunction of the inverter, the revert-to-mains unit switches the load back to the mains without interruption. When the trouble has subsided, the inverter resumes its function of supplying the load. The revert-to-mains unit is standard equipment if the mains frequency and load frequency are the same. A built-in manual bypass enables the system to be fully decoupled for service.

Where necessary an additional power supply provides an alternative standby power supply available without any limit on time. The revert-to-mains unit must not respond in this mode, as otherwise the standby power supply system would be switched directly to the safe bar, which could result in frequency deviations outside the load's tolerance range.

# 5 Public Mains

It is important to know the characteristics of the public power supply system (distribution network) when operating telecommunications power supply systems. This chapter explains the factors to be considered and the line-side conditions for which the equipment is designed.

## 5.1 Type of Voltage

Rectifiers up to about 25 A are usually built for connecting to single-phase alternating voltage (e.g. 230 V/50 Hz) and for currents from 25 to 1000 A for connecting to three-phase alternating voltage (e.g. 400 V/50 Hz). The equipment can be adapted to different alternating supply voltages or supply frequencies.

Figure 5.1 shows the voltages and the designations of the conductors with regard to the terminals at the rectifier.

The voltages of outer conductors L 1, L 2 and L 3 of the three-phase alternating voltage have a phase shift of 120°. The voltage between two outer conductors is called '*delta voltage*' (400 V), while that between an outer conductor and the neutral conductor N is called '*star voltage*' (230 V).

The amplitude of an alternating voltage is indicated by its root-mean-square (r.m.s.) value. This is lower than the peak value by a factor of 1.4.

## 5.2 Tolerances of the a.c. Supply Voltage and Supply Frequency

According to the standard VDE, power converters remain serviceable when, in the long term, the alternating voltage fluctuates between 90 and 110% of the rated voltage (e.g. 230 or 400 V).

Rectifiers are normally designed for *fluctuations in the supply* ranging from $-25$ to $+15\%$, i.e. as long as the supply voltage remains within this tolerance, the equipment's control system can keep the output constant at, for example, $\pm0.5\%$ or 1% (static tolerance range).

Apart from the above long-term voltage changes there also occur short-term, non-periodic undervoltages and overvoltages (voltage surges) in the mains supply. Atmospheric discharges can cause overvoltages up to 100 times the normal value for the mains voltage.

Such overvoltages are limited to permissible values in the low-voltage distribution system by means of surge dissipators. In addition, the varistors and TAZ

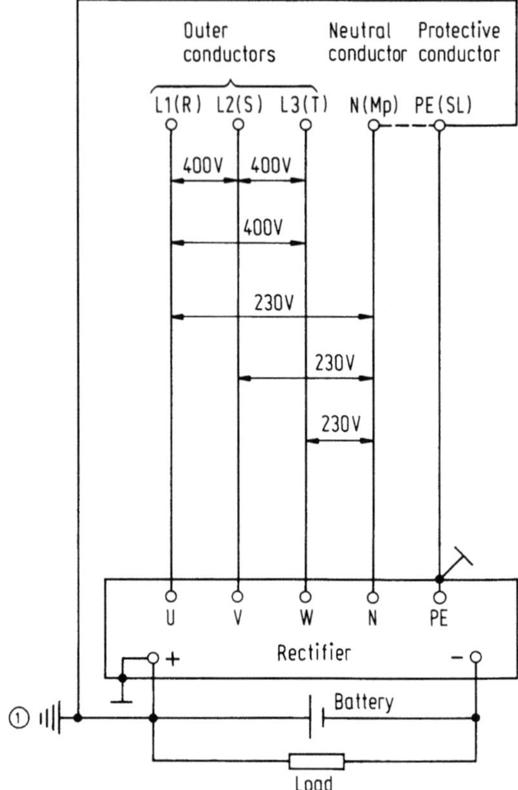


**Fig. 5.1.** Conductor designations and voltage particulars

(transient-absorption zener) diodes partly built into the equipment provide further limiting.

*Frequency fluctuations* in larger grid systems are normally slight. The control system of rectifiers, even in the case of frequency changes within the tolerance range of $\pm 5\%$ of the rated supply frequency $f_N$ (generally 50 Hz or 60 Hz), keep the output voltage constant within the range of, for example, $\pm 0.5\%$.

According to the standard VDE, power converters must remain serviceable if the frequency of the public power supply system varies by up to $\pm 1\%$ from the rated value.

## 5.3 Wave Shape and Distortion Factor of the a.c. Supply Voltage

Loads connected to the mains, e.g. rectifiers, generate harmonic oscillations and produce feedback to the mains. Harmonic waves also occur in the mains network itself. The following applies to the mains to ensure error-free working of power converters:

- the relative harmonic content of the supply voltage must be at most 10%,
- for each harmonic the limit values specified by VDE 0160 may not be exceeded and
- the largest periodic momentary value for the alternating supply voltage may not lie more than 20% above the respective peak value for the fundamental.

The *distortion factor, k*, represents a measure for the harmonic content:

$$k = \frac{\text{r.m.s. of the harmonics}}{\text{r.m.s of the alternating quantity}}$$

With ideal sinusoidal behaviour of the alternating supply voltage the distortion factor $k = 0$.

What harmonic currents occur depends on the rectifier's pulse number, $p$. They have the frequencies $v \cdot f$ with the ordinal number of the harmonic

$$v = k \cdot p \pm 1(k = 1,\ 2,\ 3,\ldots)$$

where $v > 1$.

With non-controlled power converter circuits the phase position of the harmonic currents is fixed, while with controlled ones the harmonic currents depend on the control angle, $\alpha$.

Every periodic oscillation can be represented as the sum of component sinusoidal oscillations (Fig. 5.2).

The ideal square-wave current curve, which applies to a certain load in a sinusoidal network, contains a great number of components of different frequencies. Table 5.1 contains all odd ordinal numbers $v$ from 1 to 25. The supply frequency of, for example, 50 Hz is the fundamental (basic oscillation 100%) and receives the ordinal number $v = 1$. The harmonics oscillate at integral multiples of the fundamental frequency. The higher the frequency, the lower the amplitude of a harmonic, so that its disturbing influence also decreases.

The magnitude of the harmonic current is independent of the pulse number $p$ and is the same in all circuits (if the frequency in question occurs at all in the higher-pulse circuits). For example, the eleventh harmonic (ordinal number $v = 11$) with the frequency 550 Hz has with the three pulse numbers $p = 2, 6$ and 12 the same magnitude as the ideal harmonic current, namely 9.09%. It is therefore beneficial to raise the pulse number $p$, because the strongest current harmonics with low number and frequency do not then occur.

In thyristor-controlled rectifiers rated at up to about 25 A single-phase alternating current, double-pulse converter circuits are normally used in a semi-controlled single-phase bridge circuit. In thyristor-controlled rectifiers rated from about 25 A to 200 A the fully controlled (six-pulse) three-phase bridge circuit is normally used.

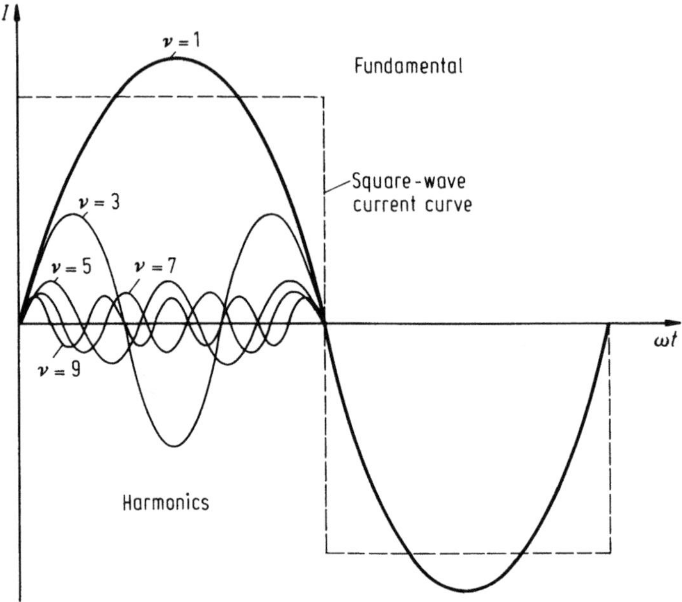


**Fig. 5.2.** Fundamental and harmonics of the mains supply. $\omega t$ Angular velocity (angular frequency × time) $2\pi f \cdot t$, $v$ Ordinal harmonic number, Fundamental: $v = 1 = 50$ Hz Harmonics: $v = 3 = 150$ Hz, $v = 5 = 250$ Hz, $v = 7 = 350$ Hz, $v = 9 = 450$ Hz

**Table 5.1.** Ideal harmonic currents on the alternating current side of power converters

| $v$ | $vf$ | Pulse number $p$ | | |
|---|---|---|---|---|
| | | 2 | 6 | 12 |
| | | $\frac{I_{vi}}{I_{L1i}}\%$ | | |
| 1 | 50 | 100 | 100 | 100 |
| 3 | 150 | 33.33 | – | – |
| 5 | 250 | 20 | 20 | – |
| 7 | 350 | 14.29 | 14.29 | – |
| 9 | 450 | 11.11 | – | – |
| 11 | 550 | 9.09 | 9.09 | 9.09 |
| 13 | 650 | 7.69 | 7.69 | 7.69 |
| 15 | 750 | 6.67 | – | – |
| 17 | 850 | 5.88 | 5.88 | – |
| 19 | 950 | 5.26 | 5.26 | – |
| 21 | 1050 | 4.76 | – | – |
| 23 | 1150 | 4.35 | 4.35 | 4.35 |
| 25 | 1250 | 4 | 4 | 4 |

In thyristor-controlled rectifiers rated at 500 and 1000 A, *two* (six-pulse) three-phase bridge circuits are used. This results in satisfactory twelve-pulse operation in each equipment.

The ratio of the vth harmonic to the fundamentals depends on the ordinal number, the control angle and the overlap. All line and thyristor side reactances cause a delay in current transfer between the commutating thyristors. The reactances flatten the current rise in the commutating circuit. The edges are then no longer ideally square but trapezoidal (Fig. 5.3). The amplitudes of the harmonics decrease with constant modulation as the influence of the reactances in the commutating circuit increases. The higher-order harmonics are attenuated more strongly by the influence of the overlap than are the lower-order ones. The larger the overlap angle $u$ (i.e. the more inductances there are present in the circuit), the longer the trapezoidal semi-oscillations become. In this way the r.m.s. values of the actual harmonics are reduced. The overlap angle depends not only on the inductances in the circuit but also on the control angle $\alpha$.

A continual change of the relative harmonic content can be assumed between the 'infinitely smoothed' direct current as in Fig. 5.3(a), and the 'pulsing direct current at the gap boundary', as in Fig. 5.3(c). The values for the fifth and sixth harmonics increase sharply as the gap boundary is approached, whilst the higher-order harmonics decrease.

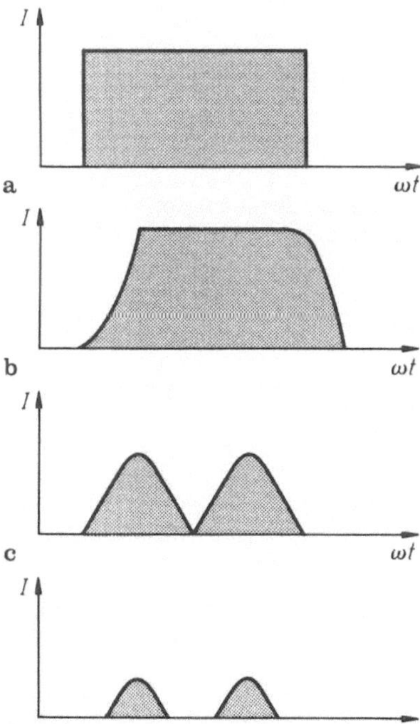

Fig. 5.3a–d. Dependence of relative harmonic content on the shape of the current. a Idealized rectangular shape; b Trapezoidally flattered by inductors; c Pulsing direct current at gap boundary; d Direct current with gaps

The ratio of the harmonic current $I_\nu$ to the basic oscillation $I_1$ current with square wave current and pulsing direct current with six-pulse fully controlled three-phase bridge circuit can be seen from Table 5.2.

The experimental values for ascertaining the maximum harmonic currents (on the supply side) to be expected from a controlled converter in fully controlled three-phase bridge circuit (six-pulse) are to be taken from Table 5.3.

The control angle $\alpha = 0$ (full modulation) is, as is well known, the same as that of an uncontrolled converter. Here with $u = u_0$ the commutation time is at its longest. The harmonics are thus particularly slight. There is the least modulation with the largest control angle $\alpha$. This does not lead, however, to the maximum values for the harmonics. The maximum harmonic current occurs when there is roughly half-modulation.

Figure 5.4 shows the dependence of the harmonic current $I_\nu$ on the control angle $\alpha$ with the individual harmonic numbers $\nu$ (Fundamentals $\nu = 1 = 100\%$). In the example a (six-pulse) fully controlled three-phase bridge circuit has been selected.

### 5.3.1 Measures for Reducing Retroactive Effects on the Mains Supply

*Suppression of Current Harmonics by Raising the Pulse Number.* There are various possibilities for reducing the occurrence of harmonics or suppressing their propagation back into the mains supply. The simplest way is to prevent harmonics occurring with the aid of a *circuit (with higher pulse number)*.

**Table 5.2.** Harmonic currents with (six-pulse) fully controlled three-phase bridge circuit

| $\nu$ | $\dfrac{I_\nu}{I_1}$ | |
|---|---|---|
| | With square-wave shape of the current | With pulsing direct current at gap boundary |
| 5 | 0.2 | 0.48 |
| 7 | 0.14 | 0.17 |
| 11 | 0.09 | 0.08 |
| 13 | 0.07 | 0.05 |

**Table 5.3.** Maximum content of harmonic currents in r.m.s. current

| $\nu$ | $\dfrac{I_{\nu\,max}}{I_L}\,[\%]$ |
|---|---|
| 5 | 30 |
| 7 | 12 |
| 11 | 6 |
| 13 | 5 |

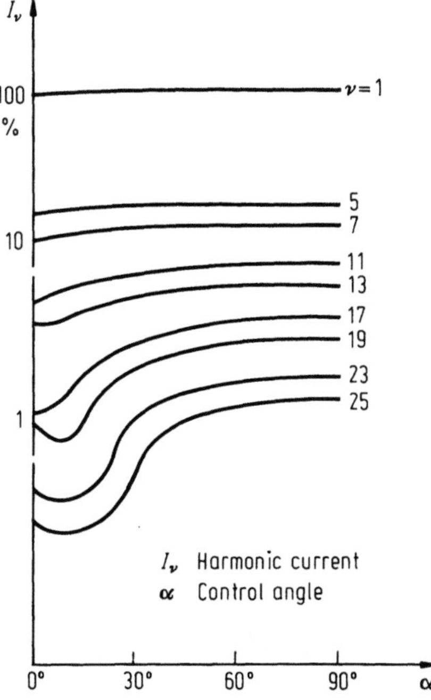


**Fig. 5.4.** Dependence of the harmonic current on the modulation

Thyristor-controlled single-phase rectifiers of up to about 25 A rated current have a two-pulse converter circuit, and three-phase rectifiers from 25 to 1000 A have a six-pulse one. In rectifiers with a rated current from 500 to 1000 A the undesirable mains disturbance is still further reduced by changing to a twelve-pulse arrangement. For this purpose, two six-pulse 250-A current converters are used in 500-A devices and two six-pulse 500-A converters in 1000-A devices (Fig. 5.5). The two circuits are operated with a phase shift of 30°, producing a twelve-pulse mains feedback instead of a six-pulse one. The output of each current converter is filtered individually before they are connected in parallel to deliver the doubled current. Another possibility is to reduce harmonics by means of special control methods (e.g. oscillator package control).

*Suppression of current harmonics by an additional boost converter.* Modern switching-mode power supplies contains an additional boost converter which draws a sinusoidal current from the mains with low conducted EMI.

*Suppression of Current Harmonics Using Filter Circuits. Screening circuits,* also called filter or absorption circuits, can be used to eliminate the harmonics on the low-voltage side caused by power converters. They are tuned to the frequency of the harmonics (Fig. 5.6).

In practice filters are normally used for the fifth, seventh, eleventh and thirteenth harmonics. A common absorption circuit, tuned to the twelfth harmonic,

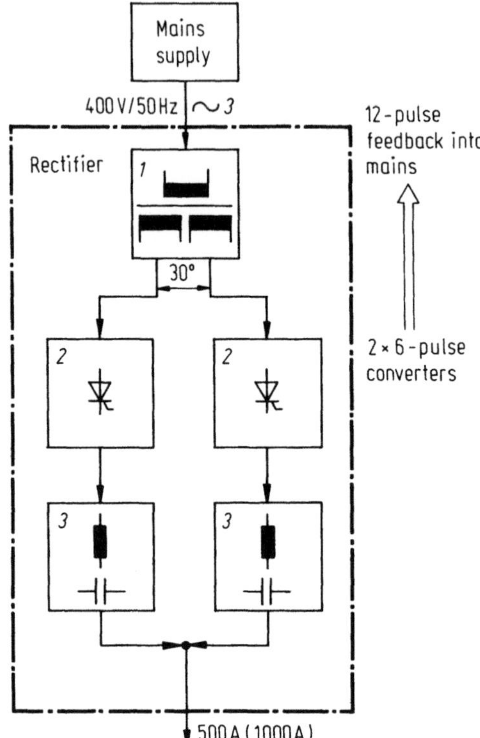



**Fig. 5.5.** Arrangement of two phase-shifted rectifiers to obtain a 500 A or 1000 A rectifier with 12-pulse feedback to the mains. 1 Main transformer, 2 Six-pulse fully controlled three-phase bridge circuit (thyristor set) for 250 A (500 A) each, 3 Output filter

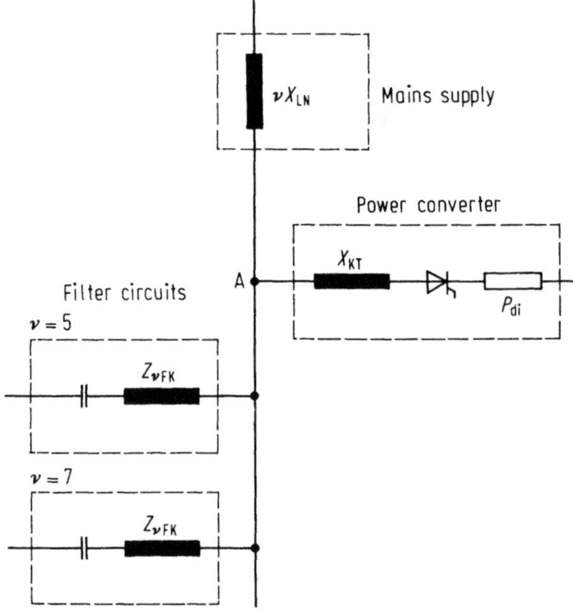



**Fig. 5.6.** Connection of the power converter and filter to the mains supply. $X_{KT}$ Short-circuit reactance at power converter transformer, $\nu X_{LN}$ effective reactance at vth harmonic, $Z_{\nu FK}$ effective impedance of the filter circuit for the vth harmonic, $P_{di}$ ideal d.c. rated power

is assigned to the eleventh and thirteenth. In this way the harmonic currents can be reduced considerably.

If all ohmic resistances are disregarded and supply capacitances are excluded, the mains supply can be replaced by an inductance $X_{LN}$, which defines the short-circuiting power $S_K$ of the supply at the converter connection point A. $vX_{LN}$ is then valid for the effective reactance at the vth harmonic.

If filter circuits are connected to converter the connection point A, thus in the immediate proximity of the converter whose inductances and capacitances are selected so that a series resonance (and thereby a short-circuit) is produced for the harmonics to be suppressed, the currents of the frequencies concerned no longer flow via the mains supply, but via the filter circuits. Below their tuning frequency the filter circuits work capacitively – for the basic oscillation too. They can therefore also be used for reactive power compensation of the system.

## 5.4 Power Failures

The reliability of public power supplies in Germany is normally very high. However, there are around 100 to 200 occurrences per year of *short-term power failures of* < 0.5 s duration. Such interruptions occur for example in the case of changeovers and automatic short-circuit reclosings.

Short-term failures in the millisecond range can possibly be bridged by the capacitors present in the rectifiers for filtering. In the event of power failures of longer duration, batteries, standby power supply systems, etc., are used.

In Germany *two to four long-term failures* lasting a few minutes to several hours occur on average each year. Of these failures 97% are of the order of less than 6 hours and only 3% are longer.

There should also be included one interruption due to a fault with the service line and one due to maintenance work or faults in the mains high- and low-voltage switchgears, so that in all some six failures a year are to be expected with each system (Table 5.4).

**Table 5.4.** Frequency of interruption in the supply of power by telecommunications power supply systems (based on statistics of 8000 d.c. telecommunications power supplies in Germany)

| Duration of the interruption in power supply | Failures in % | Number of mains failures per year and power supply system | One fault per power supply system in years |
|---|---|---|---|
| Up to 2 min | 40 | 2.4 | 0.4 |
| 2 to 30 min | 25 | 1.5 | 0.7 |
| 30 to 60 min | 14 | 0.85 | 1.2 |
| 1 to 2 h | 9 | 0.55 | 1.8 |
| 2 to 4 h | 8 | 0.5 | 2 |
| More than 4 h | 4 | 0.2 | 5 |
| Total | | 6 | |

# 6 Energy Storage

An introduction to battery systems was presented in Section 1.4.

## 6.1 Stationary Lead-Acid Batteries

### 6.1.1 Requirements

Stationary batteries (accumulators) are expected to meet the following requirements (Fig. 6.1):

- long shelf life before commissioning,
- ease of installation,
- low commissioning cost,
- simple charging technique,
- high efficiency,
- reliability (high availability of capacity at any time),
- long life,
- low maintenance cost,
- safety against explosion,
- mechanical durability,
- ability to withstand short-circuits and
- environmental acceptability.

### 6.1.2 Charging and Discharging

Figure 6.2 illustrates the discharging and charging process and shows chemical equations in lead-acid batteries (accumulators). If electrodes consisting of lead or lead compounds are immersed in a vessel filled with dilute sulphuric acid ($H_2SO_4 + H_2O$) as electrolyte, a secondary voltaic cell results. The active material lead dioxide ($PbO_2$) is produced electrochemically (formation) at the positive electrode. At the negative electrode lead oxide is pressed into a grid of lead. Through forming, this paste is converted electrochemically into finely distributed spongy lead ($Pb$), the active material. In a charged lead battery the active material at the positive electrode consists of lead dioxide ($PbO_2$) and at the negative electrode of lead ($Pb$).

If two electrodes are linked via a resistance (load, $R_L$), the current $I$ flows (Fig. 6.2a). During this process the chemical conversion of the active materials of both plates is taking place.

**Fig. 6.1.** Room with lead-acid batteries (with tubular positiv electrodes OPzS) on floor stillages. (Photo by courtesy of Varta AG)

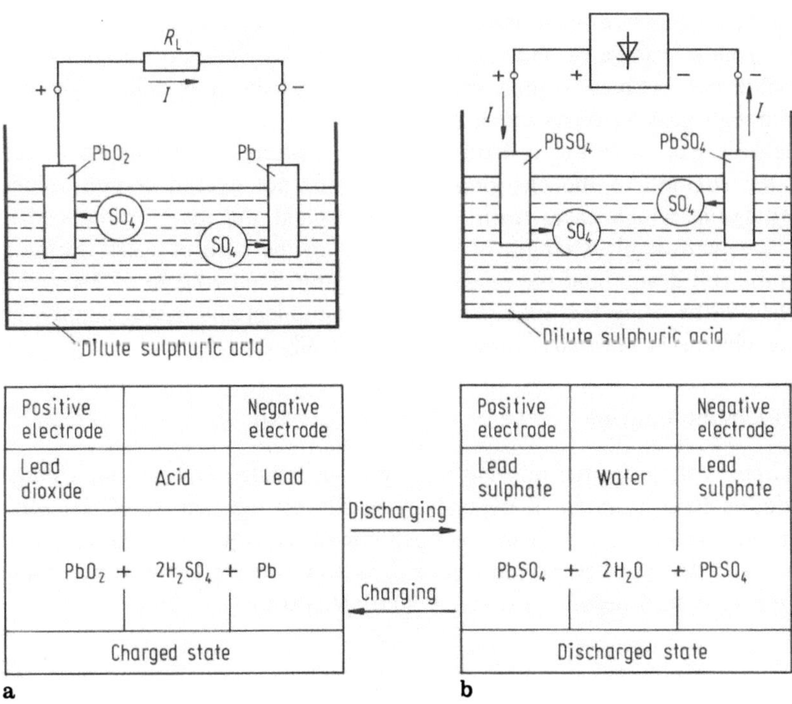


**Fig. 6.2a, b.** Discharging and charging processes in a lead-acid battery

Because of the electrochemical processes during discharging, both the lead dioxide of the positive plate as well as the lead of the negative plate are converted into lead sulphate ($PbSO_4$). Acid ($H_2SO_4$) is used up and water ($H_2O$) is formed, the concentration of the acid decreasing according to the energy drawn off. The gradually increasing internal resistance causes an initially slow, then quicker, drop in the voltage until a lower limit is reached (final discharge voltage), which is determined by the strength of the discharge current.

Conversion of the active, charged material into lead sulphate (discharging) involves a great increase in volume. During discharge the pores in the material become blocked and acid is prevented from reaching the inner particles of the material. This means an insufficient transfer of material (diffusion), a slowing down of the reaction and a decrease in conductivity, which in the end leads to a fall in the discharge voltage until it reaches the specified final discharge voltage.

The battery can be charged by connecting it to a d.c. source (e.g. rectifier), provided the voltage of the d.c. source is greater than that of the battery (Fig. 6.2b).

When charging, the active materials of the two electrodes and the sulphuric acid are restored to the original state existing before discharge. As an unwanted secondary reaction, water is decomposed into hydrogen and oxygen as soon as the cell-voltage exceeds 1.23 V. Fortunately these reactions are hindered at the lead (Pb) and lead-dioxide ($PbO_2$) surface so much that they cause only a creeping self-discharge of a few percent of capacity per month. But, if the cell voltage exceeds a certain value hydrogen and oxygen evolution can gain in volume enormously. So, from a voltage of about 2.4 V/cell the gassing rate is fast enough to produce visible gas bubbles. (Often the voltage of 2.4 V/cell is called 'gassing voltage', although gassing starts much earlier.)

As intensive gas evolution is harmful for the material of the plates in the long term, the strength of the charging current must not exceed certain values once the voltage at which gases form has been reached and must be reduced if necessary. The permissible strength of the current in the case of gases forming depends on the cell design and the method of charging. This voltage of 2.4 V/cell is not reached when using the battery in a telecommunications power supply, as the charging voltage is normally limited to 2.33 V/cell.

### 6.1.3 Open-Circuit Voltage

The open-circuit voltage of the off-load battery is also called equilibrium voltage or electromotive force (e.m.f.); it depends primarily on the density of the acid. The higher the density, the higher the open-circuit voltage. It is sufficient in practice to know that the open-circuit voltage is approximately the same as the value for the rated acid density[1] (charged state) plus 0.84.

---

[1] In kilograms per litre at a temperature of 20 °C.

### 6.1.4 Rated Voltage

The rated voltage of the lead-acid battery is 2 V/cell. It is more or less approximated when the battery is discharged at a low rate ($\sim$ 10-hour rate) (see Fig. 6.3).

### 6.1.5 Discharging Voltage

The voltage during discharging depends on the level of the discharge current and on the time. The higher the discharge current and the longer discharging lasts, the lower the voltage. Causes for this are the decrease in acid density and with it the e.m.f., and also the increasing voltage drop due to the internal cell resistance $R_i$.

For example, with OPzS batteries the acid density when discharging with the ten-hour current $I_{10}{}^2$ starts at 1.24 kg/l and falls by 0.12 to 0.13 kg/l. When discharging with even larger currents the decrease in acid density is less, corresponding to the smaller amount of energy that can be taken off down to the permissible final discharge. Taking into account the level of the discharging current, the state of the discharge of the battery can thus be ascertained from the density of the acid. It should be noted that the density of the acid also depends on the temperature.

At the start of discharging of a fully charged battery the voltage passes through a minimum of 20 to 30 mV per cell maximum (voltage dip S, Fig. 6.3).

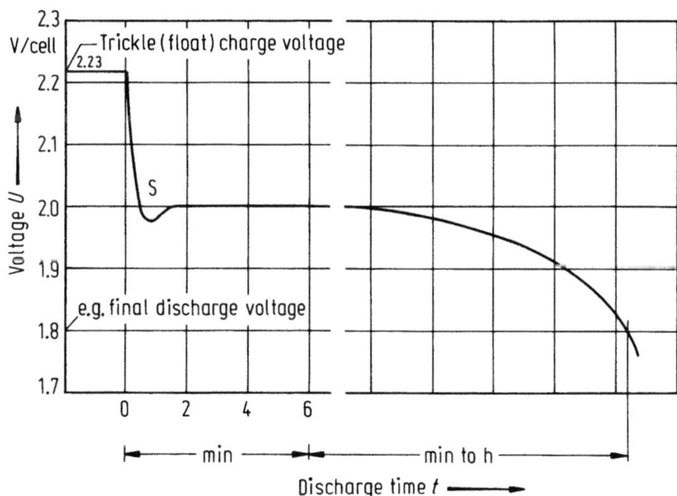


**Fig. 6.3.** Discharge voltage curve of a lead-acid battery

---

[2] $I_{10}$ designates the current, in amperes, in relation to the rated capacity, with which a battery is discharged in 10 h down to a given final discharge voltage.

One cause for this brief drop in voltage is that the production of lead sulphate ions is temporarily delayed. Because of this unsteady behaviour the value always given as the initial discharge voltage is the one measured after removal of 10% of the capacity corresponding to the respective discharge current.

The behaviour of the voltage following the voltage dip is initially proportional to the decline in acid density. The voltage thus slowly decreases in accordance with the characteristic curve of the battery in question.

Diffusion problems, depletion of active material and a decrease in conductivity also determine the behaviour of the voltage during load. This is the reason why the voltage drops very rapidly towards the end of discharging, and the higher the discharge current the less energy can be taken from a cell until the final discharge voltage is reached. Thus the consumable capacity is reduced by some 50% if there is a one-hour discharge current instead of the ten-hour one.

Figure 6.4 shows as an example discharge curves of an OPzS battery versus the available capacity. The rated capacity $K_{10}$ here is 100 Ah. $K_{10}$ accordingly indicates the amount of current in ampere-hours that, on discharging over a period of ten hours, can be consumed with the associated current $I_{10}$. Thus 10 A can be consumed for ten hours before the associated final discharge voltage $U_s$ of 1.79 V/cell is reached.

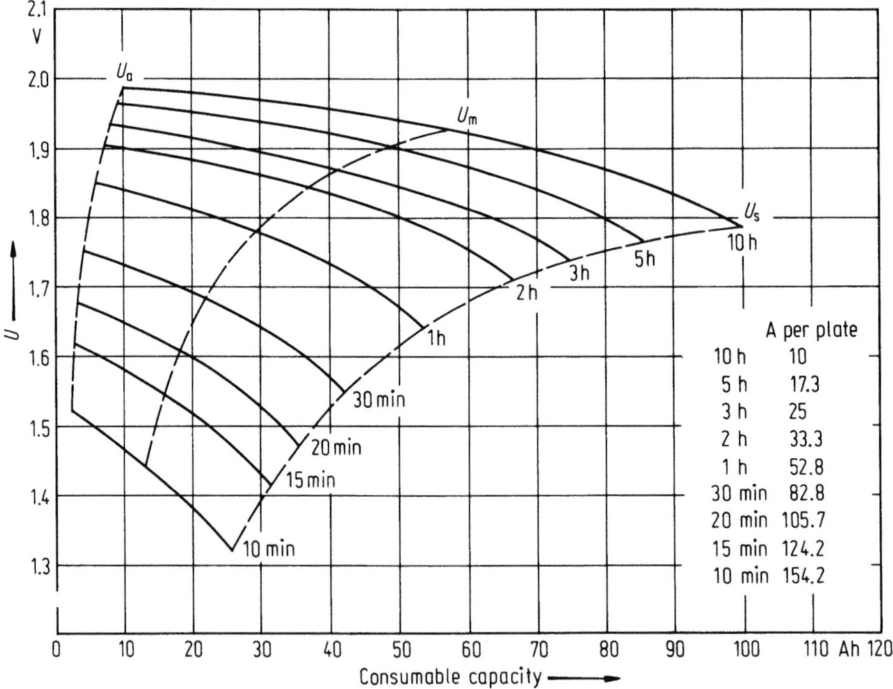


**Fig. 6.4.** Discharge voltage and available capacity of the OPzS 100 battery plate as a function of constant discharge current strengths. (*Source*: Varta AG) $U_a$ Initial discharge voltage, $U_m$ mean discharge voltage, $U_s$ final discharge voltage

When discharging with the current $I_2(\triangleq 33.3\,\text{A})$ an amount of energy of only a little over 50% of the rate capacity $(K_{10})$ could be consumed until there is a drop below the same voltage of 1.79 V/cell.

The final discharge voltage is the specified minimum voltage level when discharging with the assigned current (e.g. 1.8 V/cell). If the voltage falls below this level a 'deep discharge' occurs that stresses the active material. Repeated discharging below the final discharge voltage harms the structure of the active material in the plates on account of the changes in volume. Frequent discharges below the final discharge voltage thus leads to a marked shortening of the battery's service life.

The final discharge voltage (related to the battery) must not be confused with the minimum voltage limit for the communications system, below which it must not fall at the end of a bridging time. This voltage is called the system-conditioned final voltage $U_{\text{smin}}$.

### 6.1.6  Self-Discharging

Due to the slow production of hydrogen at the negative electrode and oxygen at the positive electrode lead-acid batteries are always subject to slight spontaneous discharging (self-discharging) even with open circuit voltage. In this case, the usual phenomena accompanying discharging can be determined, namely sulphating of the plates and a decrease in acid density. The loss in capacity occurring in modern stationary batteries is 2 to 5% per month, slightly higher at the beginning but is gradually reduced with discharge at a temperature of around 20 °C. These losses must be compensated for by a continuous supply of energy. This is one reason why the battery is constantly operated on float charging (also called trickle charging) in standby parallel and changeover modes. Self-discharging proceeds more rapidly at higher temperatures; its rate is approximately doubled per 10 °C temperature increase.

### 6.1.7  Float (Trickle) Charging and Charging Voltage

Float charging is the most common method for lead-acid batteries in those stationary applications where the battery is only used for emergency-energy supply. Permanent overcharge is achieved by the supply of constant voltage only slightly higher than the open-circuit voltage (usually about 2.23 V/cell). During float (trickle) charging[3] the battery continuously receives a small current of 10 to 40 mA per 100 Ah of rated capacity.

---

[3] The general meaning of 'trickle charging' is the charging of the lead-acid battery with a small constant current $(I_{200})$. This is to be used for example, for recharging of long-term stored batteries. In special books charging with a constant voltage is generally called 'float charging'.

If the recommendations in the operating instructions are observed, lead-acid batteries can be run for their whole lifetime with a constant float voltage of 2.23 V/cell. Recharging is also possible with this voltage. To shorten the charging time it is also possible to apply temporarily a higher voltage but this should remain below the voltage at which heaving gassing occurs.

In general a 'short-order' charging voltage of 2.33 V/cell is applied. The duration of short-order charging can be adjusted on a electronic timer. Depending on the device a time from 1 to 24 h can be set. At the end of the short-order charging, the rectifier is switched back to float charging. In systems with valve-regulated maintenance-free batteries, the characteristic 'charging at 2.33 V/cell' is blocked. The battery attains its fully charged state with the 2.23 V/cell characteristic as well, although this can take several days.

Table 6.1 shows for lead-acid battery types OPzS, OGi-Block and OGiE the capacity available again after the respective times as a percentage, taking into account a charging factor of 1.2. The previously consumed capacity is replaced by charging in accordance with $IU$ characteristic curves. A float charging voltage of 2.23 V/cell or charging voltage of 2.33 V/cell is assumed.

In a power failure, for example, 50% of the battery's rated capacity ($K_{10}$) will be consumed. The full capacity is to be restored after return of the mains

**Table 6.1.** Charging time with lead-acid batteries necessary after drawing off capacity in order to restore the desired level with $IU$ charging (2.23 or 2.33 V/cell)

| Consumption of $K_{10}$ (%) | Charging time time $t$ (h) | Capacity available after charging (%) | | | | | | | |
|---|---|---|---|---|---|---|---|---|---|
| | | 2.23 V/cell | | | | 2.33 V/cell | | | |
| | | $0.5\,I_{10}$ | $I_{10}$ | $1.5\,I_{10}$ | $2\,I_{10}$ | $0.5\,I_{10}$ | $I_{10}$ | $1.5\,I_{10}$ | $2\,I_{10}$ |
| 25 | 3 | 87.5 | 94 | 95 | 96 | 87.5 | 95 | 97 | 97 |
| | 6 | 94 | 96 | 97 | 97 | 97 | 98 | 98 | 99 |
| | 10 | 96 | 97 | 97.5 | 97.5 | 100 | 100 | 100 | 100 |
| | 20 | 97 | 98 | 98 | 98 | – | – | – | – |
| 50 | 3 | 62.5 | 75 | 82 | 85 | 62.5 | 75 | 86 | 89 |
| | 6 | 75 | 89 | 90 | 92 | 75 | 92 | 94 | 95 |
| | 10 | 90 | 93 | 93 | 94 | 90 | 96 | 96 | 96 |
| | 20 | 97 | 97 | 97 | 97 | 100 | 100 | 100 | 100 |
| 75 | 3 | 37.5 | 50 | 62.5 | 68 | 37.5 | 50 | 62.5 | 75 |
| | 6 | 50 | 73 | 80 | 83 | 50 | 74 | 86 | 89 |
| | 10 | 66 | 86 | 88 | 89 | 66 | 89 | 93 | 94 |
| | 20 | 93 | 93 | 93 | 93 | 94 | 98 | 100 | 100 |
| 100 | 3 | 12.5 | 25 | 37.5 | 47 | 12.5 | 25 | 37.5 | 50 |
| | 6 | 25 | 50 | 65 | 70 | 25 | 50 | 76 | 80 |
| | 10 | 41.5 | 73 | 78 | 80 | 41.5 | 80 | 85 | 89 |
| | 20 | 80 | 91 | 91 | 91 | 82 | 95 | 100 | 100 |

$K_{10}$ Capacity, ten hours. $I_{10}$ Current, ten hours.
*Source*: Varta AG.

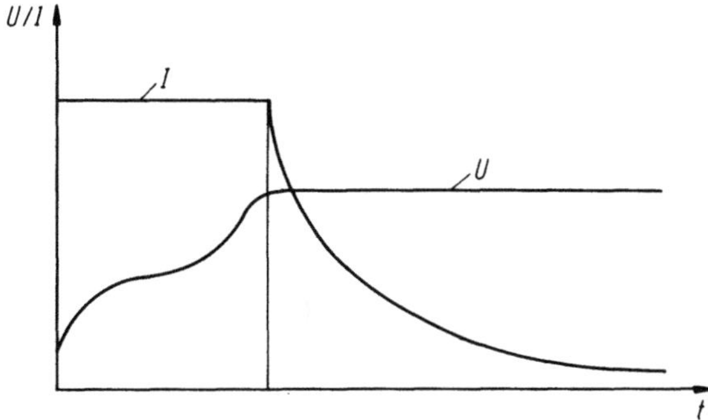

Fig. 6.5. Charging of lead-acid batteries for stationary systems in accordance with $IU$ characteristic

supply. If charging is done with 2.23 V/cell and a current of 0.5 $I_{10}$, then 90% of the capacity will have been put back after a charging time of 10 h and 97% after 20 h. This means for a 100 Ah battery ($K_{10}$) a current strength of 0.5 $I_{10}$ = 5 A. Some days will pass, however, before the battery is again 100% charged (fully charged state). If, on the other hand, charging occurs with a voltage of 2.33 V/cell and a current of 0.5 $I_{10}$, then full recharge (100%) is already reached after 20 h.

The behaviour of voltage and current when charging is finally determined by the size of the rectifiers and their characteristics. In standard DIN 41 772 letter symbols are specified for the characteristics of rectifiers and thereby for the different methods of charging, and also for changeover and disconnection processes. The meanings are:

- $U$ constant-voltage characteristic,
- $I$ constant-current characteristic,
- W taper characteristic,
- O automatic characteristic switching and
- a automatic disconnection after full charging.

Lead-acid batteries are charged in accordance with the $IU$ characteristic by the rectifiers of the telecommunications power supply system (Fig. 6.5). This runs in two sections. First the charging current remains constant with a rising charging voltage until, depending on the characteristic chosen, the voltage of 2.23 V/cell or 2.33 V/cell is reached. From this figure on, the voltage is kept constant and consequently charging proceeds with a current falling to lower values. When the preset time is reached an automatic switch back from 2.33 V/cell to 2.23 V/cell occurs.

### 6.1.8 Initial Charging Voltage (Commissioning)

Modern lead-acid batteries are supplied either filled and charged or unfilled and dry-charged. They therefore require only short initial charging. A large part of the capacity is immediately available after filling with acid.

With dry-charged cells it is possible when charging according to the *IU* characteristic to carry out the initial charging at a voltage of 2.33 V/cell or at the float voltage of 2.23 V/cell unless the batteries have been stored for a long time. In this case prolonged charging should have to be carried out. It is assumed that at normal temperature and air humidity there is still a residual capacity of some 85% available after a storage of 3 to 5 years. Instructions from the manufacturers of batteries on initial charging must be followed closely to achieve a long life for the battery.

### 6.1.9 Capacity

The capacity of a battery is a measure of its efficiency and size. It is measured in ampere hours (Ah) and indicates the amount of energy a battery can deliver in a certain time (h) when discharging with a constant current (A) until a given voltage is reached (final discharge voltage).

Rated capacity means the rated value for the capacity with a specified discharge current strength and also the specified assigned final discharge voltage and temperature. The higher the discharge current, the smaller the capacity and discharge-voltage. The rated capacity particular for all battery types is valid at an acid temperature of 20 °C. At lower or higher temperatures the capacity is reduced or increased, within a limited temperature range, but about 1% per 1 K. The full value for the rated capacity with new batteries is usually achieved after several charge/discharge cycles.

The capacity curves in Fig. 6.6 show, for lead-acid batteries of the OPzS type, the available capacities for different sizes of plate. The dependence of capacity on the discharge current strength can also be seen. Batteries with a capacity of more than 250 Ah are usually split, for operational reasons, into two parallel-connected groups, which together provide the capacity required.

### 6.1.10 Efficiency and Charging Factor

A distinction is made between *ampere-hour efficiency* and *watt-hour efficiency*. These are calculated as follows:

$$\text{Ampere-hour efficiency:} = \frac{\text{available ampere-hours in Ah}}{\text{charged ampere-hours in Ah}};$$

$$\text{Watt-hour efficiency:} = \frac{\text{available watt-hours in Wh}}{\text{charged watt-hours in Wh}}.$$

Both variables depend on the cell construction, on the temperature of the acid and on the level of charging and discharging currents. The ampere-hour efficiency is between 83% and 90%, the watt-hour efficiency between 67% and 75%.

The *charging factor* is the reciprocal value of the ampere-hour efficiency. It is more customary in practice to write it as:

$$\text{Charging factor} = \frac{\text{charged ampere-hours in Ah}}{\text{available ampere-hours in Ah}}.$$

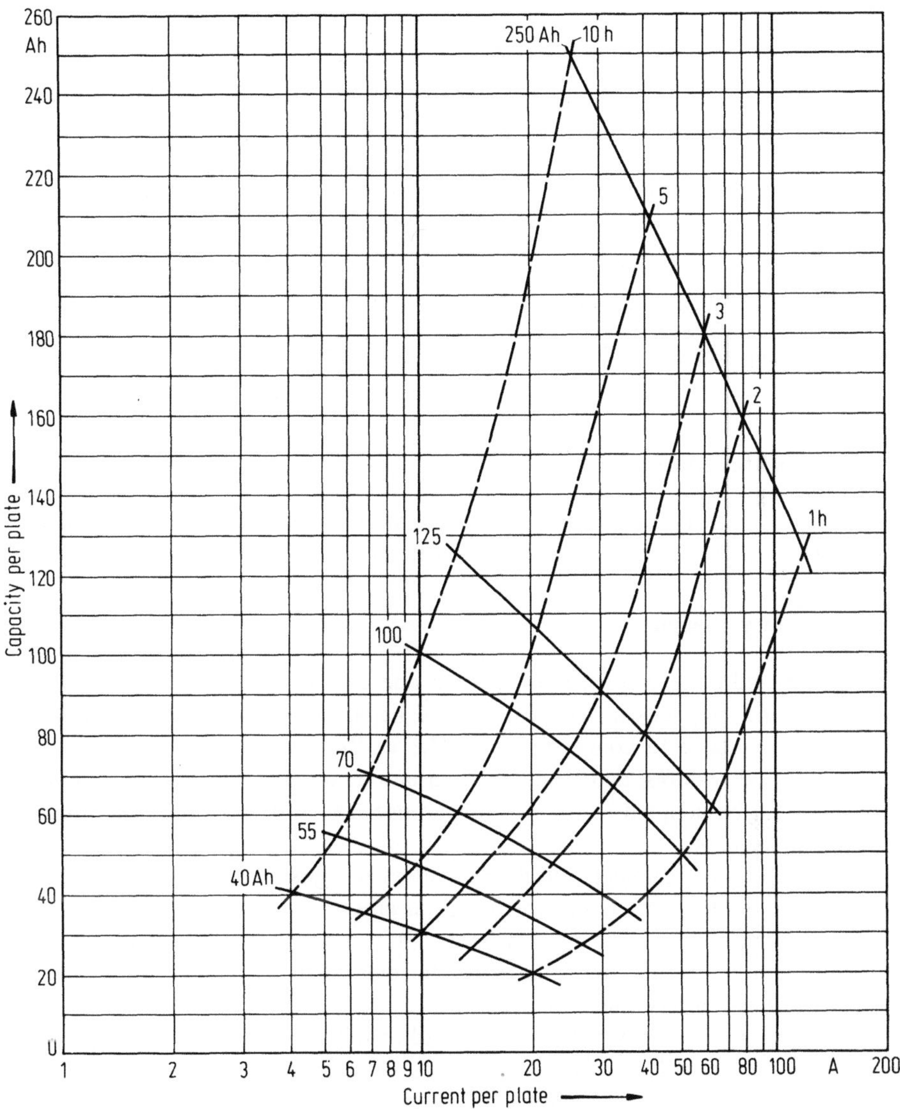


**Fig. 6.6.** Capacity curves for types OPzS lead-acid batteries (plates 40 to 250 Ah) at a temperature of 20 °C. (*Source*: Varta AG)

The charging factor is usually between 1.1 and 1.2.

*Note:* Charging efficiency and charging factor are appropriate parameters only for charging processes limited in time. They lose any meaning with prolonged charging, e.g. during float, when only the secondary reactions are maintained by the current and 'charging' no longer occurs. The charging factor would then exceed all limits and the efficiency approach zero.

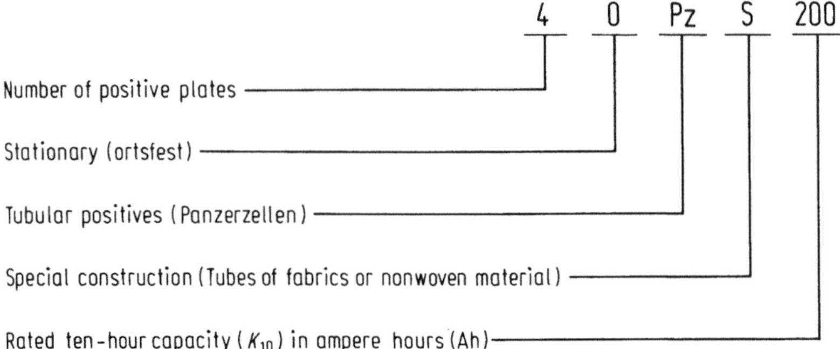



**Fig. 6.7.** Example of a type designation according to DIN

### 6.1.11 DIN Designation of Battery Type

Figure 6.7 shows as an example the designation of a tubular lead-acid battery according to DIN (Deutsche Industrie Normen).

### 6.1.12 Battery Design

Every lead-acid battery consists of at least one cell with dilute sulphuric acid as electrolyte and one set each of positive and negative plates; these are welded in alternating sequence to connector bars. The realization of this principle can be achieved in many ways, and many such variations do actually exist. Above all, a battery has to be adapted to its special purpose, and this adaptation may require special construction.

So an enormous number of different battery designs made by various methods are available on the market, and an attempt to present a comprehensive description would far exceed the scope of this book. Only a few examples can be shown; for more detailed information, the reader is referred to catalogues and brochures offered by most battery manufacturers.

The designation of a conventional lead-acid battery generally derives from the type of positive plates used. The types of vented batteries mainly used in telecommunications power supply systems are:

- pasted plates or grid plates (OGi in DIN),
- tubular positive plates (ironclad plates) (OPzS in DIN),
- Planté plates (GroE in DIN).

Smaller capacities (usually up to 200 Ah) are often made as block batteries with 2, 3 or 6 cells per block (4, 6 and 12 V/block). The following figures show some examples. In all cases the negative plates are of grid construction. These are available up to the largest capacities in *closed* containers.

Batteries of the tubular-plate design, and also the pasted-plate batteries are nowadays made with positive plates of lead containing less than 2% of antimony (previously 9 to 12%). The strength of these plates is ensured by means of special alloying additions (e.g. selenium). As a result of the reduction in the proportion of antimony in combination with design parameters maintenance intervals of up to three years are possible. Within the stipulated interval, maintenance-free operation is guaranteed, provided that the battery is used in accordance with the relevant recommendations, e.g. the battery must not be overcharged. In particular, operation in the standby paralled mode or changeover mode with continuous float (trickle) charging of the battery (at 2.23 V/cell) is a prerequisite. A further important consideration is the temperature. It is also possible to use an antimony-free alloy.

In many countries, e.g. the USA, lead-calcium alloy is used as grid material for standby batteries. They show even lower water consumption than low-antimony batteries, but the maintenance interval is also a question of battery design (reserve electrolyte).

### 6.1.13 Batteries with Tubular Positives

*Construction.* A battery with tubular positive electrodes (OPzS) is illustrated in Figs. 6.8a, b and c. Figures 6.8d and e show the internal construction of the cell. The OPzS cell embodies tubular positive plates in conjunction with negative grid plates. The positive tubular plates consist of a side-by-side arrangement of thin lead rots – again of a special low-antimony alloy – provided with centring pieces, over which are drawn woven pockets of a highly acid-permeable insulating material. The active material (lead dioxide) is contained in the space between the current conducting lead rod and the pocket. The pockets provide the mechanical support of the active material.

In the negative plates the active material (initially lead oxide) is pasted into a grid of hard lead. Through the forming process the lead oxide is converted electrochemically into finely divided spongy lead (Pb).

A microporous separator is used to insulate the plates, sometimes in conjunction with a spacer of corrugated PVC. The electrolyte is dilute sulphuric acid usually with a density of 1.24 kg/l at 20 °C.

*Design.* Individual OPzS cells are manufactured in transparent acid-proof plastic containers for capacities from 250 to 1500 Ah, in translucent acid-proof plastic containers for capacities of 2000, 2500 and 3000 Ah and in hard rubber containers for capacities from 3500 to 12 000 Ah.

Cells with capacities of from 250 to 3000 Ah are designed for bolted connections (with insulated copper links).

Cells of from 3500 to 12 000 Ah are currently available only with welded connections. The links are of hard lead, welded onto the cell terminals.

*Application.* OPzS batteries are suitable for medium and long discharge periods because tubular plates cannot be manufactured economically with diameters smaller than about 8 mm. This diameter is synonymous with the plate thickness

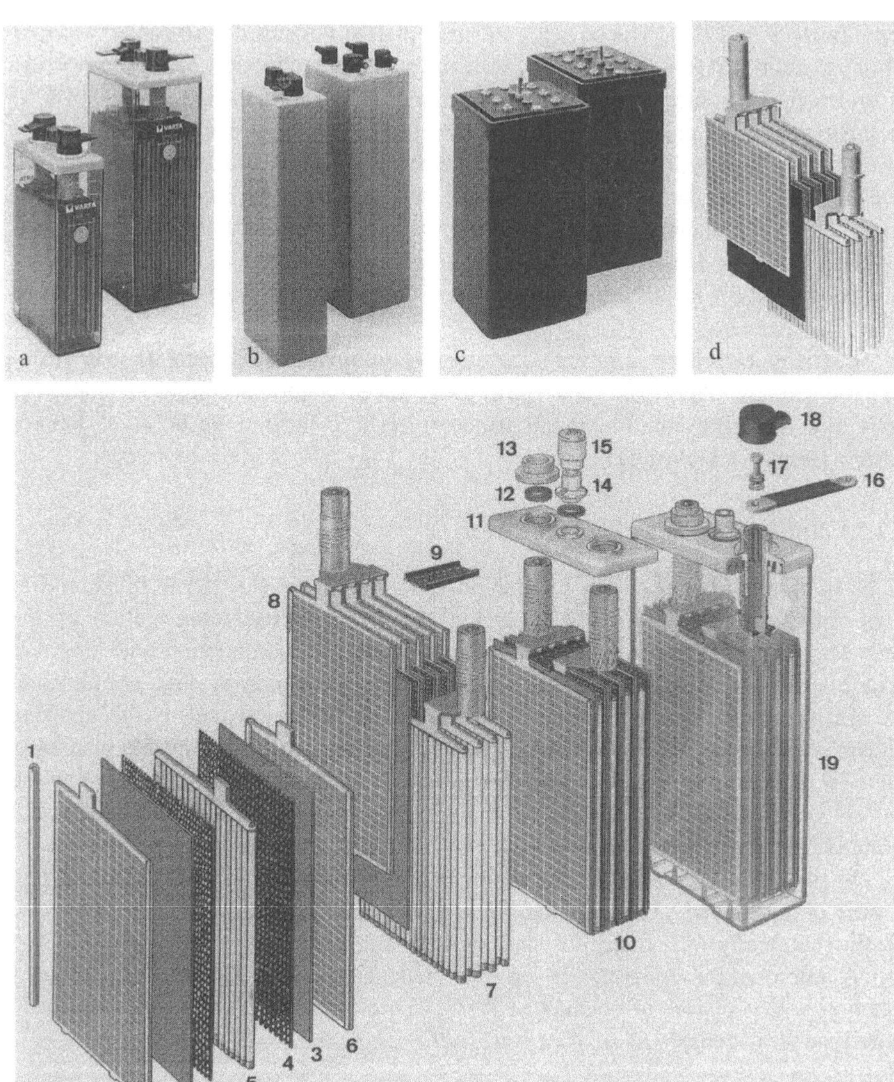



**Fig. 6.8a–e.** Lead-acid battery with tubular positive electrodes (OPzS). (Photo by courtesy of Varta AG). Legend to **e**: *1* edge insulator (enlarged), *2* negative end plate, *3* microporous separator, *4* perforated corrugated separator, *5* positive tubular plate, *6* negative centre plate, *7* positive plate set with terminal bar and safety pole, *8* negative plate set with terminal bar and safety pole, *9* cover plate, *10* plate group, *11* cell cover, *12* terminal seal, *13* cover disc, *14* filling vent with sealing ring, *15* gas drying plug, *16* cell connector, *17* bolt with locking washer, *18* terminal cap, *19* OPzS cell in transparent container

that determines the minimum resistance. They are therefore mainly installed when the battery is required to cater for a mains supply interruption of from 1 to 10 h. Other application areas, apart from telecommunications, are for example:

- alarm systems,
- signalling systems,
- fire alarm systems,
- equivalent power supply systems conforming to national and international standards.

*Installation.* OPzS batteries are installed on stillages, or in stepped racks (see Section 6.6)

*Service life.* The service life of OPzS batteries can be quoted as 12 to 15 years (for a reduction in capacity to 80%).

*Batteries with pasted plates as block batteries (OGi) and single cells (OGiE).* These batteries are rapidly gaining favour for telecommunications power supplies, particularly for small-to-medium-sized telephone systems. They are manufactured with pasted plates of varying thickness. Worthy of note are their low internal resistance, with correspondingly good voltage characteristics (even under rapid-discharge conditions with high loading), and the fully insulated construction.

*Construction.* Block batteries of the OGi type batteries are illustrated in Fig. 6.9(a). The internal construction is shown in Fig. 6.9 b) and c). The OGi-block batteries and the OGiE batteries shown in Fig. 6.9 incorporate positive rod plates in association with negative grid plates. The rod plates are made of a special low-antimony lead alloy and of a special design derived from tubular grid active material (initially lead oxide) is pasted round the rods and lead dioxide ($PbO_2$) is formed from it by an electrochemical process. To reduce shedding of active material and the consequent formation of sludge, the positive rod plate is enclosed in a highly acid-permeable nonwoven pocket. To locate the vertical bars and to ensure mechanical stability the positive plates are provided with horizontal braces.

In the negative plates the active material (initially lead oxide) is pasted into a grid of hard lead. Through the forming process the lead oxide is converted electrochemically into finely divided spongy lead (Pb). Positive and negative plates are assembled into plate stacks. Insulation is provided by a microporous separator and the electrolyte is dilute sulphuric acid with a density of 1.24 kg/l at 20 °C.

*Design.* The OGi-block batteries and the OGiE batteries are supplied in translucent plastic containers; they are in the form of three-cell blocks constituting 6 V units with capacities from 12.5 to 150 Ah, and two-cell blocks, as 4 V units, with capacities of 175 and 200 Ah (OGi-block). They are also available as individual cells with capacities from 250 to 2000 Ah (OGiE).

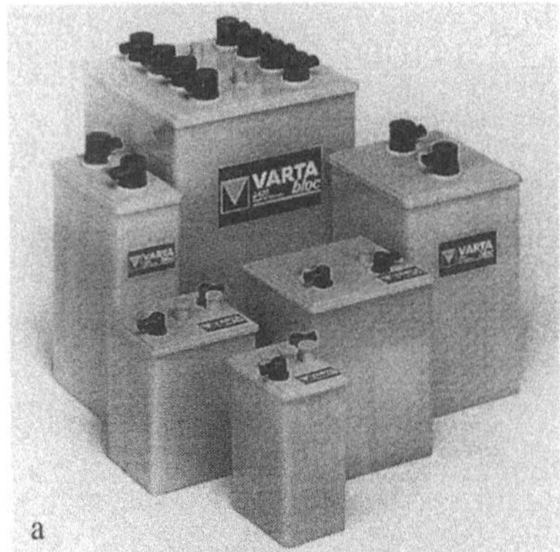
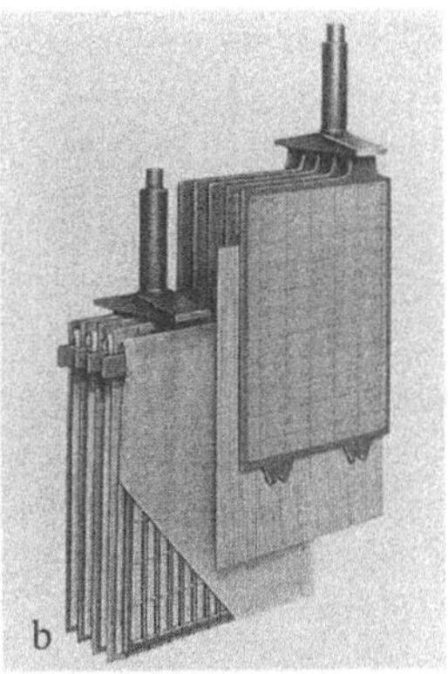
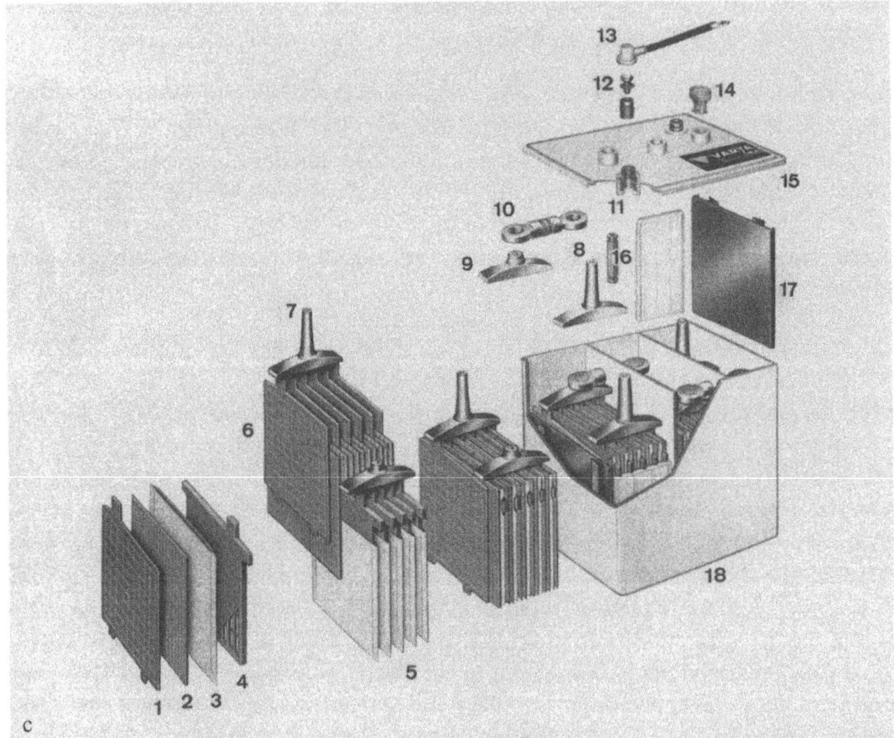

Fig. 6.9a–c. Lead-acid batteries with pasted plates as block batteries (OGi) and single cells (OGiE). (Photo by courtesy of Varta AG). Legend to c: *1* Negative grid plate, *2* microporous separator, *3* separator pocket, *4* positive rod plate, *5* positive plate set with separator pocket over each individual plate, *6* negative plate set, *7* pole section, *8* brass bushing, *9* internal pole bridge, *10* Varta con connector, *11* pole bushing, *12* pole bolt, *13* pole cap, *14* vent plugs, *15* container lid, *16* positive plate support, *17* spacer, *18* battery container

The interconnection of the block batteries or individual cells is effected by means of plastic-insulated copper links bolted to the terminals. External short-circuiting of the installed battery can be prevented by the use of additional rubber terminal caps.

*Application.* By virtue, particularly, of their previously mentioned low internal resistance the OGi-block batteries and the OGiE batteries can be used universally both for rapid-discharge operation in the range of minutes and for long discharge periods of from 1 to 10 h. As well as those indicated in connection with OPzS batteries, their applications extend to the starting of diesel generators, uninterruptible power supplies, solar generators and wind generators.

*Installation.* The block batteries are installed on stillages, or in stepped racks (see Section 6.6)

*Service life.* The service life of the OGi-block and OGiE batteries is between 12 and 15 years.

### 6.1.14 Batteries with Planté Plates (GroE)

The GroE design represents a further development of the well-known and proven Planté plate (large-surface-area plate) batteries (Gro). The Gro battery of the traditional type is no longer used, because, among other reasons, of the high water consumption associated with the design of the individual open cells. Other respects in which they fall short of the present state of battery technology are, for example, short maintenance intervals, large volume and heavy weight.

*Construction.* A GroE battery is illustrated in Fig. 6.10. The GroE cell contains positive large-surface-area plates in association with negative grid plates. Figure 6.10(b) shows the internal construction of the cell.

The positive plates consist of soft cast lead. The surface area of the plate is increased by means of a special lamellar structure, affording a large area of

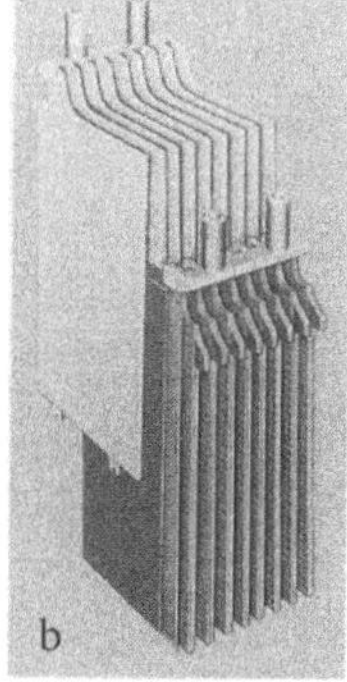

**Fig. 6.10a, b.** Lead-acid batteries with Planté plates (GroE). (Photo by courtesy of Varta AG)

contact with the acid. The active material is produced by an electrochemical process.

A microporous separator is introduced to insulate the plates. The electrolyte is dilute sulphuric acid with a density of 1.22 kg/l at 20 °C.

*Design.* GroE batteries are constructed in transparent or translucent plastic containers for capacities of from 200 to 2800 Ah.

*Application.* The principal areas of application (mainly in the short-time range $\leqq 1$ h) are, for example:

– equivalent power-supply systems in power stations,
– UPS (uninterruptible (a.c.) power supply) systems.

*Installation.* GroE batteries are installed on stillages or stepped racks (see Section 6.6)

*Service life.* The service life is between 15 and 20 years.

### 6.1.15 Comparison of Lead-Acid Batteries

*Internal resistance.* Lead-acid batteries differ significantly in their internal resistance $R_i$ according to the various kinds of plate construction (Fig. 6.11). The difference of the internal resistance is mainly caused by the positive plate design.

### 6.1.16 Vent Plugs

There are essentially three types of vent plug in stationary batteries today:

– Gas-drying plug. In this plug a filling of granules prevents the egress of acid droplets (acid mist), which can be produced by gas evolution during charging.

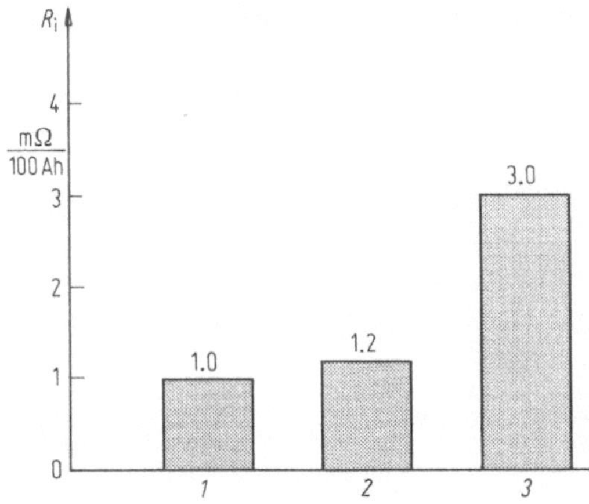

Fig. 6.11. Internal resistances $R_i$ of some stationary lead-acid batteries used in telecommunication power supplies. 1. Planté plates (GroE), 2 pasted plates (OGi-block or single cells OGiE), 3 tubular plates (OPzS)

The gas mixture itself can escape freely. These plugs are standard fittings on the OGi, OPzS and GroE cells.
– Ceramic plug. This 'flame-trap' or 'explosion-proof' plug prevents ignition of gases present in the cell due to external causes. By 'ceramic funnel plug' is understood a variant of the flame-trap plug. Topping-up water can be added through a tube, in the form of a funnel, which reaches directly into the electrolyte, without removing the plug.
– Recombination plug. The recombination plug ('recombinator') converts the explosive gas produced by charging back to water by a catalytic method, thus reducing the loss of water. This results in longer maintenance intervals for stationary batteries.

Due consideration must be given, in accordance with the manufacturer's recommendations, to the relationship between the quantity of gas given off, in the float (trickle) charging condition as well as in charging, and the capacity of the recombination plug. Under normal conditions the use of recombination plugs in stationary batteries embodying low-antimony alloys is of little interest, since maintenance intervals of up to three years are in any case achievable.

### 6.1.17 Valve-Regulated Lead-Acid Batteries

During the last 10 years, valve-regulated lead-acid batteries have gained a considerable market share in stationary applications, especially for short bridging times in UPS systems, but also in telecommunications. Batteries of this type are maintenance-free throughout their entire lifetime and mainly differ from conventional batteries in the following details:

– immobilization of the sulphuric acid electrolyte by means of a gelling process or by absorption in a felt of glass fibres (absorbent-glass-mat separator) and
– valves instead of the conventional plugs. The valves prevent the occurrence of oxygen from the surrounding air but are open to relieve excess pressure in the cells.

Additional advantageous features of valve-regulated lead-acid batteries are:

– no electrolyte leakage, even in case of container fracture,
– operation in any position,
– extremely low gassing (mainly hydrogen at the rate of self-discharge),
– no electrolyte stratification in cycle operation and
– no acid fumes.

The valve-regulated lead-acid battery is based on the internal oxygen cycle. This means that gaseous oxygen ($O_2$), generated at the positive electrode by water decomposition (oxidation of $O^{2-}$ ions) mainly during overcharging (float charging), but also as a secondary reaction during charging, is reduced to $O^{2-}$ ions when it reaches the surface of the negative electrode, and water is formed together with protons ($H^+$). Written as reaction equations:

|  at the positive electrode  |  at the negative electrode  |
|:---:|:---:|
| $2 \cdot H_2O \Rightarrow O_2 + 4 \cdot H^+ + 4 \cdot e^-$ | $O_2 + 4 \cdot H^+ + 4 \cdot e^- \Rightarrow 2 \cdot H_2O$ |

Both reactions balance each other, and the state of charge of the electrode materials is not altered. Only slight concentration gradients are caused in the electrolyte, because protons ($H^+$) that are released at the positive electrode are consumed at the negative electrode, and water is generated at the negative electrode.

*Note:* The above process is often called 'recombination', because the reduced oxygen ($O^{2-}$) together with protons ($H^+$) form water. But this is different in principle from the catalytic recombination, where gaseous oxygen and gaseous hydrogen are combined with water vapour at the surface of a catalyst. The internal oxygen cycle generates water from the negative electrode from water and protons ($H^+$ ions), but the reaction does not include gaseous hydrogen. This is important, because it emphasizes that gaseous hydrogen cannot be removed by the internal oxygen cycle.

As a prerequisite for the internal oxygen cycle, the electrolyte has to be immobilized to leave open space for the required fast gas diffusion, because the diffusion rate of gaseous oxygen is $10^7$ times (!) faster than that of oxygen solved in the electrolyte.

Two methods are used for electrolyte immobilization:

1 *Absorption* of the electrolyte *by a glass felt* of fine fibres (absorbent-glass mat). The fine fibres of these felts (diameters in the µm range) form a pore system that absorbs the electrolyte by capillary forces. Larger pores stay void when a corresponding small volume of electrolyte is filled into the battery, or after a certain share of the water was lost by electrolysis. The empty pores and the empty space beside the separators provide the gaseous phase for fast oxygen transport.

2 *Gelled electrolyte by addition of SiO$_2$*. Gelling is another way to immobilize the electrolyte. Highly dispersed mixtures of sulphuric acid with specially prepared silicon dioxide (siliceous acid, $SiO_2$) are used. They form a stable gel over a long term, even to the powerful oxidizing effects of the positive electrode. Shrinkage during solidification forms cracks that run through the solid electrolyte and render fast transport of gaseous oxygen.

Real pores cannot be observed in the gelled electrolyte, as in a felt. But the gelled sulphuric acid behaves in the same way as in a felt with a pore structure one order of magnitude finer than that of the usual microglass-mat separator.

Both ways are used to immobilize the electrolyte. A common feature of both methods is that the battery does not contain liquid electrolyte, and the advantages mentioned apply in general to both designs of valve-regulated lead-acid batteries.

Figure 6.12 illustrates this development of water loss during float charging of a battery with gelled electrolyte. At the beginning, this battery still contained liquid electrolyte which limits the efficiency of the internal oxygen cycle and

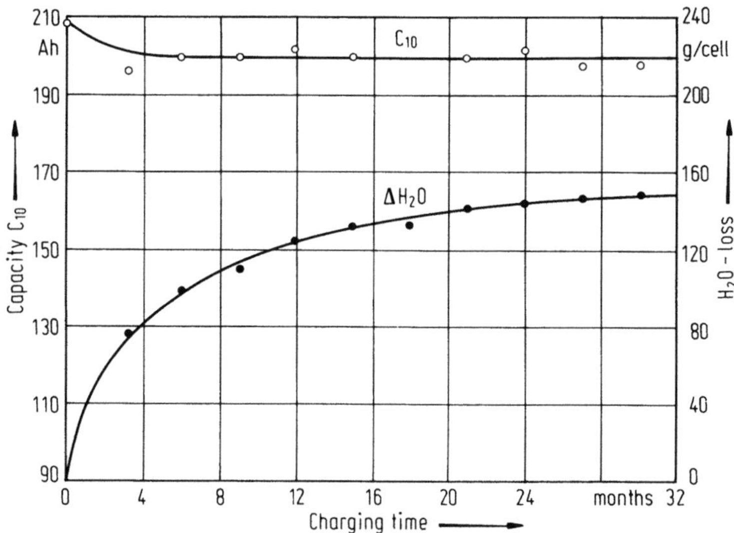


**Fig. 6.12.** Water loss and capacity development of a valve-regulated lead-acid battery with gelled electrolyte at beginning of float charge. Sonnenschein A600, 12 V 4/180 float voltage 2.23 V/cell (*Source*: Accumulatorenfabrik Sonnenschein GmbH)

causes water loss by electrolysis. But, after less than one year of float charging the electrolyte volume is so reduced that the internal oxygen cycle balances the float current, and water loss in nearly reduced to the unavoidable hydrogen evolution rate due to self-discharge. Maximum efficiency of the internal oxygen cycle is thus attained automatically. However, a gradual water loss can never be avoided in valve-regulated lead-acid batteries because:

- Hydrogen is evolved at the negative electrode, at least at the rate of self-discharge at open cell voltage ($Pb + H_2SO_4 \Rightarrow PbSO_4 + H_2$), which results in a minimum $H_2$-evolution of 1 to 2 cm$^3$ per hour and 100 Ah of nominal capacity.
- Oxygen is consumed by grid corrosion in the positive electrode ($Pb + 2 \cdot H_2O \Rightarrow PbO_2 + 4 \cdot H^+$) at a rate comparable to the above hydrogen evolution.

In total both reactions result in a gradual water loss which does not limit service life under normal operational conditions. But hydrogen evolution and grid corrosion are temperature dependent and their rate is approximately doubled per each 10 °C. At high temperatures, a dramatic reduction of service life can result. For this reason, valve-regulated lead-acid batteries are more sensitive to operational conditions than vented ones.

The slight decrease of the 10-hour capacity (upper curve in Fig. 6.12) is also caused by water loss. The utilization of the active material is slightly reduced with immobilized electrolyte, because ion transport is only possible by diffusion, convection of the electrolyte is no longer possible. Thus, valve-regulated lead-acid batteries must contain 10 to 15% more active material to reach the same capacity compared to their flooded design.

The low hydrogen evolution rate has already been taken into account in VDE 0510, where the ventilation needed for battery rooms equipped with such batteries is only 25% of that for conventional batteries.

Because the electrolyte of such batteries is leak-proof, no special precautions against acid spillage are required for the battery rooms.

When cycling batteries with liquid electrolytes, acid stratifaction occurs if charging is such that no gas is formed as a result of electrolysis of water, which would act to stir the electrolyte. If special precautions are not taken to prevent acid layering, the result is premature battery failure, especially due to corrosion and sulphate formation in the lower plate regions.

The resulting acid stratification in the same cell design but with different electrolyte immobilization is shown in Fig. 6.13.

- in the flooded cell, after 10 cycles the acid density increased at the bottom to $1.29\,g/cm^3$, while an acid density of $1.17\,g/cm^3$ is observed at the top,
- in the cell with absorbent-glass-mat separation some stratification is to be seen, although much smaller compared to the flooded cell. Furthermore, the grade stratification depends on the filling grade of the glass mat. Less acid volume results in less stratification,
- in the gelled electrolyte, acid stratification can be neglected.

Figure 6.13 indicates that there are marginal differences between the two methods of electrolyte immobilization. The stronger immobilization of gelled electrolyte has the advantage that even in tall cells stratification is no matter. For batteries

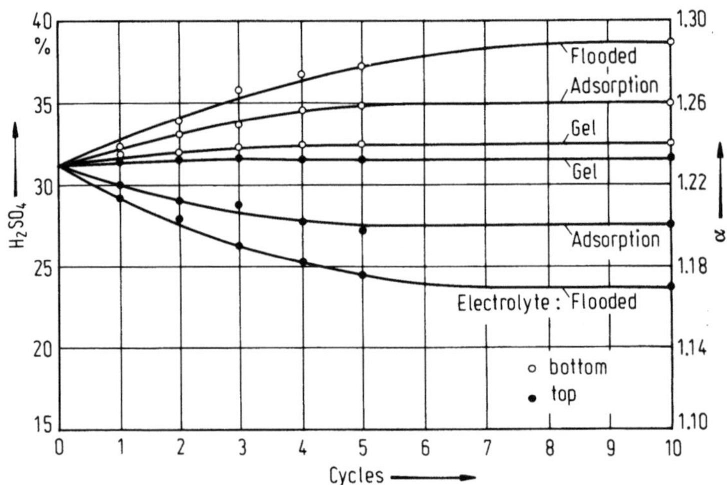


**Fig. 6.13.** Electrolyte stratification with different technology. OPzS design, nom. capacity 350 Ah. (*Source*: Accumulatorenfabrik Sonnenschein GmbH)

**Fig. 6.14.** Valve-regulated-block battery with pasted plates and gelled electrolyte (OGiV). (Photo by courtesy of Accumulatorenfabrik Sonnenschein GmbH)

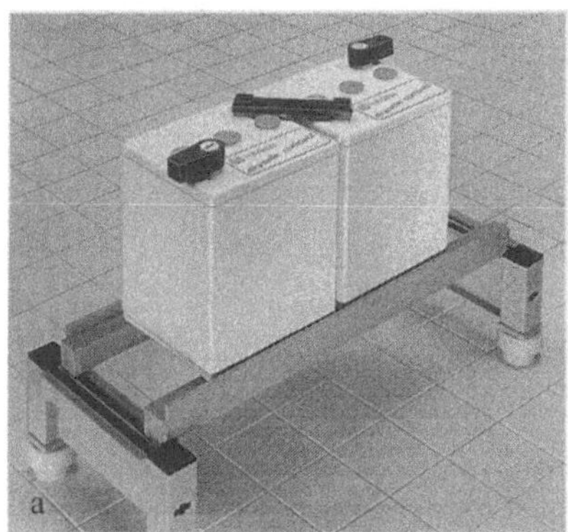

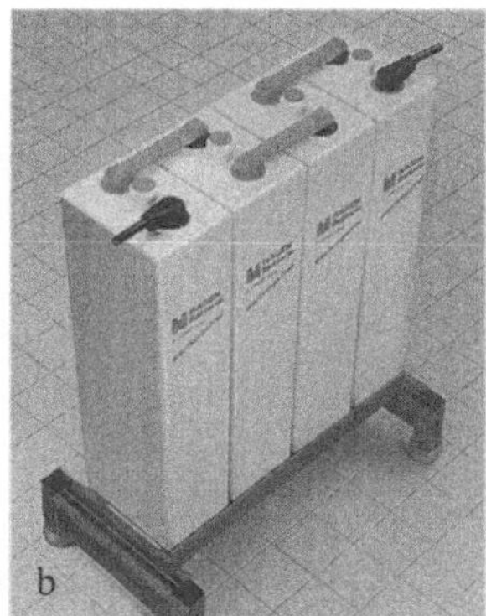

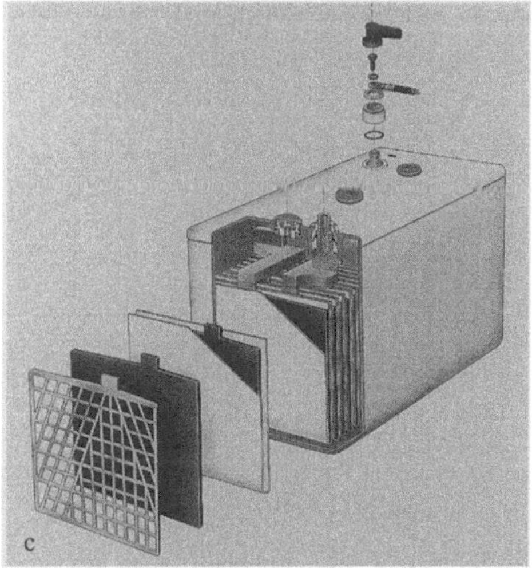

**Fig. 6.15a–c.** Valve-regulated-block battery with pasted plates and a absorbent-glass-mat separator (OGiV). (Photo by courtesy of Hagen Batteries AG)

with absorbent-glass-mat separation the cell height is usually restricted to about 30 cm; taller cells are recommended to be operated in the horizontal position. Furthermore the larger pore sizes in absorbent-glass-mat separators cause the effect that water loss occurs only at the expense of the separator, while the fine pores in the active material remain filled. As a result, the electrolyte resistance is more increased in unfavourable operational conditions.

On the other hand, absorbent-glass-mats do not require an additional separator to prevent the growth of lead dendrites which may penetrate the separator and cause short circuits. Thus, valve-regulated lead-acid batteries with absorbent-glass-mat separators can be designed with a lower internal resistance.

Valve-regulated lead-acid batteries are manufactured as OGiV with positive grid plates in 4- or 6-V modules as block batteries in the capacity range from 12.5 Ah to 200 Ah (with an expected service life of between 4 and 10 years) (Figs. 6.14 and 6.15).

**Fig. 6.16.** Valve-regulated-single-cell battery with tubular plates and gelled electrolyte (OPzV). (Photo by courtesy of Accumulatorenfabrik Sonnenschein GmbH)

**Fig. 6.17a, b.** Valve-regulated-block battery with pasted plates and absorbent-glass-mat separator (OGiV) **(a)** and valve-regulated-block battery with pasted plates and gelled electrolyte (OGiV) **(b)**. (Photo by courtesy of Varta AG)

**Fig. 6.18.** Valve regulated-single-cell battery with tubular plates and gelled electrolyte (OPzV) in horizontal installation. (Photo by courtesy of Accumulatorenfabrik Sonnenschein GmbH)

On the market they are also supplied as single-cell OPzV in 2-V units from 200 Ah to 3000 Ah (with a expected service life of about 10 years) which employ positive tubular cells and negative grid plates (Fig. 6.16).

Figure 6.17a and b show other types of modern valve-regulated lead-acid batteries.

For certain special applications, such as building into shelters or battery racks, it is possible to install batteries horizontally (Fig. 6.18). There are high-power-high-current versions for use in UPS systems, which are especially suited to high current drain over short periods of time.

## 6.2 Stationary Alkaline Batteries

*Nickel-cadmium Batteries.* The nickel-cadmium battery is based on nickel hydroxide (NiOOH) and cadmium (Cd) as positive and negative electrode material. The electrochemical reactions during discharge are:

- the oxidation of cadmium $Cd \Rightarrow Cd^{2+} + 2 \cdot e^-$ at the negative electrode,
- the reduction of trivalent nickel to the bivalent ion $Ni^{3+} + e^- \Rightarrow Ni^{2+}$ at the positive electrode.

During charging both reactions are reversed.

As in the lead-acid battery, all reactants are solids of low solubility and the reactions are fairly reversible.

The advantages are:

- potassium hydroxide (KOH) does not (practically) take part in the electrochemical reaction. Only water that represents the gross of the electrolyte is consumed (or released). Thus, only small concentration changes occur.

The consequences for battery practice are:

- spacing between electrodes can be minimized. This means low electrolyte resistance and suitability for extremely high loads;
- the freezing point of the electrolyte hardly changes, so there are no problems even at very low temperatures ($-40\,°C$);
- electrolyte stratification does not occur, boost charging is always possible.
- corrosion of metallic parts is rarely observed (plate supports and conducting elements are made of nickel). The consequence is:
  - nickel-cadmium batteries can be stored for practically unlimited periods without refreshing charges.
- hydrogen evolution does not occur at the equilibrium potential of the cadmium electrode. The consequence is:
  - nickel-cadmium batteries can be made 'sealed'.

The disadvantages of nickel-cadmium batteries are:

- the cell voltage of only 1.2 V means that about 66% more cells are required, compared to a lead-acid battery of the same voltage. The larger number of cells causes increased maintenance expenditures;
- the charge efficiency of nickel-cadmium batteries is inferior to that of lead-acid batteries;
- nickel-cadmium batteries are about twice as expensive as lead-acid batteries of similar nominal capacities, but this is compensated for high load applications by superior high-rate performance. Thus, smaller capacities can be used with nickel-cadmium batteries.

In telecommunication power supplies most of their advantages have no or little relevance. So, they are employed only seldom and shall be treated only briefly.


*Electrode Design.* A considerable variety of electrode design characterizes nickel-cadmium batteries. The most important plate types are:

- *Pocket plates* which consist of strips of active material (about 1 cm wide and 3 mm thick) that are wrapped in perforated nickel sheets (or nickel plated steel sheets), called 'pockets'. These pockets are clamped together by U-formed nickel strips at both ends, and form an electrode. Graphite is employed as a conducting aid that is added to the active material.
- *Sintered plates* which are based on a spongy substrate of sintered-nickel powder. The material is sintered on both sides of a perforated nickel (or nickel-plated) sheet, that later represents the current conductor and lug of the plate. The active material is deposited from a dissolved state by chemical or electrochemical reactions into the fine pores of the sintered nickel and forms a thin layer with excellent electrical connection to the substrate. Sintered plates are characterized by their outstanding high load performance.
- In *fibre plates* (FNC electrodes) the substrate is a fleece of nickel-plated fibres. The organic or carbon fibres are left underneath the coating or removed by

thermal decomposition. The active material is precipitated from the dissovled state or filled into the fleece as a slurry. The electrochemical behaviour corresponds to that of sintered plates.
- *Plastics-bonded cadmium electrodes* represent a variety of pressed electrodes used as negative electrodes in sealed batteries. They were developed to replace the more expensive sintered electrodes.

### 6.2.1 Nickel-Cadmium Batteries in Telecommunications

For normal telecommunication nickel-cadmium batteries with pocket plates are mainly employed.

*Construction.* TP and T nickel-cadmium type batteries are illustrated in Fig. 6.19 (a) and (b). Fig. 6.19 (c) shows their internal construction, with pocket plates. Both the positive and the negative electrodes are pocket plates. The plates are insulated by a plastic mesh or a perforated corrugated separator. The electrolyte is dilute potassium hydroxide with a density of 1.20 kg/l.

*Design.* Individual cells type TP are supplied in impact-resisting, temperature-stable, translucent plastic containers with capacities of from 10 to 315 Ah, and those of type T in steel containers with capacities of from 380 to 1250 Ah (capacities 5-hour rated).

*Application.* Ni-Cd pocket-plate cells types TP and T are very suitable for long discharge periods; they are preferably applied, therefore, in cases where mains supply interruptions of 1 to 10 h are required.

*Charging.* In general, an initial charging current of $1.5 \times I_5$ up to a voltage of 1.55 V/cell is recommended. The maximum continuous voltage permitted in a

**Fig. 6.19a–c.** Nickel-Cadmium type batteries TP (**a**) and T (**b**). (Photo by courtesy of Varta AG)

telephone system is for example 58 V. A battery of 38 cells can therefore be charged only up to 1.52 V/cell.

TP cells should be floated at 1.4 V/cell to maintain their charge. Under this condition they draw a current of 30 to 60 mA/100 Ah ($C_5$). The permissible operating temperature range is from $-20$ to $+45\,°C$. An equalizing charge is necessary, in accordance with the manufacturer's instructions, in order to maintain full capacity.

*Characteristic data.*

- nominal voltage: 1.2 V/cell,
- nominal capacity ($C_5$), related to a nominal charging current $I_5$ of $0.2 \times C_5$ (A) up to a final discharge voltage of 1.0 V/cell at $20\,°C$,
- charging voltage: 1.55 V/cell,
- float charging voltage: 1.4 V/cell,
- charging factor: 1.4 (in cycle applications),
- electrolyte: potassium hydroxide with a density of $1.20 \pm 0.02$ kg/l and
- internal resistance: approximately $1.4\,m\Omega/100$ Ah.

*Installation.* Blocks of multi-cell arrangements of TP batteries and individual T-type cells are installed on stillages, or in tiered or stepped racks.

*Service life.* The life of TP and T-type batteries is between 10 and 15 years.


## 6.3 Lithium Primary Batteries

The use of primary batteries for standby applications is of interest when the required energy is so low that the energy content is sufficient for the whole service life and recharging is thus not needed. Lead-acid and nickel-cadmium batteries are not suited for such applications, because their self-discharge would limit service life too early.

The lithium primary battery with organic electrolyte represents a battery with a self-discharge rate of less than 1% per year. For this reason it is suitable (and often used) for memory backup. It represents light-weight energy sources supplying long-term, reliable power under a wide range of ambient conditions.

The lithium battery can be incorporated as a permanent component in an electronic circuit and displays the following features:

- high cell voltage: rated voltage 3 V, open-circuit voltage up to 3.67 V;
- broad temperature range: $-55\,°C$ to $+75\,°C$;
- high energy density: 630 Wh/kg or $1240$ Wh/dm$^3$;
- long storage life and high reliability: 10 years' storage life at room temperature, capacity loss $< 1\%$ per year;
- life: 10 years;
- safety: no internal pressure, no gassing.

## 6.4 Design Pointers

### 6.4.1 Standby Time (Backup Time)

The following indications apply exclusively to lead-acid batteries. In d.c. telecommunications power supply systems the standby time (backup time) specified may range from minutes to hours. A period of 1–4 hours is typical. A standby time of 6 hours is generally provided for the German Postal Administration Telekom Network. In installations which include a stationary standby generator, a period of 4 hours is allowed. In unmanned local exchanges a battery reserve of up to 72 hours is provided.

### 6.4.2 Number of Cells

#### D.C. Telecommunications Power Supply System

The number of battery cells is determined by the upper and lower voltage limits of the communications system. In 48 V systems, lead-acid batteries of 24, 25 or 26 cells are used. In 60 V systems, it is customary to use either 30 or 31 cells. Either 25 cells or 26 may be chosen, or 31 in preference to 30. This enables the battery to be more deeply discharged, and thus increases its usable capacity. It must be borne in mind, however, that at the beginning of the discharge the voltage applied to the load is higher by the voltage of one cell.

Normally 25 cells are used in 48 V systems. The system conditioned for the final voltage $U_{smin}$ is 1.83 V/cell.

Alternatively it is possible to choose, for example, 24 rather than 25 cells. A lead-acid battery of 24 cells can be used in the parallel mode if the permissible upper limit of continuous voltage is not exceeded in either the float (trickle) charging conditions (2.23 V $\times$ 24 = 53.52 V) or the charging condition (2.33 V $\times$ 24 = 55.92 V).

24-cell batteries (8 $\times$ 6 V blocks) are used for example in small to medium size telecommunications systems. In this case the battery can also be installed in the power supply cabinet, in the cabinet of the telecommunications system or in a battery cabinet. The system conditioned for the final voltage $U_{smin}$ in this case is 1.87 V/cell.

#### UPS Systems

In UPS systems the nominal voltage may be, for example, 220 or 440 V, and the corresponding number of cells 110 or 220. The capacity is normally based upon standby times of from 5 to 60 minutes.

### 6.4.3 Capacity

The projected capacity depends upon the required standby time and the permitted voltage limits for the communications system. It can be arrived at approximately as follows:

$$\text{Nominal capacity} = \text{standby time} \times I_{HV} \times \text{mean correction factor}$$

where standby time is the period for which the battery is required to supply the communications system when necessary.

$I_{HV}$ is the mean current drawn by the communications system during the busy period.

The mean correction factor depends upon the specified standby time, the system conditioned for the final voltage and the battery construction; it is given in Table 6.2 for OPzS batteries with commonly adopted standby times and system conditioned final voltages.

The mean current $I_{HV}$ in a communications system during the busy period is determined mainly by the amount of traffic and the number of connected lines (subscribers' or junction lines). This value is obtained by multiplying the current consumption of the individual circuit elements by the traffic intensity (erlang) with which these circuit elements are loaded. The values can usually be obtained from the trunking diagram of the communications system.

*Example of the determination of battery capacity for a digital electronic switching system (EWSD)*
Given:

– mean current during the busy period $I_{HV} = 330\,\text{A}$ (typical for 5000 subscribers);
– required standby time: 2h;

**Table 6.2.** Mean correction factor for the determination of tubular (OPzS) battery capacity (accurate for 100 Ah plates)

| Communications system | ESK 10 000 E and EWSA | EMS or EWSD and KN system | EMD Export 48 V | EMD Export 60 V | Direct loads |
|---|---|---|---|---|---|
| System-conditioned final voltage $U_{Smin}$ (V) | 46.6 | 44 or 45.8 | 46 | 58 | 57.4 |
| System-conditioned final voltage $U_{Smin}$/cell (V) | 1.79 | 1.83 | 1.84 | 1.87 | 1.92 |
| Number of cells | 26 | 24 or 25 | 25 | 31 | 30 |
| Standby time (h) | Mean correction factor | | | | |
| 1 | 2.41 | 2.63 | 2.87 | 3.10 | 4.31 |
| 2 | 1.70 | 1.84 | 1.94 | 2.11 | 2.89 |
| 3 | 1.44 | 1.54 | 1.61 | 1.73 | 2.25 |
| 4 | 1.30 | 1.38 | 1.42 | 1.52 | 1.92 |
| 5 | 1.19 | 1.28 | 1.31 | 1.41 | 1.75 |
| 6 | 1.13 | 1.22 | 1.23 | 1.35 | 1.67 |

– nominal system voltage: 48 V;
– permissible continuous tolerance on operating voltage, measured at the assemblies:
  (a) for apparatus connected directly to the operating voltage (direct loads): 44 to 58 V,
  (b) for d.c./d.c. converters and inverters: 40 to 75 V;
– maximum permitted voltage drop in distribution system for example 1.8 V[4]

Required:

– battery capacity.

Since both the direct loads and the d.c./d.c. converters and inverters are fed from a *single* power supply system, the significant voltage limits for the dimensioning of the battery are the narrower ones for the direct loads. Taking the voltage drops into account, the system-conditioned final voltage is

$$U_{\text{Smin}} = 44\,\text{V} + 1.8\,\text{V} = 45.8\,\text{V}.$$

In terms of one battery cell, this becomes 45.8 V/25 $\approx$ 1.83 V/cell.

Having regard to the required standby time and $I_{\text{HV}}$, the capacity in ampere-hours can now be read from the battery manufacturer's data.

The discharge current for determining the battery size is calculated as follows, taking the mean battery discharge voltage into account:

$$I_{\text{E}} = \frac{I_{\text{A}} \cdot U_{\text{N}}}{U_{\text{E}}}$$

where $I_{\text{E}}$ Discharge current, $I_{\text{A}}$ load current at rated system voltage, $U_{\text{N}}$ rated system voltage, $U_{\text{E}}$ mean discharge voltage approx. 46 V for 25-cell battery.

Table 6.3 shows, as an example, the range of types and the selection guide for the OPzS individual-cell batteries. In addition, the variation in capacity and final discharge voltage $U_{\text{S}}$ (relative to periods of 1, 3, 5, 8 and 10 h) is indicated for each battery type.

The discharge currents given in Table 6.4 correspond to the system-conditioned final voltage $U_{\text{Smin}}$ of 1.83 V/cell (for a 25-cell lead-acid battery) derived above for the example of a communications system, based upon standby periods of from 0.5 to 10 h. In the example, an $I_{\text{HV}}$ of 330 A was assumed. Here a discharge current of 330 A is found for a standby time of 2 h. If the corresponding value of nominal capacity $C_{10}$ is traced to the left of the table, it is found to be 1200 Ah (at the ten-hour rate).

In power supply systems with battery capacities greater than 250 Ah e.g. for public communications systems, the battery will be divided into two groups for safety reasons. Instead of one battery of the calculated capacity, two batteries are selected, each of half the capacity.

---

[4] or 2.7 V (depending on the system configuration).

**Table 6.3** Types of tubular (OPzS) individual-cell batteries with data (approximate values). (Table by courtesy of Varta AG)

| Type designation | Type-No. | Capacity [Ah] | | | | | Cell dimensions [mm] | | | Cell dimensions [inch] | | | Weight [kg] | | Weight [lbs] | |
|---|---|---|---|---|---|---|---|---|---|---|---|---|---|---|---|---|
| | | $K_{10}$ | $K_8$ | $K_5$ | $K_3$ | $K_1$ | L | W | H | L | W | H | Cell wt w.elec. | Electrol. Wt | Cell wt w.elec. | Electrol. wt |
| | | Final voltage $U_s/V$ | | | | | | | | | | | | | | |
| | | 1.8 | 1.75 | 1.78 | 1.765 | 1.71 | | | | | | | | | | |
| 4 OPzS 200 | 1705004175 | 200 | 199 | 172 | 150 | 108 | 103 | 206 | 405 | 4.06 | 8.11 | 15.94 | 17.3 | 4.9 | 38.1 | 10.8 |
| 5 OPzS 250 | 1705005175 | 250 | 249 | 215 | 188 | 135 | 124 | 206 | 405 | 4.88 | 8.11 | 15.94 | 20.8 | 6.0 | 45.8 | 13.2 |
| 6 OPzS 300 | 1705006175 | 300 | 298 | 258 | 225 | 162 | 145 | 206 | 405 | 5.71 | 8.11 | 15.94 | 24.3 | 7.2 | 53.6 | 15.9 |
| | | Final voltage $U_s/V$ | | | | | | | | | | | | | | |
| | | 1.795 | 1.75 | 1.765 | 1.745 | 1.675 | | | | | | | | | | |
| 5 OPzS 350 | 1707005175 | 350 | 345 | 303 | 268 | 231 | 124 | 206 | 520 | 4.88 | 8.11 | 20.47 | 26.9 | 7.9 | 59.3 | 17.4 |
| 6 OPzS 420 | 1707006175 | 420 | 414 | 363 | 321 | 270 | 145 | 206 | 520 | 5.71 | 8.11 | 20.47 | 31.5 | 9.4 | 69.4 | 20.7 |
| 7 OPzS 490 | 1707007175 | 490 | 483 | 424 | 375 | 324 | 166 | 206 | 520 | 6.54 | 8.11 | 20.47 | 36.1 | 10.9 | 79.6 | 24.0 |
| | | Final voltage $U_s/V$ | | | | | | | | | | | | | | |
| | | 1.79 | 1.75 | 1.77 | 1.745 | 1.67 | | | | | | | | | | |
| 6 OPzS 600 | 1710006175 | 600 | 549 | 519 | 450 | 324 | 145 | 206 | 695 | 5.71 | 8.11 | 27.36 | 44.8 | 12.9 | 98.7 | 28.4 |
| 8 OPzS 800 | 1710008175 | 800 | 732 | 692 | 600 | 432 | 210 | 191 | 695 | 8.27 | 7.52 | 27.36 | 61.3 | 16.9 | 135.1 | 37.2 |
| 10 OPzS 1000 | 1710010175 | 1000 | 915 | 865 | 750 | 540 | 210 | 233 | 695 | 8.27 | 9.17 | 27.36 | 74.6 | 21.1 | 164.4 | 46.5 |
| 12 OPzS 1200 | 1710012175 | 1200 | 1098 | 1038 | 900 | 648 | 210 | 275 | 695 | 8.27 | 10.83 | 27.36 | 88.0 | 25.2 | 194.0 | 55.5 |
| | | Final voltage $U_s/V$ | | | | | | | | | | | | | | |
| | | 1.79 | 1.75 | 1.76 | 1.735 | 1.64 | | | | | | | | | | |
| 12 OPzS 1500 | 1712512175 | 1500 | 1512 | 1350 | 1188 | 828 | 210 | 275 | 845 | 8.27 | 10.83 | 33.27 | 114.4 | 34.2 | 252.1 | 75.4 |
| 15 OPzS 1875 | 1712515175 | 1875 | 1890 | 1688 | 1485 | 1035 | 212 | 397 | 822 | 8.35 | 15.63 | 32.36 | 145.0 | 51.0 | 319.6 | 112.4 |
| 16 OPzS 2000 | 1712516175 | 2000 | 2016 | 1800 | 1584 | 1104 | 212 | 397 | 822 | 8.35 | 15.63 | 32.36 | 153.0 | 48.0 | 337.2 | 105.8 |
| 20 OPzS 2500 | 1712520375 | 2500 | 2520 | 2250 | 1980 | 1380 | 212 | 487 | 822 | 8.35 | 19.17 | 32.36 | 190.0 | 60.0 | 418.8 | 132.2 |
| 24 OPzS 3000 | 1712524375 | 3000 | 3024 | 2700 | 2376 | 1656 | 212 | 576 | 822 | 8.35 | 22.68 | 32.36 | 225.0 | 72.0 | 495.9 | 158.7 |

| | | | | Final voltage $U_\mathrm{s}/V$ | | | | | | | | | | | | | | |
|---|---|---|---|---|---|---|---|---|---|---|---|---|---|---|---|---|---|---|
| | | | | 1.79 | 1.75 | 1.775 | 1.73 | 1.63 | | | | | | | | | | |
| 14 | OPzS | 3500 | 1725014491 | 3500 | 3542 | 3150 | 2772 | 1932 | 430 | 412 | 890 | 16.93 | 16.22 | 35.04 | 312 | 92 | 687.8 | 202.8 |
| 16 | OPzS | 4000 | 1725016491 | 4000 | 4048 | 3600 | 3168 | 2208 | 430 | 456 | 890 | 16.93 | 17.95 | 35.04 | 356 | 104 | 784.8 | 229.3 |
| 18 | OPzS | 4500 | 1725018491 | 4500 | 4554 | 4050 | 3564 | 2484 | 430 | 500 | 890 | 16.93 | 19.69 | 35.04 | 399 | 115 | 879.6 | 253.5 |
| 20 | OPzS | 5000 | 1725020491 | 5000 | 5060 | 4500 | 3960 | 2760 | 430 | 554 | 890 | 16.93 | 21.81 | 35.04 | 442 | 127 | 974.4 | 280.0 |
| 22 | OPzS | 5500 | 1725022491 | 5500 | 5566 | 4950 | 4356 | 3036 | 430 | 598 | 890 | 16.93 | 23.54 | 35.04 | 478 | 138 | 1053.8 | 304.2 |
| 24 | OPzS | 6000 | 1725024491 | 6000 | 6072 | 5400 | 4752 | 3312 | 430 | 642 | 890 | 16.93 | 25.28 | 35.04 | 507 | 157 | 1117.7 | 346.1 |
| 28 | OPzS | 7000 | 1725028491 | 7000 | 7084 | 6300 | 5544 | 3864 | 430 | 882 | 890 | 16.93 | 34.72 | 35.04 | 644 | 214 | 1419.8 | 471.8 |
| 32 | OPzS | 8000 | 1725032491 | 8000 | 8096 | 7200 | 6336 | 4416 | 430 | 882 | 890 | 16.93 | 34.72 | 35.04 | 680 | 210 | 1499.1 | 463.0 |
| 36 | OPzS | 9000 | 1725036491 | 9000 | 9108 | 8100 | 7128 | 4968 | 430 | 1078 | 890 | 16.93 | 42.44 | 35.04 | 816 | 263 | 1799.0 | 579.8 |
| 40 | OPzS | 10000 | 1725040491 | 10000 | 10120 | 9000 | 7920 | 5520 | 430 | 1078 | 890 | 16.93 | 42.44 | 35.04 | 855 | 260 | 1884.9 | 573.2 |
| 44 | OPzS | 11000 | 1725044491 | 11000 | 11132 | 9900 | 8712 | 6072 | 430 | 1254 | 890 | 16.93 | 49.37 | 35.04 | 961 | 309 | 2118.6 | 681.2 |
| 48 | OPzS | 12000 | 1725048491 | 12000 | 12144 | 10800 | 9504 | 6624 | 430 | 1254 | 890 | 16.93 | 49.37 | 35.04 | 1000 | 306 | 2204.6 | 674.6 |

| | | | | Final voltage $U_\mathrm{s}/V$ | | | | | | | | | | | | | | |
|---|---|---|---|---|---|---|---|---|---|---|---|---|---|---|---|---|---|---|
| | | | | 1.80 | 1.75 | 1.780 | 1.765 | 1.710 | | | | | | | | | | |
| [a] 6V 3 OPzS | | 150 | 1405030305 | 150 | 149 | 129 | 113 | 81 | 226 | 220 | 380 | 8.90 | 8.66 | 14.96 | 36.7 | 5.2 | 80.9 | 11.5 |

[a] 6V-bloc batteries

The capacity values shown in the tables relate to loadings from a fully charged condition and an ambient temperature of 20 °C.
Connector losses are taken into account.

**Table 6.4.** Project planning data, e.g. tubular range 4 OPzS 200 to 48 OPzS 12 000-discharge currents for a system-conditioned final discharge voltage $U_{smin}$ of 1.83 V/cell (approximate values). (Table by courtesy of Varta AG)

| $U_s = 1.83$ V/cell discharge currents (A) | | | | | | | | | | |
|---|---|---|---|---|---|---|---|---|---|---|
| Typ | | 30′ | 1 h | 2 h | 3 h | 4 h | 5 h | 6 h | 8 h | 10 h |
| [a]6V3 OPzS | 150 | 87.3 | 64.8 | 43.5 | 33.6 | 27.6 | 23.7 | 20.9 | 17.0 | 14.1 |
| 4 OPzS | 200 | 116 | 86.4 | 58.0 | 44.8 | 36.8 | 31.6 | 27.8 | 22.6 | 18.8 |
| 5 OPzS | 250 | 145 | 108 | 72.5 | 56.0 | 46.0 | 39.5 | 34.8 | 28.3 | 23.5 |
| 6 OPzS | 300 | 174 | 129 | 87.0 | 67.2 | 55.2 | 47.4 | 41.7 | 33.9 | 28.2 |
| 5 OPzS | 350 | 180 | 140 | 99.0 | 77.5 | 63.5 | 55.0 | 48.5 | 39.5 | 33.0 |
| 6 OPzS | 420 | 216 | 168 | 118 | 93.0 | 76.2 | 66.0 | 58.2 | 47.4 | 39.6 |
| 7 OPzS | 490 | 252 | 196 | 138 | 108 | 88.9 | 77.0 | 67.9 | 55.3 | 46.2 |
| 6 OPzS | 600 | 293 | 229 | 165 | 130 | 108 | 93.6 | 82.2 | 66.6 | 56.1 |
| 8 OPzS | 800 | 391 | 306 | 220 | 173 | 144 | 124 | 109 | 88.8 | 74.8 |
| 10 OPzS | 1000 | 489 | 383 | 275 | 217 | 181 | 156 | 137 | 111 | 93.5 |
| 12 OPzS | 1200 | 586 | 459 | 330 | 260 | 217 | 187 | 164 | 133 | 112 |
| 12 OPzS | 1500 | 672 | 552 | 420 | 342 | 291 | 252 | 222 | 180 | 150 |
| 15 OPzS | 1875 | 840 | 690 | 525 | 427 | 364 | 315 | 277 | 225 | 187 |
| 16 OPzS | 2000 | 896 | 736 | 560 | 456 | 388 | 336 | 296 | 240 | 200 |
| 20 OPzS | 2500 | 1120 | 920 | 700 | 570 | 486 | 420 | 370 | 300 | 250 |
| 24 OPzS | 3000 | 1344 | 1104 | 840 | 684 | 583 | 504 | 444 | 360 | 300 |
| 14 OPzS | 3500 | 1358 | 1148 | 910 | 756 | 651 | 567 | 504 | 413 | 343 |
| 16 OPzS | 4000 | 1552 | 1312 | 1040 | 864 | 744 | 648 | 576 | 472 | 392 |
| 18 OPzS | 4500 | 1746 | 1476 | 1170 | 972 | 837 | 729 | 648 | 531 | 441 |
| 20 OPzS | 5000 | 1940 | 1640 | 1300 | 1080 | 930 | 810 | 720 | 590 | 490 |
| 22 OPzS | 5500 | 2134 | 1804 | 1430 | 1188 | 1023 | 891 | 792 | 649 | 539 |
| 24 OPzS | 6000 | 2328 | 1968 | 1560 | 1296 | 1116 | 972 | 864 | 708 | 588 |
| 28 OPzS | 7000 | 2716 | 2296 | 1820 | 1512 | 1302 | 1134 | 1008 | 826 | 686 |
| 32 OPzS | 8000 | 3104 | 2624 | 2080 | 1728 | 1488 | 1296 | 1152 | 944 | 784 |
| 36 OPzS | 9000 | 3492 | 2952 | 2340 | 1944 | 1674 | 1458 | 1296 | 1062 | 882 |
| 40 OPzS | 10000 | 3880 | 3280 | 2600 | 2160 | 1860 | 1620 | 1440 | 1180 | 980 |
| 44 OPzS | 11000 | 4268 | 3608 | 2860 | 2376 | 2046 | 1782 | 1584 | 1298 | 1078 |
| 48 OPzS | 12000 | 4656 | 3936 | 3120 | 2592 | 2232 | 1944 | 1728 | 1416 | 1176 |

[a] 6V-bloc batteries

The current levels shown in the tables relate to loadings from a fully charged condition and an ambient temperatures of 20 °C.

Connector losses are taken into account.

If the battery is divided into two 600 Ah groups of 25 cells each, the battery type, with its specification, is as follows:

- OPzS individual-cell battery in transparent plastic containers.
- 6 OPzS 600,
- nominal capacity $C_{10}$ 600 Ah (at the ten-hour rate),
- discharge current 165 A for 2 h standby time and a system-conditional final voltage $U_{Smin}$ of 1.83 V/cell.

**Table 6.5.** Project planning data, e.g. tubular range 4 OPzS 200 to 48 OPzS 12 000-discharge currents for a system-conditioned final discharge voltage $U_{smin}$ of 1.87 V/cell (approximate values). (Table by courtesy of Varta AG)

$U_s = 1.87$ V/cell discharge currents (A)

| Typ | | 30′ | 1 h | 2 h | 3 h | 4 h | 5 h | 6 h | 8 h | 10 h |
|---|---|---|---|---|---|---|---|---|---|---|
| [a]6V3 OPzS | 150 | 73.8 | 56.3 | 39.0 | 30.3 | 25.1 | 21.8 | 19.2 | 15.5 | 12.9 |
| 4 OPzS | 200 | 98.4 | 75.0 | 52.0 | 40.4 | 33.4 | 29.0 | 25.6 | 20.6 | 17.2 |
| 5 OPzS | 250 | 123 | 93.8 | 65.0 | 50.5 | 41.8 | 36.3 | 32.0 | 25.8 | 21.6 |
| 6 OPzS | 300 | 147 | 112 | 78.0 | 60.6 | 50.1 | 43.5 | 38.4 | 31.0 | 25.9 |
| 5 OPzS | 350 | 148 | 117 | 83.8 | 67.0 | 56.5 | 49.5 | 44.0 | 35.8 | 29.5 |
| 6 OPzS | 420 | 177 | 140 | 100 | 80.4 | 67.8 | 59.4 | 52.8 | 42.9 | 35.4 |
| 7 OPzS | 490 | 207 | 163 | 117 | 93.8 | 79.1 | 69.3 | 61.6 | 50.1 | 41.3 |
| 6 OPzS | 600 | 240 | 192 | 142 | 115 | 97.8 | 84.6 | 74.4 | 60.6 | 51.6 |
| 8 OPzS | 800 | 320 | 256 | 189 | 153 | 130 | 112 | 99.2 | 80.8 | 68.8 |
| 10 OPzS | 1000 | 400 | 320 | 237 | 192 | 163 | 141 | 124 | 101 | 86.0 |
| 12 OPzS | 1200 | 480 | 384 | 284 | 230 | 195 | 169 | 148 | 121 | 103 |
| 12 OPzS | 1500 | 540 | 456 | 354 | 294 | 252 | 222 | 198 | 162 | 138 |
| 15 OPzS | 1875 | 675 | 570 | 442 | 367 | 315 | 277 | 247 | 202 | 172 |
| 16 OPzS | 2000 | 720 | 608 | 472 | 392 | 336 | 296 | 264 | 216 | 184 |
| 20 OPzS | 2500 | 900 | 760 | 590 | 490 | 420 | 370 | 330 | 270 | 230 |
| 24 OPzS | 3000 | 1080 | 912 | 708 | 588 | 504 | 444 | 396 | 324 | 276 |
| 14 OPzS | 3500 | 1106 | 952 | 756 | 630 | 560 | 497 | 448 | 378 | 315 |
| 16 OPzS | 4000 | 1264 | 1088 | 864 | 720 | 640 | 568 | 512 | 432 | 360 |
| 18 OPzS | 4500 | 1422 | 1224 | 972 | 810 | 720 | 639 | 576 | 486 | 405 |
| 20 OPzS | 5000 | 1580 | 1360 | 1080 | 900 | 800 | 710 | 640 | 540 | 450 |
| 22 OPzS | 5500 | 1738 | 1496 | 1188 | 990 | 880 | 781 | 704 | 594 | 495 |
| 24 OPzS | 6000 | 1896 | 1632 | 1296 | 1080 | 960 | 852 | 768 | 648 | 540 |
| 28 OPzS | 7000 | 2212 | 1904 | 1512 | 1260 | 1120 | 994 | 896 | 756 | 630 |
| 32 OPzS | 8000 | 2528 | 2176 | 1728 | 1440 | 1280 | 1136 | 1024 | 864 | 720 |
| 36 OPzS | 9000 | 2844 | 2448 | 1944 | 1620 | 1440 | 1278 | 1152 | 972 | 810 |
| 40 OPzS | 10000 | 3160 | 2720 | 2160 | 1800 | 1600 | 1420 | 1280 | 1080 | 900 |
| 44 OPzS | 11000 | 3476 | 2992 | 2376 | 1980 | 1760 | 1562 | 1408 | 1188 | 990 |
| 48 OPzS | 12000 | 3792 | 3264 | 2592 | 2160 | 1920 | 1704 | 1536 | 1296 | 1080 |

[a] 6V-bloc batteries

The current levels shown in the tables relate to loadings from a fully charged condition and an ambient temperatures of 20 °C.

Connector losses are taken into account.

Tables 6.5 to 6.7 shows other different examples of project planning data-discharge currents and discharge power for $U_S$ 1.83 and $U_S$ 1.87 V/cell.

## 6.5 Maintenance

Surveillance of stationary batteries during float charge is important to ensure that the battery is kept fully charged and will deliver its expected capacity in an

**Table 6.6.** Project planning data, e.g. tubular range 4 OPzS 200 to 48 OPzS 12 000-discharge power for a system-conditioned final discharge voltage $U_{smin}$ of 1.83 V/cell (approximate values). (Table by courtesy of Varta AG)

$U_s = 1.83$ V/cell discharge power (W/cell)

| Typ | | 30′ | 1 h | 2 h | 3 h | 4 h | 5 h | 6 h | 8 h | 10 h |
|---|---|---|---|---|---|---|---|---|---|---|
| [a]6V3 OPzS | 150 | 162 | 121 | 82.5 | 64.2 | 52.8 | 45.9 | 40.2 | 32.4 | 27.6 |
| 4 OPzS | 200 | 216 | 162 | 110 | 85.6 | 70.4 | 61.2 | 53.6 | 43.2 | 36.8 |
| 5 OPzS | 250 | 270 | 202 | 137 | 107 | 88.0 | 76.5 | 67.0 | 54.0 | 46.0 |
| 6 OPzS | 300 | 324 | 243 | 165 | 128 | 105 | 91.8 | 80.4 | 64.8 | 55.2 |
| 5 OPzS | 350 | 332 | 260 | 186 | 146 | 120 | 105 | 93.0 | 75.5 | 63.5 |
| 6 OPzS | 420 | 399 | 312 | 223 | 175 | 144 | 126 | 111 | 90.6 | 76.2 |
| 7 OPzS | 490 | 465 | 364 | 261 | 205 | 168 | 147 | 130 | 105 | 88.9 |
| 6 OPzS | 600 | 540 | 426 | 309 | 249 | 206 | 180 | 158 | 127 | 108 |
| 8 OPzS | 800 | 720 | 568 | 412 | 332 | 275 | 240 | 211 | 170 | 144 |
| 10 OPzS | 1000 | 900 | 710 | 515 | 415 | 344 | 300 | 264 | 213 | 181 |
| 12 OPzS | 1200 | 1080 | 852 | 618 | 498 | 412 | 360 | 316 | 255 | 217 |
| 12 OPzS | 1500 | 1140 | 948 | 732 | 612 | 516 | 462 | 408 | 334 | 282 |
| 15 OPzS | 1875 | 1425 | 1185 | 915 | 765 | 645 | 577.5 | 510 | 418 | 352 |
| 16 OPzS | 2000 | 1520 | 1264 | 976 | 816 | 688 | 616 | 544 | 446 | 376 |
| 20 OPzS | 2500 | 1900 | 1580 | 1220 | 1020 | 860 | 770 | 680 | 558 | 470 |
| 24 OPzS | 3000 | 2280 | 1896 | 1464 | 1224 | 1032 | 924 | 816 | 669 | 564 |
| 14 OPzS | 3500 | 2562 | 2142 | 1736 | 1442 | 1232 | 1092 | 966 | 791 | 672 |
| 16 OPzS | 4000 | 2928 | 2448 | 1984 | 1648 | 1408 | 1248 | 1104 | 904 | 768 |
| 18 OPzS | 4500 | 3294 | 2754 | 2232 | 1854 | 1584 | 1404 | 1242 | 1017 | 864 |
| 20 OPzS | 5000 | 3660 | 3060 | 2480 | 2060 | 1760 | 1560 | 1380 | 1130 | 960 |
| 22 OPzS | 5500 | 4026 | 3366 | 2728 | 2266 | 1936 | 1716 | 1518 | 1243 | 1056 |
| 24 OPzS | 6000 | 4392 | 3672 | 2976 | 2472 | 2112 | 1872 | 1656 | 1356 | 1152 |
| 28 OPzS | 7000 | 5124 | 4284 | 3472 | 2884 | 2464 | 2184 | 1932 | 1582 | 1344 |
| 32 OPzS | 8000 | 5856 | 4896 | 3968 | 3296 | 2816 | 2496 | 2208 | 1808 | 1536 |
| 36 OPzS | 9000 | 6588 | 5508 | 4464 | 3708 | 3168 | 2808 | 2484 | 2034 | 1728 |
| 40 OPzS | 10000 | 7320 | 6120 | 4960 | 4120 | 3520 | 3120 | 2760 | 2260 | 1920 |
| 44 OPzS | 11000 | 8052 | 6732 | 5456 | 4532 | 3872 | 3432 | 3036 | 2486 | 2112 |
| 48 OPzS | 12000 | 8784 | 7344 | 5952 | 4944 | 4224 | 3744 | 3312 | 2712 | 2304 |

[a] 6V-bloc batteries
The power values shown in the tables relate to loadings from a fully charged condition and an ambient temperatures of 20 °C.
Connector losses are taken into account.

emergency. Two problems may arise:

– the battery as a whole may lose unnoticed charge,
– individual cells of the battery may suffer failure (short-circuits, interrupted current connections). In an emergency, these cells will limit the capacity of the whole battery.

To prevent such failures or secure their early discovery, lead acid batteries, for example in power stations or telephone exchanges, are subjected to regular repeated checking patrols, looking for irregularities such as leakage, strong

**Table 6.7.** Project planning data, e.g. tubular range 4 OPzS 200 to 48 OPzS 12 000-discharge power for a system-conditioned final discharge voltage $U_{smin}$ of 1.87 V/cell (approximate values). (Table by courtesy of Varta AG)

$U_s = 1.87$ V/cell discharge power (W/cell)

| Typ | | 30′ | 1 h | 2 h | 3 h | 4 h | 5 h | 6 h | 8 h | 10 h |
|---|---|---|---|---|---|---|---|---|---|---|
| [a]6V3 OPzS | 150 | 139 | 106 | 74.1 | 58.2 | 48.6 | 42.0 | 37.2 | 30.0 | 25.2 |
| 4 OPzS | 200 | 186 | 142 | 98.8 | 77.6 | 64.8 | 56.0 | 49.6 | 40.0 | 33.6 |
| 5 OPzS | 250 | 232 | 177 | 123 | 97.0 | 81.0 | 70.0 | 62.0 | 50.0 | 42.0 |
| 6 OPzS | 300 | 279 | 213 | 148 | 116 | 97.2 | 84.0 | 74.4 | 60.0 | 50.4 |
| 5 OPzS | 350 | 275 | 220 | 160 | 127 | 108 | 95.0 | 84.5 | 69.0 | 57.5 |
| 6 OPzS | 420 | 330 | 264 | 192 | 152 | 130 | 114 | 101 | 82.8 | 69.0 |
| 7 OPzS | 490 | 385 | 308 | 224 | 177 | 151 | 133 | 118 | 96.6 | 80.5 |
| 6 OPzS | 600 | 450 | 360 | 273 | 219 | 186 | 162 | 144 | 117 | 100 |
| 8 OPzS | 800 | 600 | 480 | 364 | 292 | 248 | 216 | 192 | 156 | 133 |
| 10 OPzS | 1000 | 750 | 600 | 455 | 365 | 310 | 270 | 240 | 195 | 167 |
| 12 OPzS | 1200 | 900 | 720 | 546 | 438 | 372 | 324 | 288 | 234 | 200 |
| 12 OPzS | 1500 | 1020 | 864 | 674 | 558 | 486 | 432 | 387 | 320 | 270 |
| 15 OPzS | 1875 | 1275 | 1080 | 843 | 697 | 607 | 540 | 484 | 400 | 337 |
| 16 OPzS | 2000 | 1360 | 1152 | 899 | 744 | 648 | 576 | 516 | 427 | 360 |
| 20 OPzS | 2500 | 1700 | 1440 | 1124 | 930 | 810 | 720 | 646 | 534 | 450 |
| 24 OPzS | 3000 | 2040 | 1728 | 1348 | 1116 | 972 | 864 | 775 | 640 | 540 |
| 14 OPzS | 3500 | 2128 | 1834 | 1442 | 1204 | 1064 | 966 | 868 | 721 | 616 |
| 16 OPzS | 4000 | 2432 | 2096 | 1648 | 1376 | 1216 | 1104 | 992 | 824 | 704 |
| 18 OPzS | 4500 | 2736 | 2358 | 1854 | 1548 | 1368 | 1242 | 1116 | 927 | 792 |
| 20 OPzS | 5000 | 3040 | 2620 | 2060 | 1720 | 1520 | 1380 | 1240 | 1030 | 880 |
| 22 OPzS | 5500 | 3344 | 2882 | 2266 | 1892 | 1672 | 1518 | 1364 | 1133 | 968 |
| 24 OPzS | 6000 | 3648 | 3144 | 2472 | 2064 | 1824 | 1656 | 1488 | 1236 | 1056 |
| 28 OPzS | 7000 | 4256 | 3668 | 2884 | 2408 | 2128 | 1932 | 1736 | 1442 | 1232 |
| 32 OPzS | 8000 | 4864 | 4192 | 3296 | 2752 | 2432 | 2208 | 1984 | 1648 | 1408 |
| 36 OPzS | 9000 | 5472 | 4716 | 3708 | 3096 | 2736 | 2484 | 2232 | 1854 | 1584 |
| 40 OPzS | 10000 | 6080 | 5240 | 4120 | 3440 | 3040 | 2760 | 2480 | 2060 | 1760 |
| 44 OPzS | 11000 | 6688 | 5764 | 4532 | 3784 | 3344 | 3036 | 2728 | 2266 | 1936 |
| 48 OPzS | 12000 | 7296 | 6288 | 4944 | 4128 | 3648 | 3312 | 2976 | 2472 | 2112 |

[a] 6V-bloc batteries

The power values shown in the tables relate to loadings from a fully charged condition and an ambient temperatures of 20 °C.

Connector losses are taken into account.

gassing actions, low electrolyte level and increased temperature (especially in valve-regulated types).

## 6.5.1 Surveillance of Vented Batteries

Quarterly voltage and density readings are nowadays usual with vented lead-acid batteries.

*Float-voltage check.* Voltage checks can provide considerable information. If, for example a number of cells show float-voltages below or above a tolerated limit, which usually covers the average cell voltage $\pm$ 50 m V (cf. the corresponding DIN or IEC publications). This may indicate:

– the battery might not be charged properly, or
– a number of cells suffer internal short circuits.

*Recommendation.* This battery should be floated for a further period, e.g. 14 days, and then the voltage reading repeated. Three different results can be expected:

1. All cell voltages are now within the desired range. *This indicates that the battery is in the proper state and was not fully charged, when the first voltage reading took place.*
2. The situation is principally not changed, but different or even more cells show too low or high voltage. *This indicates that the charge/discharge balance of the battery is out of order, e.g. the battery might have been used unnoticed for peak shaving.*
3. The situation has not changed, the same cells show too low voltages. This indicates that the 'low voltage cells' are defect. *For these weak cells acid density readings should be taken to check whether the state of charge is not yet affected. If the weak cells show reduced acid density, they suffer a serious fault, and the manufacturer should be informed.*

*Note: Correlation cannot be expected between float voltage and cell capacity, because float voltage is caused by oxygen and hydrogen evolution and is not directly related to the discharge reaction.* A too low float voltage only indicates that side reactions consume a portion of the current that is no longer available for gas evolution. The capacity of the cell is not affected, as long as these side reactions (internal short-circuit or increased hydrogen evolution at the negative electrode) only decrease gas evolution. Thus, cells with low float voltage may deliver high capacity. Only if the rate of the side reaction exceeds the float current, gradual discharge of the cell begins which can be detected by acid density readings.

*Acid-Density Checks.* Acid density checks are required only to obtain supplementary data for 'low-voltage cells', because the acid density can never fall below the tolerated range as long as the correct cell voltage is observed, presuming that the battery has properly been put into service, including accurate adjustment of the initial acid concentration. As long as the correct or nearly correct acid density is observed, the cell will deliver full capacity, in spite of its low cell voltage. A continuous and significant decline of acid density, however, is alarming, because it suggests that either the battery is charged insufficiently or the cell concerned is gradually discharged by an internal short circuit.

*Note:* It is recommended to refrain from repeated and superfluous density readings for two reasons:

1. when acid samples are taken from the battery, these may be acid spillage, which is not vapourized but remains on the cell for time. The wet surface gets dirty by absorbing dust which may reduce the insulation, and corrosion occurs if the acid reaches metallic surfaces.
2. Usually acid sampling is possible only above the plates. But, after recharge or water refill this part of the electrolyte remains diluted due to stratification for a certain period until the acid concentration is equalized by diffusion. Density readings during this period are misleading.

*Résumé.* Regularly checked lead acid batteries of vented design require capacity tests only after about two thirds of their expected service life (increase of the internal resistance due to corrosion of the positive grid may then be noticeable). This means that capacity measurements should be started in long-life stationary types, like Varta OPzS or Varta bloc, after about 10 years after installation. Ageing effects that are not indicated by voltage readings, e.g. increase of the internal resistance due to grid corrosion, may then reduce the rated capacity.

### 6.5.2 Surveillance of Valve-Regulated Lead-Acid Batteries

The checks described can in principle be transferred to valve-regulated lead-acid batteries. However, a number of influences reduce the evidence of the test:

– the voltage deviation of the individual cell is increased compared to the vented design on account of the more effective internal oxygen cycle that depends largely on the (gradually changing) electrolyte-filling level.
– valve-regulated lead-acid batteries are often used as block batteries with three or six cells per block. So only block voltages can be measured during float charging. In a block battery; the standard deviation per block amounts to $\sqrt{n} \cdot \sigma$, when $\sigma$ represents the standard deviation of the single cell. Consequently the standard deviation amounts to 1.73 times $\sigma$ for a 6 volt block (3 cells) and to 2.45 times $\sigma$ for a 12 volt block (6 cells). So the evidence of float-voltage readings is reduced.
– acid density readings, which provide important complementary parameters to determine the state of charge in vented lead-acid batteries, are not possible with the valve regulated design.
– an additional ageing process of the valve-regulated design is gradual water loss, especially in unfavourable operational conditions (increased temperature). The extent of this loss cannot be determined, because neither the remaining electrolyte volume nor its density can be measured. Even battery weight is not a suitable quantity, because oxygen, consumed by grid corrosion, remains within the cell. Water loss can only be evaluated from gas measurements of high accuracy (*special attention has to be paid to fugacity of hydrogen that requires metallic tubing*) which is impossible in normal battery service.

Water loss causes increased internal resistance that reduces the discharge voltage and thereby the available capacity, especially with high loads. This predominantly concerns batteries with absorbent-glass-mat separators, because pore sizes in absorbent-glass-mat separators are coarse, compared to those of the active material of both electrodes. Thus, capillary forces of the active material are higher, water loss occurs mainly at the expense of the water content in the separator, and so increases the electrolyte resistance. In gelled electrolyte a real pore structure cannot be ascertained by direct measurements, but the electrolyte behaves in the same way as in a felt with a pore system one order of magnitude finer than that of the absorbent-glass-mat separator. Thus, 'capillary competition' between the active material and immobilized electrolyte is not relevant.

Changes of the internal resistance are not detectable by float-voltage readings, because of the weak float current. Impedance measurements are recommended and suitable devices are offered. But this means additional surveillance expenditures, and most devices require that the float charge is switched off during impedance measurements.

An approximate method, applied in practice to pick out poor cells, makes use of the a.c.-drop per cell, caused by the ripple that is superimposed to the float current. It is presumed that the ripples does not change during measurement, a diverging a.c.-resistance indicates that the internal resistance of the cell concerned obviously differs significantly from the bulk of the battery.

*Résumé*. Capacity measurements are required more often with valve-regulated lead-acid batteries, when the state of the battery has to be known exactly.

Valve-regulated lead-acid batteries are especially sensitive to unfavourable operational conditions that increase water loss. At too high an operational temperature, this battery design is prone to dry out. Regular checks of battery temperature and the observance of correct operational conditions are strictly recommended.

### 6.5.3 Battery Monitoring

Battery monitoring means continual supervision of the battery (Fig. 6.20). Perfect monitoring does not only control the present situation, but also registers (and evaluates) changes with time and abnormal situations of the battery (e.g., significant discharges). Thus, not only the actual 'state of charge' is determined but also information is acquired concerning the 'state of health' that describes more or less the ageing of the battery. Thus, monitoring gives confidence that the battery will deliver sufficient energy, and for a reasonable period of time.

A monitoring system is expensive, but saves capacity test, especially in valve-regulated lead-acid batteries.

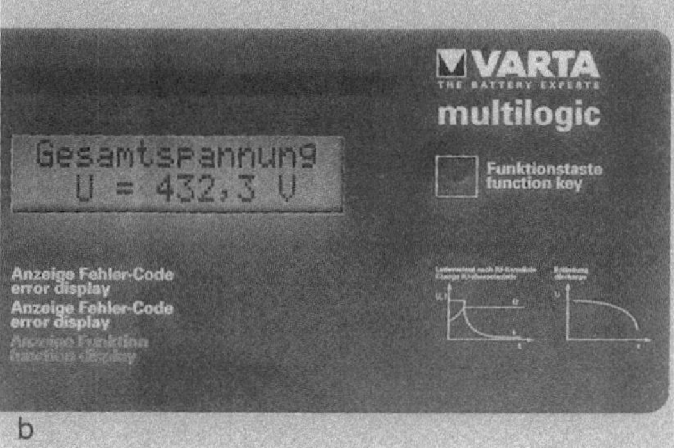

**Fig. 6.20a, b.** Battery-multilogic-measuring and monitoring unit **(a)** with display panel **(b)**. (Photo by courtesy of Varta AG)

## Application

The multilogic (Varta AG) was conceived for the monitoring of uninterruptible power supply equipment with batteries in standby parallel mode. It continuously monitors the availability for use of the battery and the power supply equipment. This monitoring takes place both in float charge mode and also during charging and discharging of the battery. Its areas of application are battery installations in power stations, transformer stations, UPS and industrial installations, in traffic control and in safety power supply equipment to DIN VDE 0108.

## Description of Operation

The multilogic measuring and monitoring equipment periodically measures:

– the total voltage of the battery,
– the voltage of cells, bloc batteries or battery sections,
– the battery current,
– the superimposed a.c. current and
– the temperature of the battery at two selected points.

The measurements are compared with data which is stored in the microprocessor and checked for plausibility. If the specified values are exceeded a fault message will be displayed. At the same time the fault which has occurred will be stored in the memory with its date and time.

This fault details in the memory can be read out or transferred to a central control. In addition there is the facility to record and display the latest measurement data on the battery through the interface.

**Features**

- *Economical monitoring* system for battery installations,
- Information *available at all times* on the availability for use of the battery,
- The possibility of *carrying out measurements* on the battery installation from a *central point*,
- The power supply for the measuring and monitoring equipment comes from the battery,
- Faults which occur are sorted into critical and less critical faults,
- *Can be used for all vented and valve regulated batteries* and
- Possible to check maintenance-free, valve regulated batteries.


### 6.5.4 A Statement of the Principle of Battery Service

In today's modern technology batteries in stationary applications are often a single component in a complete system. They are usually the last link in a chain onto which, in the event of a breakdown of the general power supplies, falls the task of supplying important safety equipment with power (Fig. 6.21).

To be able to guarantee these high requirements at any time it is not sufficient simply to produce and install high grade long-life batteries, but also requires that the availability of the complete system be checked by an expert Service Department and if necessary guaranteed by suitable action.

The most important prerequisites for this are both trained and experienced personnel and also modern tools which are in line with todays level of technology (Fig. 6.22).

**Fig. 6.21.** UPS-system with distribution, measuring instruments, monitoring and battery system. (Photo by courtesy of Varta AG and Siemens AG)

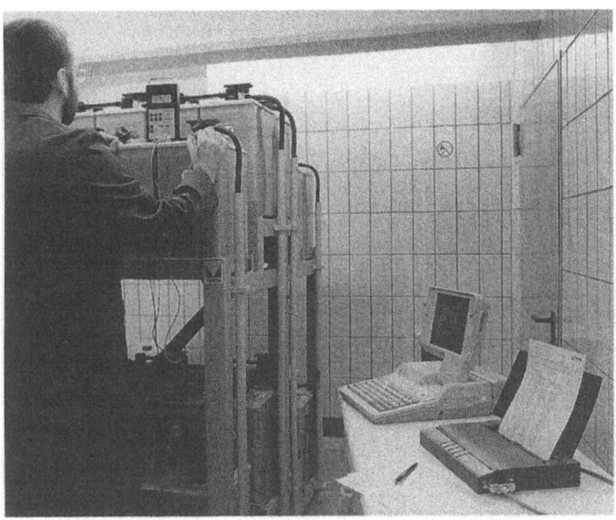

**Fig. 6.22.** Measuring and testing of a battery system for UPS. (Photo by courtesy of Varta AG)

As well as the necessary knowledge in dealing with batteries it is particularly important in the servicing of batteries and battery systems to evaluate and judge correctly the measurements and findings which occur. This requires a well-founded knowledge of the functioning of systems and of the technology of the batteries as well as practical experience on the characteristics of battery systems.

In addition to the conventional methods of testing batteries, the service personnel have at their disposal today measurement recording systems which allow the measured values to be analysed in depth with the aid of a computer supported analysis software. This offers the battery user the advantage of having meaningful test records and easily understood graphical representations of the test results.

This significant data is achieved by the use of two different data capture systems:

- mobile capacity measuring equipment with automatic recording of measurements from up to 110 measurement points simultaneously (Fig. 6.23).
- microprocessor controlled digital multimeter TCM with software and an internal memory is used to record the measurements, along with the voltage figures the device records the date and time of every measurement (Fig. 6.24).

### 6.5.5 Battery Inspections

Regular inspections of the battery are recommended not only by the manufacturers but also, in specific applications, are controlled by Standards and Regulations. This also applies in general to valve-regulated batteries. They serve to establish the condition of the equipment at that point in time with the aim of ensuring

**Fig. 6.23.** Mobile capacity measuring equipment. (Photo by courtesy of Varta AG)

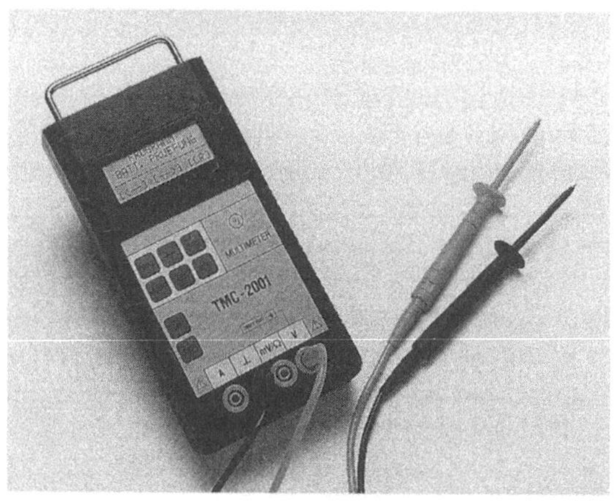

**Fig. 6.24.** Microprocessor controlled digital multimeter TCM. (Photo by courtesy of Varta AG)

the availability of the battery, discovering any deficiencies and initiating action in the event of faults.

In addition to the exterior visual impression, the essential parameters in determining the condition of the battery are traditionally the individual cell voltage in float charge mode and the specific gravity of the electrolyte in lead acid batteries.

Since more and more valve-regulated battery systems are coming into use, where there is no opportunity to make an assessment based on the electrolyte, the voltage of the individual cells assumes an ever greater importance. It has taken this development into account by equipping its stationary battery Service for with an intelligent battery measurement system (Fig. 6.25). Compare also with Fig. 6.24 (digital multimeter TCM).

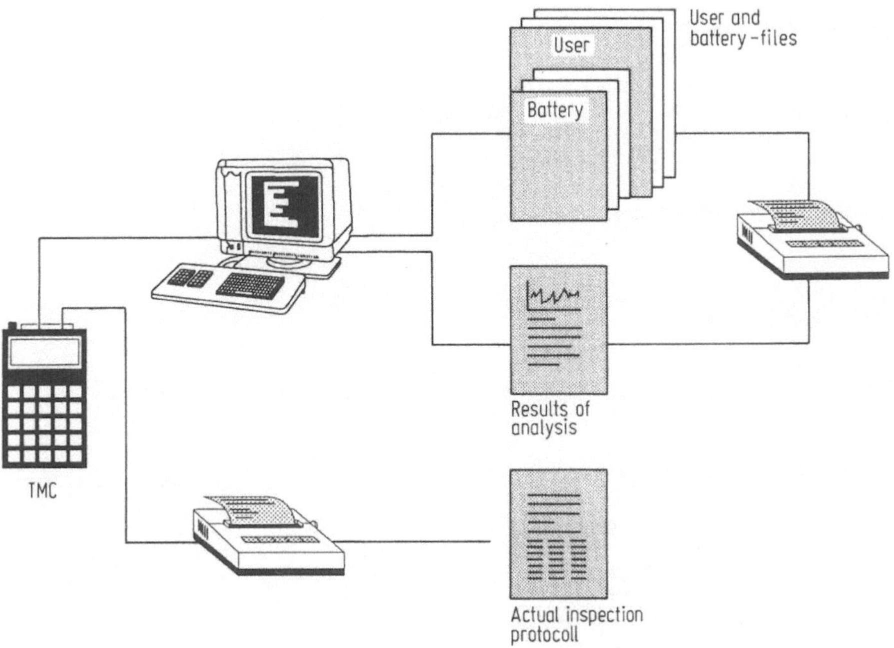


**Fig. 6.25.** Battery measuring system. (By courtesy of Varta AG)

The device has a printer connection so that a record of the test can be printed on site on a printer which is carried with it.

Its practically-oriented software permits a computer supported evaluation and analysis of the results. In this the values of all measurements taken in the course of the operation of a battery are compared with each other. This gives an overall picture of the progress of the electrical values over the whole working life of the battery. The results can be printed out as a records or as clear graphics (Figs. 6.26 and 6.27).

In addition the software makes it possible to keep battery files. This offers the advantage, particularly for users with several battery installations, of being able to co-ordinate the maintenance dates and the inspection intervals and thus reduce costs (Fig. 6.28).

In addition to the figures taken and stored by the digital multimeter, further details of the battery such as the specific gravity and the temperature of the electrolyte can be entered manually through the keyboard when a PC is used. These then form another criterion for assessing the condition of a battery. Battery inspections which are carried out, analysed and assessed by specialists in accordance with the methods described here, help the users of batteries to make a judgement on the reliability of their equipment and allow any limitations on its availability to be recognised in good time.

```
TEST RECORD:

Measurement No. 1 on 24.02.1993          08:51:04
Tester                 :   Wustlich
Operator               :   FA2DORT
Battery No.            :   1              Battery ID  :  001
Battery Type           :   12 OPzS 1500   Cell No.    :  30
Remarks on measurement :

Remarks on battery     :   30 cells 12 OPzS 1500, Index 9 / 87

All Measurements:

Battery Voltage        :   66.97 V
Lowest cell voltage    :   2.223 V cell No. 4
Highest cell voltage   :   2.246 V cell No. 18
Mean of all cells      :   2.232 V
Remarks on electrolyte :   max

Cell:           Voltage:        Temperature:        S.G.
   1 :          2.226 V         17.00 ° C           0.000  kg/l
   2 :          2.233 V          0.00 ° C           0.000  kg/l
   3 :          2.232 V          0.00 ° C           0.000  kg/l
   4 :          2.223 V          0.00 ° C           0.000  kg/l
   5 :          2.230 V          0.00 ° C           0.000  kg/l
   6 :          2.229 V          0.00 ° C           0.000  kg/l
   7 :          2.224 V          0.00 ° C           0.000  kg/l
   8 :          2.223 V          0.00 ° C           0.000  kg/l
   9 :          2.232 V          0.00 ° C           0.000  kg/l
  10 :          2.233 V          0.00 ° C           0.000  kg/l
  11 :          2.239 V          0.00 ° C           0.000  kg/l
  12 :          2.235 V          0.00 ° C           0.000  kg/l
  13 :          2.235 V          0.00 ° C           0.000  kg/l
  14 :          2.229 V          0.00 ° C           0.000  kg/l
  15 :          2.229 V          0.00 ° C           0.000  kg/l
  16 :          2.239 V          0.00 ° C           0.000  kg/l
  17 :          2.235 V          0.00 ° C           0.000  kg/l
  18 :          2.246 V          0.00 ° C           0.000  kg/l
  19 :          2.229 V          0.00 ° C           0.000  kg/l
  20 :          2.229 V          0.00 ° C           0.000  kg/l
  21 :          2.237 V          0.00 ° C           0.000  kg/l
  22 :          2.234 V          0.00 ° C           0.000  kg/l
  23 :          2.232 V          0.00 ° C           0.000  kg/l
  24 :          2.229 V          0.00 ° C           0.000  kg/l
  25 :          2.241 V          0.00 ° C           0.000  kg/l
  26 :          2.230 V          0.00 ° C           0.000  kg/l
  27 :          2.234 V          0.00 ° C           0.000  kg/l
  28 :          2.238 V          0.00 ° C           0.000  kg/l
  29 :          2.230 V          0.00 ° C           0.000  kg/l
  30 :          2.237 V          0.00 ° C           0.000  kg/l
```

**Fig. 6.26.** Test record for a 30 cell battery. (By courtesy of Varta AG)

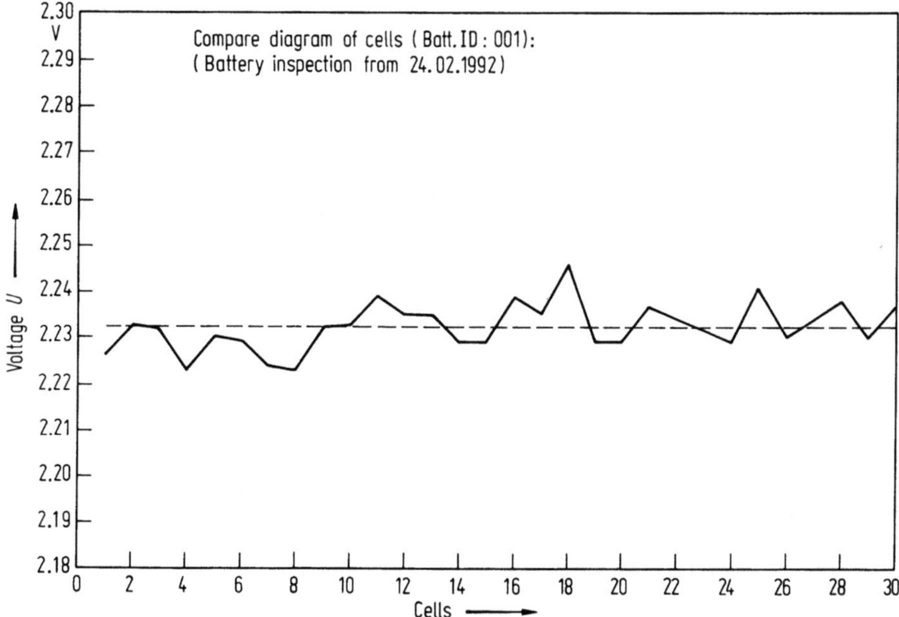


**Fig. 6.27.** Graphical representation of the cell voltage in float charge mode. (By courtesy of Varta AG)

### 6.5.6 Capacity Checks

Although there are various types of concept on the market today for checking the availability for use of a battery, the most reliable method is still a load test conducted in accordance with specified parameters. This can be carried out either with current and voltage figures specific to the installation or by means of a capacity check in accordance with the relevant Standards and Regulations.

Since the application for batteries in the last few years has moved very strongly in the direction of many cells for short back-up periods, the requirement for taking measurements in load tests has also increased.

The usefulness of such tests depends on the precision with which the measurements are taken and the possibilities for analysing the results of these measurements. In addition to this large customers, such as for example the energy supply firms, more and more frequently require qualified and comparable documentation of the tests carried out.

There are opportunities to carry out load and capacity tests with a mobile computerised test unit which is virtually up to laboratory standard in taking and evaluation of measurements.

The measurement equipment consists of

- a data capture unit with connections for
  - 110 voltage measurement points,

**Details of identification:**

| | |
|---|---|
| User | : VARTA |
| Battery I.D. | : 999 |
| Battery Number | : UPS 1 |
| Price for a check | : 0.00 DM |

**Details of the physical set up and application:**

| | |
|---|---|
| Building | : Administration |
| Floor | : Basement |
| Room | : 16 |
| Report to | : Mr. Hass |
| Application | : Siemens UPS |
| Cost Centre | : |

**General battery details:**

| | |
|---|---|
| Supply date | : 30.06.1984 |
| Commissioning date | : 20.09.1984 |
| Manufacturer | : VARTA |
| Battery Model | : Vb 2307 |
| Battery Type | : Pb |

**Technical battery details:**

| | | | |
|---|---|---|---|
| Battery voltage | : 368.00V | | |
| Target capacity | : 504.00 A | 0.25 h | (126.00 Ah) |
| Rated capacity | : 35.00 A | 10.00 h | (350.00 Ah) |
| Cells | : 184 | | |
| Min. Cell voltage | : 2.180 V | | |
| Max. Cell voltage | : 2.280 V | | |
| Min. Battery voltage | : 406.00 V | | |
| Max. Battery voltage | : 416.00 V | | |

**Additional notes on battery:**

partially with recos

**Fig. 6.28.** Battery file. (By courtesy of Varta AG)

    – shunt voltage for current measurement, and
    – temperature sensors;
– a personal computer with the appropriate software;
– a printer for test records and graphics.

All the cells of the battery being checked (up to 110) are connected to the data capture unit. The measurement leads are protected by fuses immediately after the cell take-off. Within the measurement system all the individual cell voltages are galvanically separated from each other.

After starting the measurement the system queries at regular intervals during the course of the test all cell voltages, the actual battery current and – if the temperature sensors are connected – the temperatures of the electrolyte. The whole measuring cycle lasts a maximum of 5 seconds. The frequency of the measuring intervals depends on the total duration of the test for a 5 hour capacity trial, for instance, a measuring run takes place every 30 seconds.

All measurements taken are put directly into the computer's memory and processed. The v.d.u. continually displays the following updated values

- all individual cell voltages,
- total voltage,
- cell with the highest and lowest individual voltage,
- the discharge current flowing at that instant,
- the total amount of current taken out, and
- the length of the period of discharge so far.

The graphics program in the software can display various diagrams of the measurement process in a clear form on the v.d.u. or output these to a printer when the test is completed (Fig. 6.29).

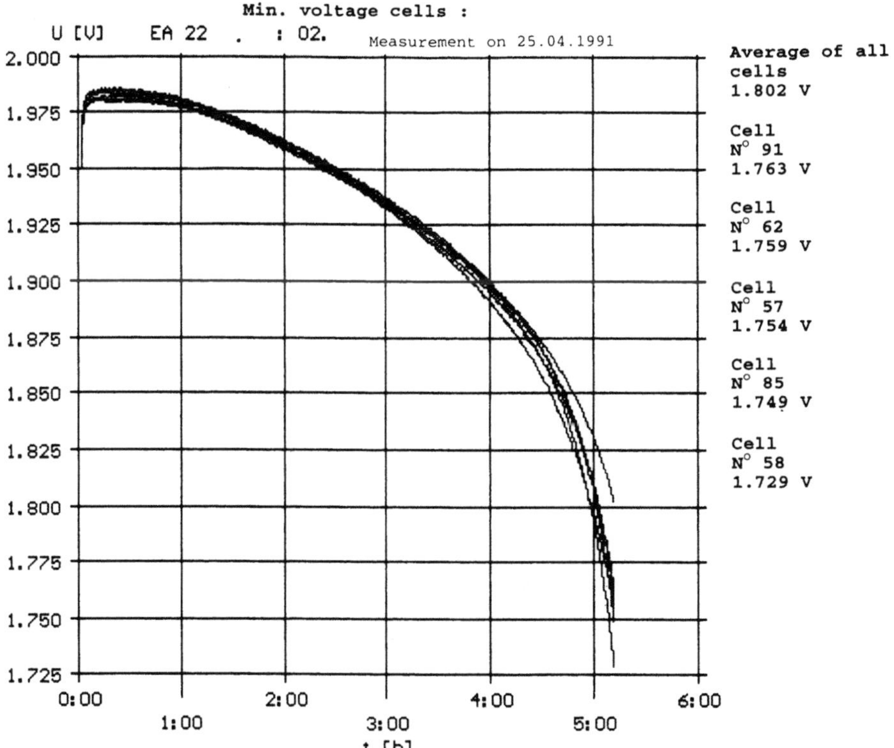


**Fig. 6.29.** Graphical analysis of the progress of individual cell voltages during discharge. Test No.: 910425.1, Battery No.: EA 22, 2nd Measurement dated 04/25/1991. (By courtesy of Varta AG)

**Test Report:**

---

**User:**

| | | |
|---|---|---|
| Company | : | KKI 2 |
| Street | : | Dammstr. |
| Town | : | 8307 Essenbach |
| | | |
| Remarks | : | 1st Capacity Check |
| | | VARTA Order No. 107880314 |
| | | 13 cells 12 GroE-H 300 |
| | | |
| Test Number | : | BTF 21.2/1 |
| | | |
| Battery Number | : | BTF 11 |
| Battery model | : | 12 GroE-H 300 |
| Number of cells | : | 13 |
| Final discharged | | |
| voltage of cell | : | 1.800 V |
| In operation since | : | 15.11.1988 |
| Rated voltage | : | 26.00 V |
| Rated capacity | : | 300.00 Ah |
| Target capacity | : | 270.00 Ah |
| Target discharge time | : | 05:00        <hh:mm> |
| | | |
| Checked on | : | 01.10.1991 |
| Checked by | : | Sudmeier |
| | | |
| Signature | : | _______________________ |

---

**Data from the test run:**

| | | |
|---|---|---|
| Battery current | : | 54.20 A |
| Duration of test | : | 05:30:16        <hh:mm:ss> |
| Measuring interval | : | 33 s |
| Capacity <Kmeas> | : | 299.52 Ah |
| Capacity <Ktest> | : | 296.47 Ah |
| Work done by current | : | 7.65 kWh |

---

**Test Report:**

Test corresponded to 109.80 % of the target capacity.

| | | |
|---|---|---|
| Lowest cell voltage | : | 1.854 V on cell 4 |
| Final voltage of battery | : | 24.40 V |

When the measurements are finished a specified test record (Fig. 6.30) is printed out on the connected printer. This record contains all the necessary data on the battery, the target figures for the test and the measured values actually achieved.

The measurement system described thus makes it possible to carry out realistic measurements which cannot often be done using conventional methods for certain applications. An example of this would be the auxiliary power batteries for UPS systems. The required back-up time normally only amounts to a few minutes. With conventional methods, whilst the total voltage and battery current can certainly be recorded precisely, a statement on the behaviour of the individual components could not be made because of the number of cells used.

In addition to the use of the measuring equipment, auxiliary loads for capacity trials are necessary.

## 6.6 Installation of Batteries

– Floor stillages (cf. Figs. 6.1, 6.15 and 6.18).
– Stepped racks (Fig. 6.31).
– Battery cabinets (Fig. 6.32).

In stepped racks sufficient separation must be allowed between the rows of cells. If the batteries are mounted in two or more vertical tiers, a space of at least 50 mm must be left free in front of and behind the racks for air circulation. The rows of batteries must be accessible from suitable service gangways. The

Fig. 6.31. Batteries with pasted plates as block batteries (OGi) on a stepped rack. (Photo by courtesy of Varta AG)

Fig. 6.30. Testrecord. (By courtesy of Varta AG)

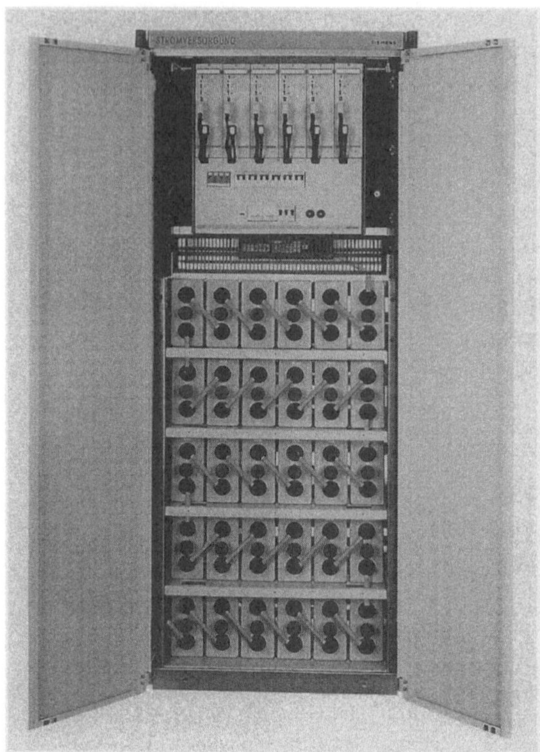

**Fig. 6.32.** Power supply connecting cabinet AS400 with valve-regulated lead-acid-single cell batteries (OPzV), see Fig. 6.16, above: Power supply unit SVE40. (Photo by courtesy of Accumulatorenfabrik Sonnenschein GmbH and Siemens AG)

width of the gangways should be appropriate to the size of the batteries, but not less than 0.5 m. Electrolyte-resistant insulation must be provided between the individual cells and earth or frame.

Recently plastic-insulated racks have been used for mounting (also stillages).

# 7 Basic Circuits and Process Control

## 7.1 Rectifier Circuits

### 7.1.1 Half-wave Rectifier

With the half-wave rectifier (Fig. 7.1) only the positive half-waves of an alternating current are passed. This circuit is suitable for *low* output powers. *More complex* filtering is necessary to suppress interference voltages.

### 7.1.2 Centre-tap Circuit

Two transformer windings are to be connected in series for the centre-tap circuit (full-wave rectifier), Fig. 7.2. Here both half-waves of the alternating voltage are used. Since only one half of the transformer's secondary winding is carrying a current at a given time, the transformer is not being operated to its full extent. The centre-tap circuit is therefore suitable only for *low* to *medium* power outputs and *low* voltages. Less complex filtering is necessary with the centre-tap circuit than with the half-wave circuit.

### 7.1.3 Bridge Circuit

With the bridge circuit (Fig. 7.3) there is current flowing through the two diodes simultaneously at each half wave, while the other two diodes are polarized in the

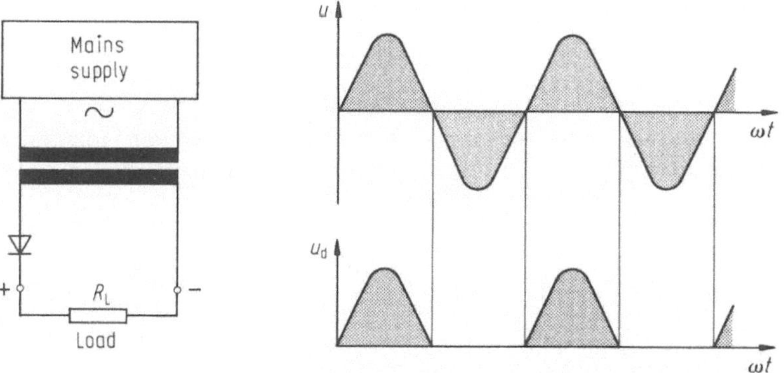


**Fig. 7.1.** Half-wave rectifier (One-way circuit). $u$ Single-phase alternating voltage, $u_d$ direct voltage, $\omega t$ angular velocity (angular frequency) multiply time $t$

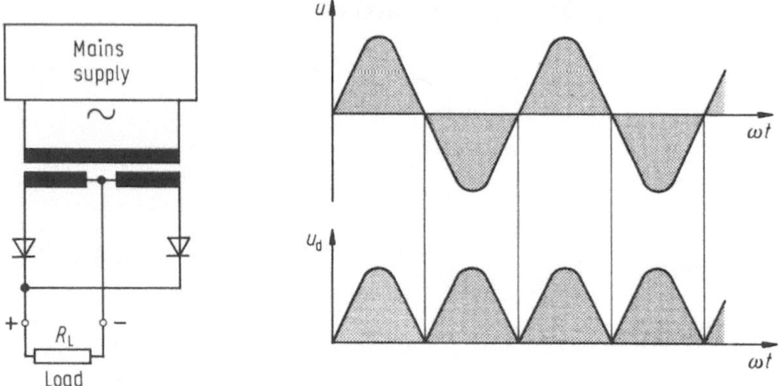

**Fig. 7.2.** Centre-tap (two-way) circuit (full-wave rectifier)

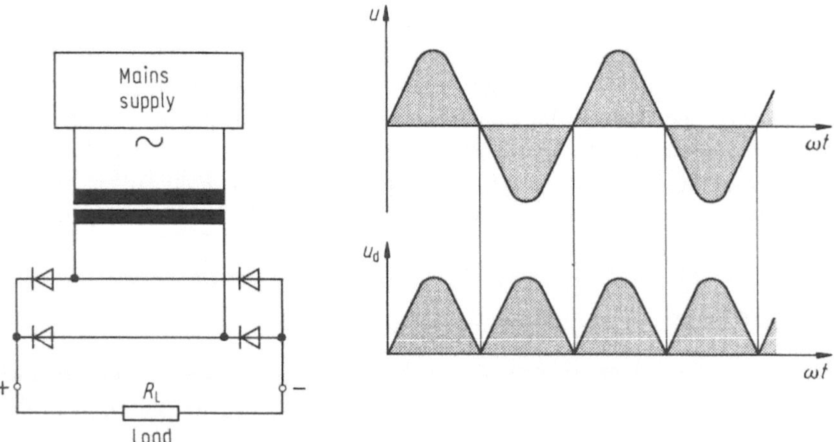

**Fig. 7.3.** Bridge circuit

reverse direction. Here both half waves are used, as with the centre-tap circuit. The bridge circuit is suitable for *medium* power levels and *medium* voltages. It is similar to the centre-tap circuit as regards filtering requirements.

### 7.1.4 Three-phase Bridge Circuit

The three-phase bridge circuit (Fig. 7.4) uses all three alternating voltages shifted by 120°. It is suitable for *high* voltage and power outputs and requires a minimum of filtering.

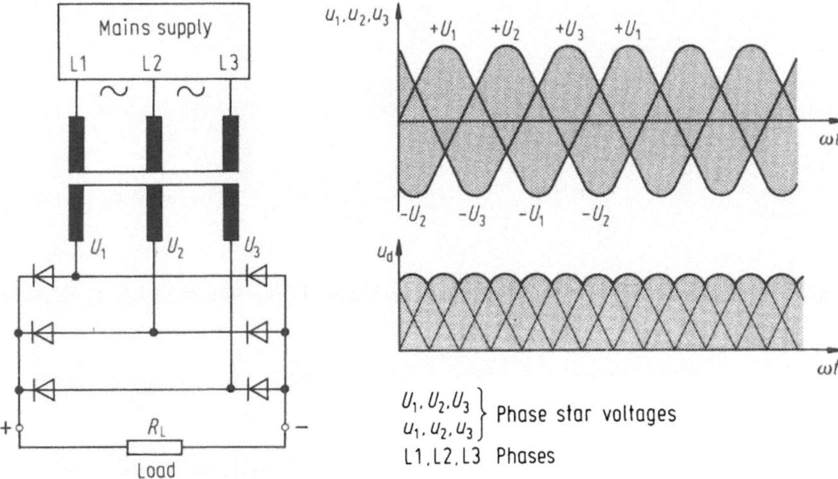


**Fig. 7.4.** Three-phase bridge circuit

## 7.2 Stabilizing and Control Procedures

Very low voltages (up to about 3 V) can be stabilized using *diodes* and dropping resistors. Voltages above this range can be stabilized using *Zener diodes*.

The method of pulse-sequence control (also called frequency control or frequency modulation) is very rarely used in telecommunications power supplies. On the other hand, pulse-width control (modulation) is very often used in all types of power converters. Phase-angle control is chiefly used in thyristor-controlled systems. The objectives of the various methods are to convert, stabilize and control voltages.

### 7.2.1 Pulse-Sequence Control

In pulse-sequence (frequency) control, the on-time $T_e$ of the final control element (power transistor or thyristor) remains constant. The output voltage is determined by varying the pulse frequency or cycle duration $T$ (Fig. 7.5). In example (a) the d.c. voltage is higher than in example (Fig. 7.5b).

### 7.2.2 Pulse-Width Control

Pulse-width control involves the modification of the duty ratio (pulse interval) by varying the on-time $T_e$ of the final power control element with the period (cycle duration) $T$ constant (Fig. 7.6). In example (Fig. 7.6a) the voltage is higher than in example (Fig. 7.6b) since the mean value of the direct current is greater.

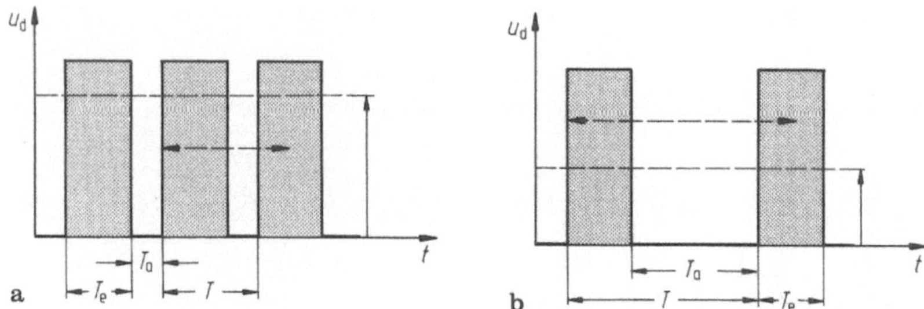

**Fig. 7.5a, b.** Pulse-sequence control. $T$ Period (varying), $T_e$ on-time (const.), $T_a$ off-time

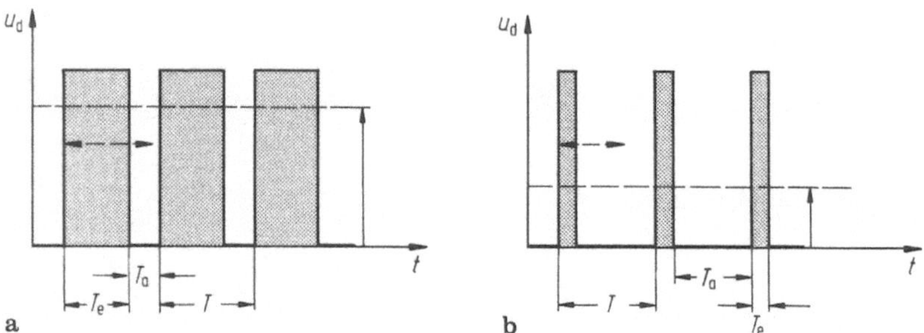

**Fig. 7.6a, b.** Pulse width control. $T$ Period (const.), $T_e$ on-time (varying), $T_a$ off-time

For completeness the *two-point current control*, in which both the on-time $T_e$ and the period $T$ are varied, should be mentioned here, too.

### 7.2.3 Phase Angle Control

In phase angle control, which is very important in practice, the onset in time of the trigger pulses (control pulses, firing pulses) is shifted in relation to the sine half-wave. Phase angle control can be explained with the aid of the single-pulse half-wave rectifier (one-way) circuit (Fig. 7.7). Multi-pulse circuits work in a similar way.

The thyristor is supplied from an alternating current source. If at a certain point in time the anode is negative with respect to the cathode, then the thyrisor is non-conducting. The firing process can only be initiated if the anode is positive with respect to the cathode. For this, the *trigger set* supplies pulses. The extent to which the positive half-waves are gated depends on the position of the trigger pulses, which are shifted by the control angle $\alpha$ (firing delay angle). In this way it is possible to regulate the voltage at the load or to keep it constant.

The control angle $\alpha$ is calculated from the 'natural' firing time of the thyristor. This angle is measured from the point at which the current waveform passes through zero, i.e. the point at which, in a rectifier circuit with diodes, the next diode takes over carrying the current (commutation).

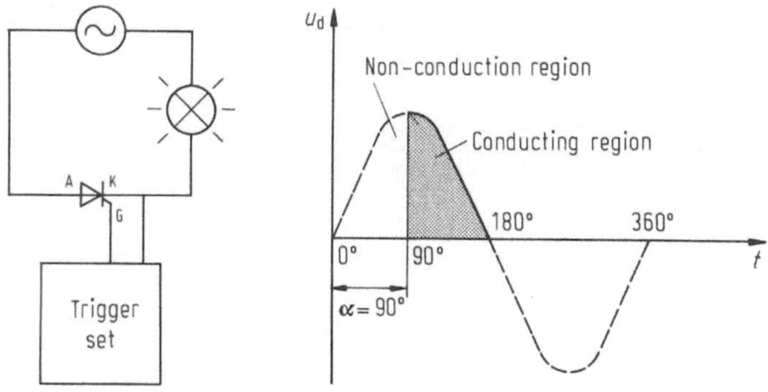



**Fig. 7.7a, b.** Thyristor in single-pulse half-wave circuit. A Anode, K cathode, G gate

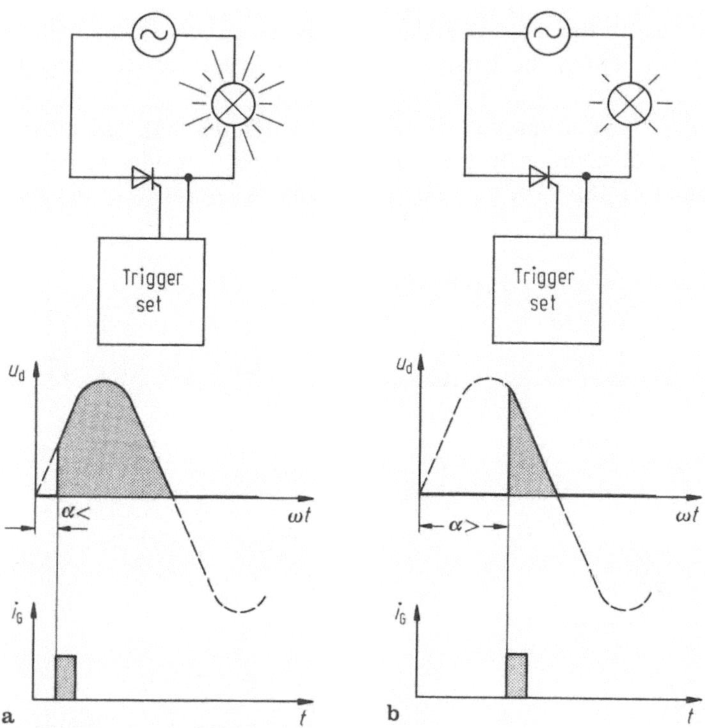



**Fig. 7.8a, b.** Time shift of trigger pulses

A trigger pulse at this point produces the highest possible direct voltage at the output of the rectifier, hence the term "full rectifier modulation" ($\alpha = 0°$).

Figure 7.7b shows a control angle of 90°. Within the range from 0° to 90° the thyristor is non-conducting, and form 90° to 180° conducting.

Figure 7.8a shows a relatively early onset of the trigger pulse ($\alpha$ small). The earlier the onset, the greater the effective voltage-time area becomes. A large voltage-time area means a high voltage, so that the lamp lights brightly.

A larger control angle means a smaller voltage-time area (Fig. 7.8b).

If no trigger pulse reaches the gate, the thyristor remains non-conducting. Accordingly, the lamp does not light.

### 7.2.4 Distribution and Design of Trigger Pulses

The trigger pulses, which pass from the trigger set via the pulse transformer[1] (Fig. 7.9e) to the thyristors of the thyristor set, must have a certain shape. These are as spikes (Fig. 7.9a), rectangular pulses (Fig. 7.9b) or 'combined pulses' (Fig. 7.9c).

With thyristor-controlled rectifiers each trigger pulse is normally cycled (7–kHz cycle, Fig. 7.9d). An advantage here is that the pulse transformer can be designed for a particularly low power. Figure 7.9f shows an ideal trigger pulse. If the gate current lies within the hatched area, the thyristor can *certainly* be fired.

The trigger pulses have a duration of 180° (at 50 Hz, 10 ms). Pulse over-coupling is used here in addition to pulse overlapping. This results in control pulses with a duration of $180° + 60° = 240° - \alpha$. This is important at the time

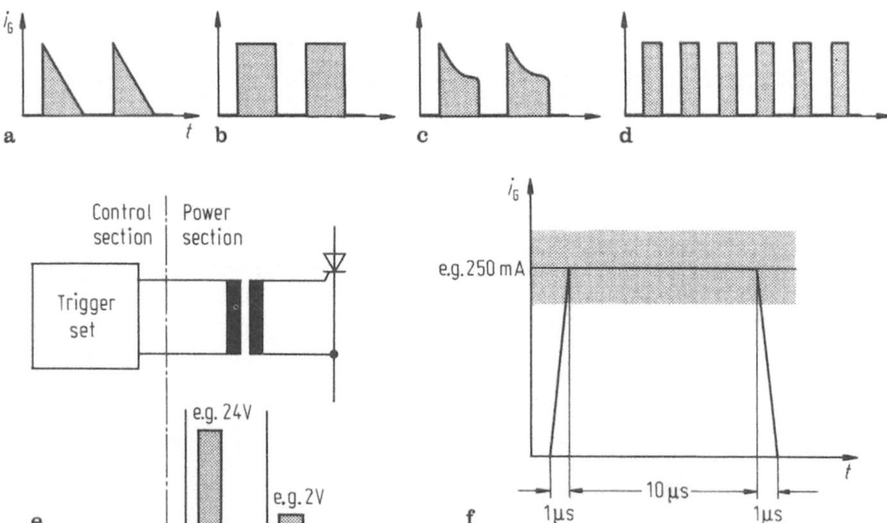


**Fig. 7.9a–f.** Trigger pulses

---

[1] The pulse transformer is used for electrical separation and adaptation of the trigger set pulses to the thyristors.

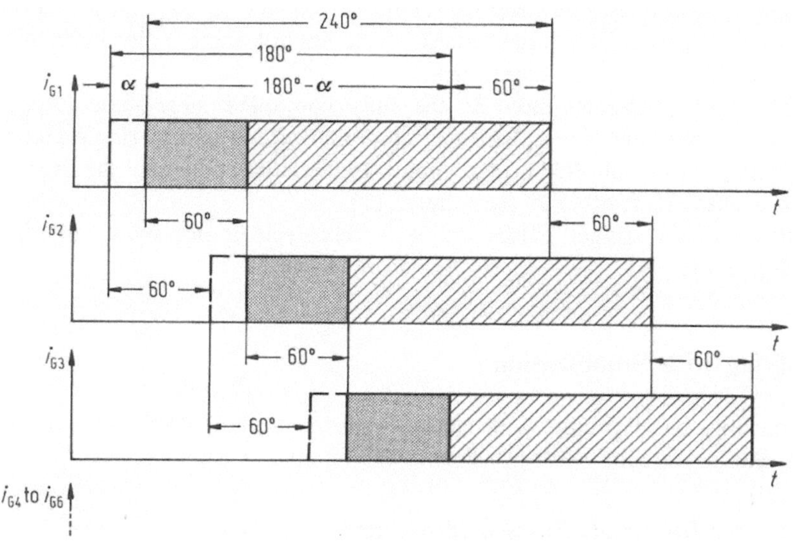

**Fig. 7.10.** 180° (240°) − α overlapping long trigger pulses

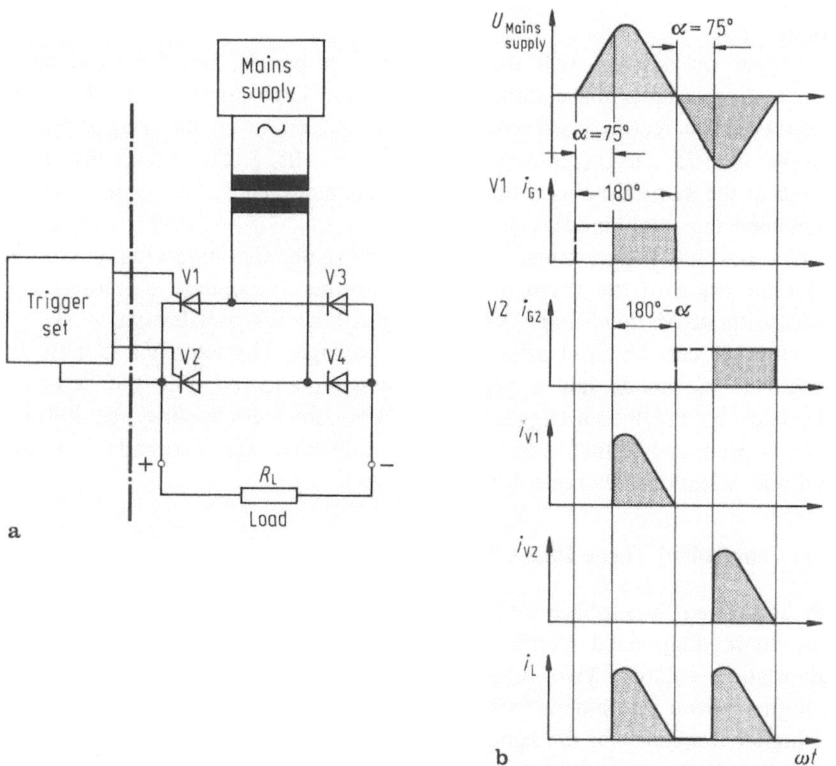

**Fig. 7.11a, b.** Semi-controlled single-phase bridge circuit

of switching on, so that the converter can begin to conduct current, but it is also advantageous with particularly large values of the control angle $\alpha$, in order that the current should not become intermittent.

$180°(240°) - \alpha$ pulses are used in the fully controlled three-phase bridge circuit, the fully controlled three-phase a.c. controller circuit and in single-phase equipment with a semi-controlled bridge circuit. In the latter equipment the trigger pulses have a phase shift of $180°$ (see Fig. 7.11).

Breakdown of the trigger pulses into a 7–kHz cycle is not shown in Fig. 7.10 (see Fig. 7.9d) .


## 7.3 Rectifying with Stabilization

At this point only basic circuits with (phase-angle controlled) thyristors will be considered. Further circuits, with transistors, are found in Chapter 8.


### 7.3.1 Semi-controlled Single-Phase Bridge Circuit

Two thyristors and two diodes are required in the semi-controlled single-phase bridge circuit (two-pulse circuit, Figure 7.11) which is normally used for rectifiers up to about 25 A.

The trigger set delivers two trigger pulses per period, one for each half-wave. They are mutually phase-shifted by $180°$ and have duration of $180° - \alpha$. Both half-waves are rectified and *phase-angled*, depending on the trigger pulses (for example $\alpha = 75°$, trigger pulses $180° - 75° = 105°$). The circuit acts as a rectifier and at the same time as a final control element. Thus, the output voltage is held constant (i.e. stabilized).

When a positive trigger pulse arrives at the gate, the following circuit is produced after firing of the thyristor V1: transformer (secondary side)/thyristor V1 (anode)/cathode V1/load/diode V4 (anode)/cathode V4/transformer.

Thyristor V2 can be fired after the zero passage. Thyristor V1 is now in the off state as the anode has a negative voltage compared with the cathode. When thyristor V2 receives a trigger pulse and becomes conducting, the following circuit is produced: transformer(secondary side)/thyristor V2(anode)/cathode V2/load/diode V3(anode)/cathode V3/transformer.


### 7.3.2 Fully controlled Three-Phase Bridge Circuit

The fully controlled three-phase bridge circuit requires six thyristors (Fig. 7.12). It is used for medium-sized rectifiers (e.g. 25, 50, 100 and 200 A) with three-phase alternating voltage. Two fully controlled three-phase bridge circuits are used in higher power equipment (500 A and 1000 A).

The higher frequency of the superimposed alternating voltage (300 Hz) compared with the single-phase bridge circuit permits the use of smaller filters. Also the distortion of the mains by harmonics is normally reduced.

The three-phase alternating voltage is stepped down by a transformer and fed to thyristors V1 to V6 (Fig. 7.12a). The trigger set must deliver six trigger pulses at intervals of 60° for each supply cycle. The trigger pulses have a duration of 180° − α respectively 240° − α. The sequence in which the thyristors V1 to V6 must receive their pulses from the trigger set is derived from the firing sequence and arrangement of the thyristors. The thyristor set (Fig. 7.13) combines rectification and the final control element function.

Fig. 7.12(b) shows the assignment of the individual voltages to thyristors V1 to V6 and Fig. 7.12(c) as an example the phase shift with a control angle of 30°.

Simultaneous firing of thyristors V6 and V1 (time A) produces the following circuit: $U_1$/thyristor V1/load/thyristor V6/$U_2$.

After a phase shift of 60° (after V1 has fired) a pulse arrives at the gate of thyristor V2, which thereby becomes conducting. The current commutates from thyristor V6 to V2. The following circuit (time B) is produced: $U_1$/thyristor V1/load/thyristor V2/$U_3$.

Thyristor V1 receives simultaneously as a second pulse the same trigger pulse which thyristor V2 received as a main pulse (not shown in Fig. 7.12, see Fig. 7.10). Thyristor V3 receives the main pulse shifted by a further 60° which also reaches thyristor V2 as a second pulse. Thyristor V3 thus becomes conducting and the current commutates from thyristor V1 to V3. The following circuit is now produced (time C): $U_2$/thyristor V3/load/thyristor V2/$U_3$.

Figure 7.14 shows the behaviour of the rectified voltage at control angles of α = 0°, 30°, 60° and 90°. As can be easily seen, with an increasing control angle

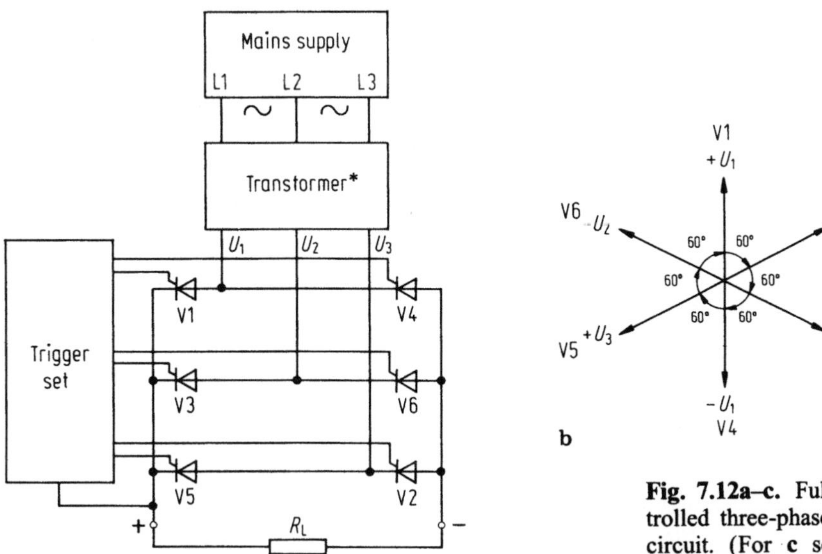


**Fig. 7.12a–c.** Fully controlled three-phase bridge circuit. (For c see page 138)

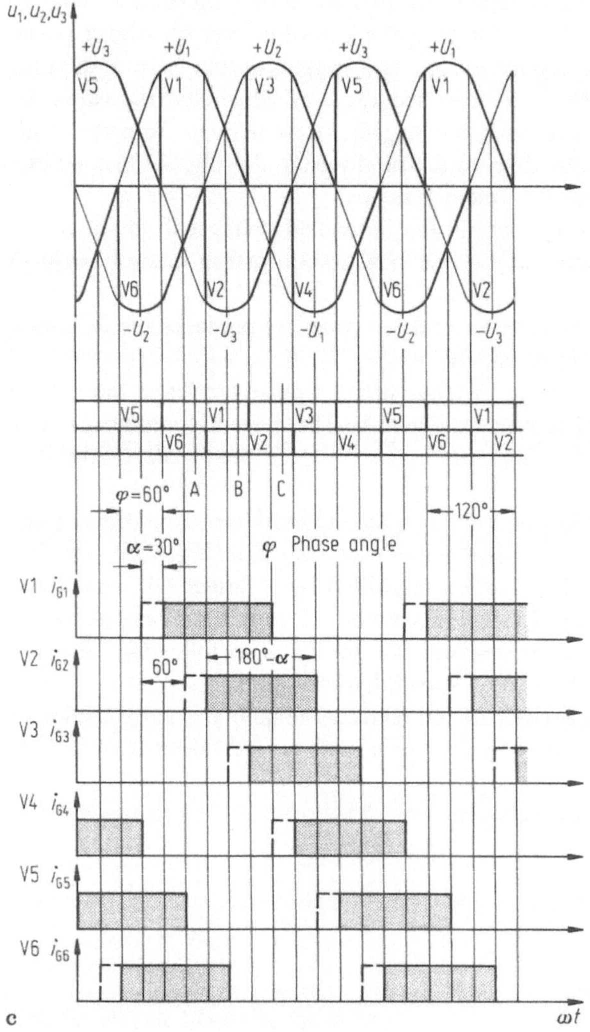


**Fig. 7.12a–c.** (*Continued*)

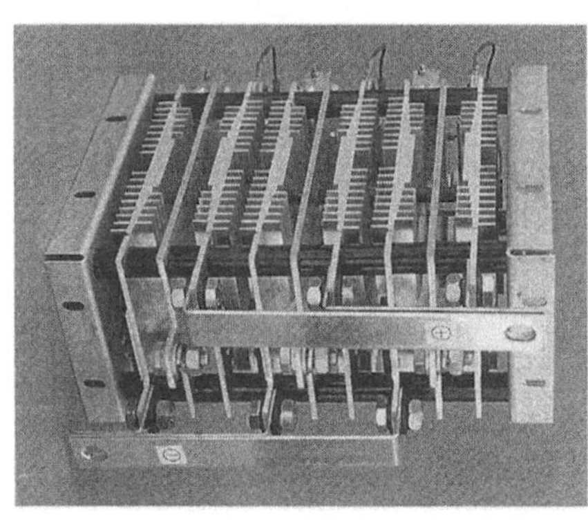

**Fig. 7.13.** Thyristor set in disc-design. (Photo by courtesy of Siemens AG)

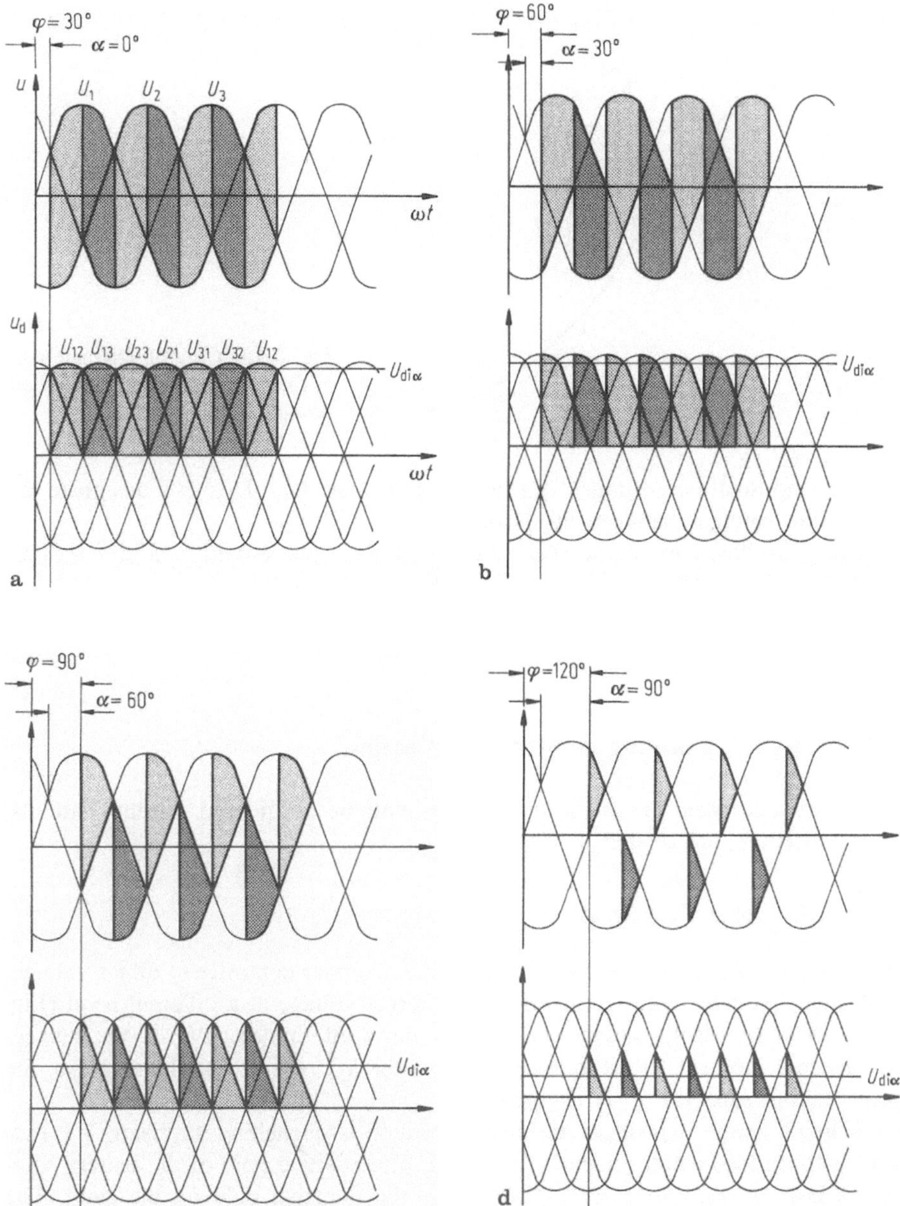



**Fig. 7.14a–d.** Fully controlled three-phase bridge circuit; representation of phase-angle control and rectification at different control angles α

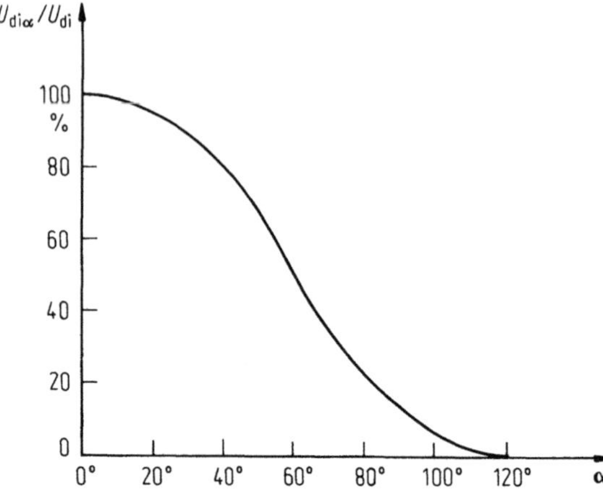

**Fig. 7.15.** Control characteristic of the fully controlled three-phase bridge circuit

the ideal no-load d.c. voltage $U_{di}\alpha$ becomes smaller. $U_{12}$, $U_{13}$, $U_{23}$ designate the linked line voltages between the phases.
Figure 7.15 illustrates at the fully controlled three-phase bridge circuit the dependence of $U_{di}\alpha/U_{di}$ on the control angle $\alpha$.

At $\alpha = 0°$ the d.c. output voltage reaches its maximum value; it drops to 0V at $\alpha = 120°$.

## 7.4 D.C./D.C. Conversion with Stabilization

Here only one basic circuit with thyristors will be considered. Further circuits, with transistors, are treated in Chapter 8.

### 7.4.1 D.C. Controller Circuit

With the aid of the d.c. controller circuit (d.c. chopper controllers) direct voltage, e.g. battery voltage, can be converted into direct voltage at a different level (Fig. 7.16). A commutating device specifically turns off thyristor V1 at the respective desired moment. If the quenching thyristor V3 receives a trigger pulse, the commutating capacitor $C$ is charged.

Circuit: Battery positive terminal/capacitor $C$/quenching thyristor V3 (anode/cathode)/load/battery negative terminal. The positive pole of the battery voltage is now on the left at capacitor C and the negative pole on the right. The main thyristor V1 is then fired by the trigger set. The battery voltage is now applied to the load and a 'circuit in reverse' is formed, i.e. the capacitor C is inversely charged.

Circuit: Capacitor $C$/main thyristor V1/ (anode/cathode)/reverse diode V4/ reverse choke $L1$/capacitor $C$.

The characters without brackets at the capacitor (plus/minus) apply to the time before reversing, those with brackets to the time after reversing. When there

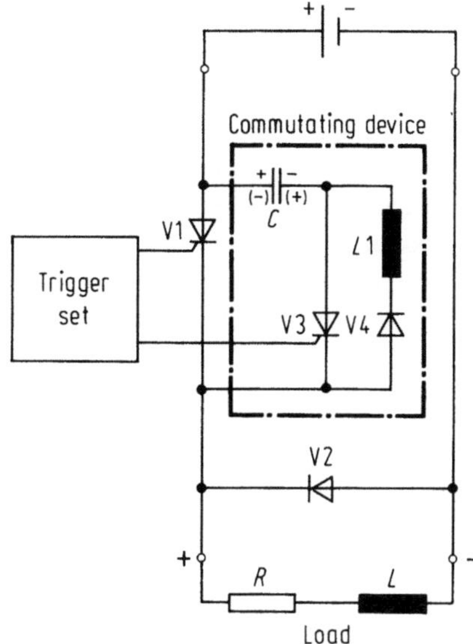


**Fig. 7.16.** D.C. controller circuit

is minus on the left and plus on the right, i.e. when the reversal process is ended the capacitor $C$ is 'prepared' to quench the main thyristor V1.

Quenching thyristor V3 is then refired by the trigger set so that the capacitor voltage is in parallel with the main thyristor V1 (minus at anode and plus at cathode). Thyristor V1 is now in the off condition and the current continues flowing via the free-wheeling diode V2. A mean value for the direct voltage $U_{di}\alpha$ is produced by the continuous alternate switching on and off of V1. Direct current (chopper) controllers make use of pulse-width control and/or pulse-sequence control (see sections 7.2.1 and 7.2.2).

## 7.5 Inverting with Stabilization

At this point only basic circuits with thyristors will be considered. Further circuits, with transistors, can be found in Chapter 8.

The inverter circuit is used to convert a direct voltage, it receives, for example, from a battery or rectifier, into alternating voltage.
The requirements of the inverter circuit in performing this function are:

- to regulate the a.c. output voltage within a specified tolerance band, so that the effects of variations in load and in the d.c. input voltage can be reduced or eliminated, and
- to provide a substantially sinusoidal output voltage waveform.

Voltage control in inverters is often effected by the phase-angle method or the pulse method (pulse-width modulation or, less frequently, pulse-frequency control).

In the *phase-angle method*, the outputs of two independent unregulated inverter units are connected in series to form a single inverter equipment, so that their voltages are added. By varying the phase of its thyristor trigger pulses, one of the inverter units is controlled in such a way that its output voltage is shifted in phase compared to that of the other unit. The larger the mutual phase displacement, the smaller is the sum of the two inverter-unit voltages. In this way the output voltage can be held constant, regardless of any disturbing influences.

In the *pulse method* (burst fire method), the conducting thyristors are triggered and turned off again repeatedly during each cycle of the a.c. output voltage, so that the individual half-cycles of the output voltage are formed from a number of pulses. By varying the pulse width (or the pulse repetition frequency) it is possible to hold the output voltage constant, regardless of any disturbing influences.

Inverters may be used in three modes of operation:

- passive (cold) standby operation (starting mode),
- active (hot) standby operation (joint mode) and
- continuous operation.

The periodic reversal of the direct voltage (inversion) can be achieved in principle by means of two circuits:

- centre-tap circuit and
- bridge circuit (see Chapter 8).

As an example the operation of the centre-tap circuit is described below.


### 7.5.1 Centre-tap Circuit

Fig. 7.17 shows an inverter in a centre-tap circuit. The trigger set supplies the two thyristors V1 and V2 alternately with pulses. In this way current from the battery flows alternately through the transformer's two partial windings T1 and T2. A square-wave alternating voltage can be taken from the secondary side of the transformer, which for practical purposes is frequently converted into an approximately sinusoidal voltage by means of suitable arrangements. While it is well known that in rectifiers the thyristors are directly connected to alternating voltage and that turning off takes place at zero passage when the current drops below the hold current, in the case of a d.c./a.c. inverter a capacitor ($C1$) has to be provided for the turning off process.

If the thyristor V1 receives a trigger pulse from the trigger set, its resistance becomes low.

Circuit: Battery positive terminal/partial winding T1/V3/V1/choke $L1$/battery negative terminal.

A voltage is induced in the secondary winding T3 and the capacitor C1 is charged via T2 and V4. The next trigger pulse is forwarded to thyristor V2. Now *both* thyristors V1 and V2 are conducting and the capacitor discharges. The

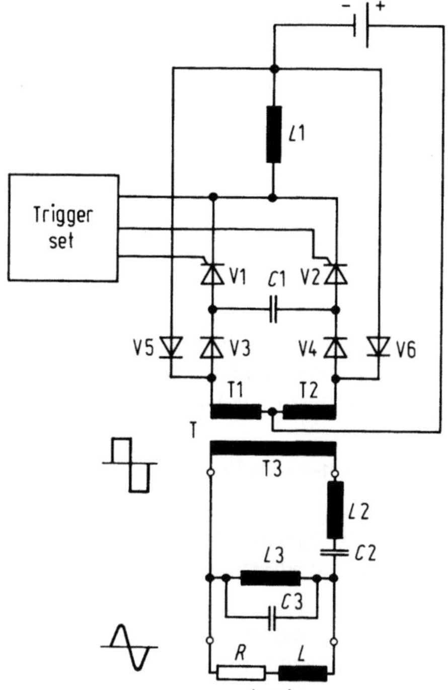


**Fig. 7.17.** Inverter in centre-tap circuit

discharge current via V2 and V1 acts in thyristor V1 against the previous flow of current. V1 thereby becomes non-conducting, while V2 remains conducting.

Circuit: Battery positive terminal/partial winding T2/V4/$L$1/battery negative pole. A voltage is again induced in the secondary winding T3. The capacitor $C$1 is charged, though now via T1 and V3 with reversed polarity, etc.

Choke $L$1 limits the current on commutation. The diodes V3 and V4 prevent the quenching capacitor $C$1 from partially discharging prematurely via the primary windings of the transformer, while diodes V5 and V6 permit the connection of reactances. Part of the reactive current returns to the battery via these diodes.

# 8 Applications of Control Systems in Power Supply Devices

## 8.1 General

This chapter treats the application of control engineering in modern power supply equipment in more detail. Some elementary correlations and terms in open-loop control and closed-loop control are presented for those readers for whom this part of the subject is not so familiar.

With rectifiers a distinction is made between non-controlled, phase-controlled and controlled equipment.

With *non-controlled rectifiers* the d.c. output voltage falls when the load increases. It is also dependent on fluctuations in the alternating supply voltage and supply frequency.

*Phase-controlled rectifiers* supply a constant d.c. output voltage with changes in load and fluctuations in the a.c. supply voltage; however, the d.c. output voltage does change in the case of fluctuations in the supply frequency.

*Controlled rectifiers* supply a constant d.c. voltage within the tolerance band of, for example, $\pm$ 0.5% or $\pm$ 1%, regardless of any change in load or fluctuation in supply voltage and frequency. They receive their energy either directly from the mains supply or from standby power supply systems and must meet both the requirements of the communications system as well as those of the battery.

This group of rectifiers is classified into equipment with a transductor power section (magnetically controlled rectifiers), thyristor power section (thyristor-controlled rectifiers), and transistor power sections (transistor-controlled rectifiers). Thyristor-controlled rectifiers generally use phase-angle control. Rectifiers with a transistor power section can be further divided into devices with a linear regulator, devices with a switching regulator, and switching-mode power supplies.

Another important group comprises d.c./d.c. converters; these are supplied with direct current and can form independent subassemblies and equipment and also constitute parts of the switching-mode power supplies or inverters.

D.C./A.C. *inverters* should also be mentioned at this point (see Sect. 8.6.5).

## 8.2 The Operation of Open- and Closed-loop Control Systems

This section starts with a comparison of the ways in which open-loop and closed-loop control systems work (Fig. 8.1). Characteristic of the open-loop control is the open control chain, whereas, that of the closed-loop controls the feedback loop.

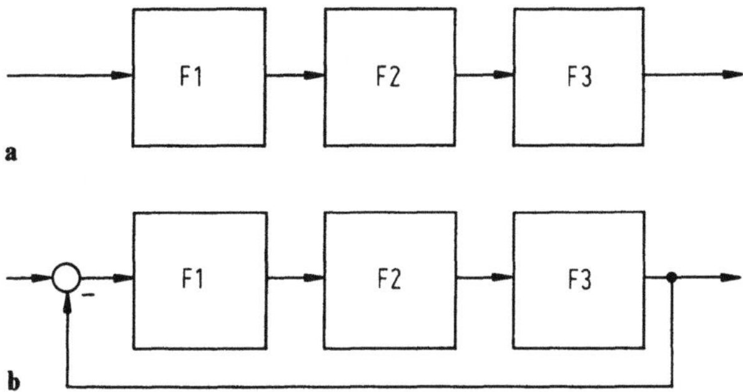

**Fig. 8.1a, b.** Operation of open-loop (**a**) and closed-loop control (**b**)

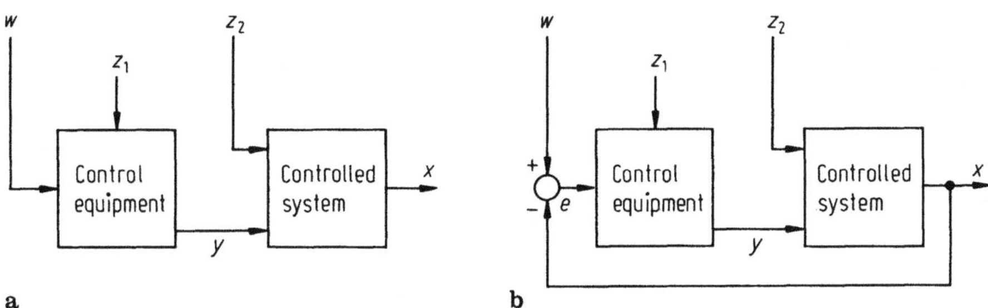

**Fig. 8.2a, b.** Basic working of open-loop control (**a**) and closed-loop control (**b**)

In *open-loop control* one or more input variables to a closed system influence the output. The output provides no feedback to the input variables. Such a control system consists of the control equipment and the controlled system (Fig. 8.2a).

The command variable $w$ arrives at the control equipment from an outside source. Depending on this command variable, the control equipment generates a correcting variable $y$, which exerts a controlling influence on the controlled system. The controlled system represents that part of an installation the tasks of which are controlled by the control equipment.

It is also called the controlled object, in which the variable to be controlled is produced. Disturbances $z$ (e.g. supply voltage fluctuations or changes in load) interfere with the control equipment (disturbance $z_1$) and also with the controlled system (disturbance $z_2$) and impair the working of the open-loop control system. Disturbances acting from outside can result in the controlled variable deviating considerably from the command variable. If the command variable $w$ changes, this causes a change in the correcting variable $y$ and thereby also in the output variable for the controlled system (control variable $x$).

Every disturbance thus influences the output variable, which can only follow the command variable to a limited extent. If the disturbance remains constant, this can be compensated for by pre-adjusting the command variable accordingly.

However, if the disturbances are constantly varying, the deficiencies of the open-loop control principle mentioned above make it necessary to use a closed-loop control system.

Despite disturbing influences, the *closed-loop control* is able to match the controlled variable as closely as possible to the value preset by the command variable. A closed-loop control system consists of control equipment and a controlled system.

With closed-loop control (Fig. 8.2b) the variable to be controlled (i.e. the controlled variable or actual value) is, if necessary, continuously monitored, compared with the constant command variable (i.e. the reference value/set-point value), and depending on the deviation, adjusted to agree with the command variable. The result is an operation within a closed circuit which is called the closed-loop control circuit. The controlled variable $x$ (actual value) is fed back to the input of the control equipment where it is compared with a fixed command variable $w$ (reference value). The difference between the reference value $w$ and the controlled variable $x$ is the error $e$. The aim of feedback control processes is always to keep the discrepancy between the reference value and controlled variable as small as possible. As long as there is a difference between the reference value and the controlled variable, the control equipment reduces the difference between them by providing an appropriate correcting variable $y$.

As with the open-loop control system, disturbances $z_1$ and $z_2$ can also affect the closed-loop control circuit, though here these effects, e.g. on the d.c. output voltage, can be corrected.

The closed-loop control system reacts to changes in the command variable or disturbances with a control process, i.e. the regulator produces a correcting variable, which brings the controlled variable to the new reference value, or holds it at the fixed reference value. This control process is always accompanied by a transient condition. This means that the regulator does not react until an error has occurred. Only then is the controlled variable brought into line with the new reference value either gradually by approximation or aperiodically.


## 8.3 Components of the Closed-loop Control System

Figure 8.3 shows the units making up the closed-loop control circuit.

### 8.3.1 Final Control Element

The unregulated voltage of the input supply is present at the input of the final control element. In the case of rectifiers, for example, this is the thyristor set made up of thyristors together with its trigger set or the power transistor unit with its driving transistors. The final control element has the task, according to the correcting variable supplied by the control equipment, of controlling the flow of energy so that it is adapted to the controlled object. The final control element thus carries out the instructions of the control equipment.

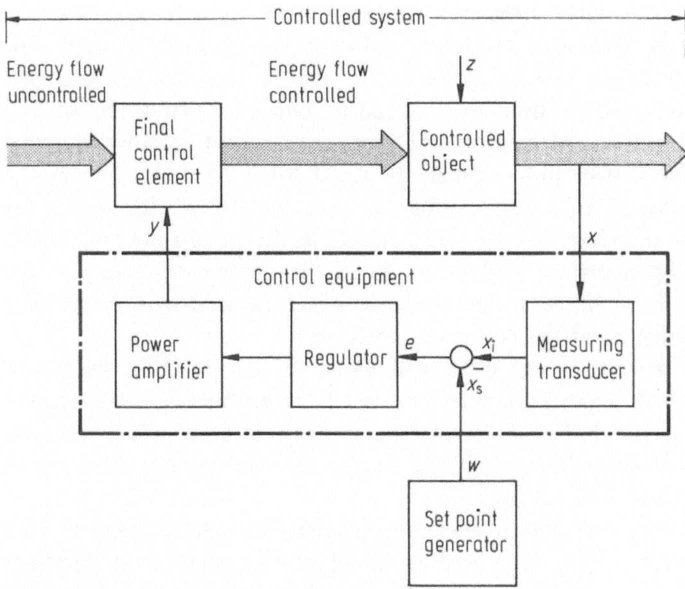


**Fig. 8.3.** Units of the closed-loop control circuit

### 8.3.2 Controlled Object

The controlled object is the main part of the controlled system influenced by the final control element. With rectifiers, for example, the d.c. output acts as the controlled object. This is influenced by disturbances $z$, e.g. changes in load. The output variable of the controlled system is picked up by the measuring transducer as a controlled variable and fed to the comparator in the regulator.

### 8.3.3 Control Equipment

The control equipment consists of the measuring transducer, the regulator and usually the power amplifier. It generates the correcting variable from the reference/actual comparison and with it influences the final control element. The error is represented by 'command variable minus controlled variable': if the controlled variable $x$ is too small and the error thus positive, the control equipment must raise the controlled variable by changing the correcting variable $y$.

If, on the other hand, the controlled variable $x$ is too large, the error is negative and the control equipment must reduce the controlled variable $x$ by changing the correcting variable $y$.

The condition is said to be corrected when the error has reached a value as close as possible to zero. How accurately the controlled variable assumes the value of the command variable, i.e. how closely the error approaches zero, depends on the regulator selected.

The *measuring transducer* measures the controlled variable and converts it so that it can be processed as an actual value in the regulator's input. For rectifiers the transducers are usually made up of sensors and measuring amplifiers. Measuring shunts, d.c. instrument transducers, current transformers of hall generators are called sensors. Measuring amplifiers are required in order to arrive at an actual value that can be processed in the regulator.

The *regulator* consists of a comparator and error amplifier with wiring for proportional and time response. Voltage- and current limiter regulators are usually used as integral components of rectifier units. In the comparator the possibly varying actual value $x_i$ of the controlled variable $x$ is compared with the usually constant reference value $x_s$ of the command variable $w$.

The command variable is set to a fixed value in the set point generator, normally with the aid of a potentiometer, and fed from an outside source to the control equipment. The stability of the command value is especially important. For this reason the potentiometer is supplied from a constant voltage source with the least possible temperature variation.

The operational amplifier, has to generate the desired control response with the aid of its feedback wiring. The output signal of the operational amplifier is usually in the range of $\pm$ 10 V. There are final control elements which do not respond to this voltage with a current of 5 or 10 mA, in which case *power amplifiers* are also required.


## 8.4 Operational Amplifiers in the Regulator

The operational amplifier (control amplifier) in the regulator must contain a feedback wiring and, at its input in a comparator, it has to compare the command variable (reference value) with the controlled variable (actual value) to determine the error (Fig. 8.4).

A set reference voltage $U_s$ usually fixed at a constant value by the command variable $w$ reaches the comparison point, at which the error $e$ is determined, via

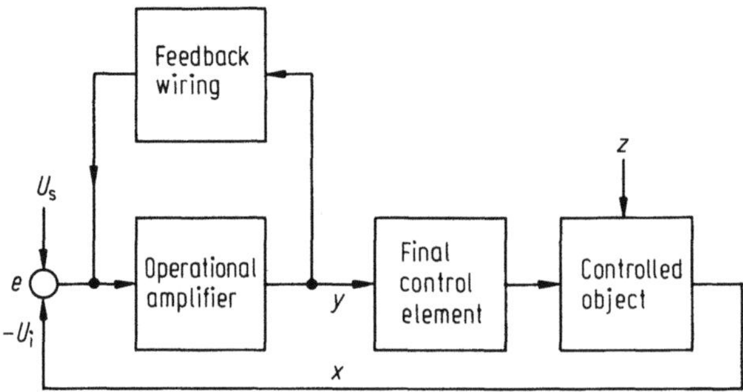


**Fig. 8.4.** Working of closed-loop control circuit with operational amplifier in the regulator

the set point channel. The reference voltage is compared with the actual voltage $-U_i$ of the controlled variable $x$. The actual voltage $-U_i$ is picked up at the output of the controlled object and thus at the controlled system.

The regulator must hold the controlled variable at the normally fixed set value for the command variable. The feedback wiring (here negative feedback) is provided to assure an optimal 'transient response' in the control procedure.

If an error other than zero occurs, the regulator delivers a modified correcting variable $y$ to its output. This is the instruction for the final control element to set its output so that the controlled variable is corrected as quickly as possible to the value specified by the command variable. To obtain the simplest possible control circuit the inverting input of the operational amplifier is used for connecting the control difference. This is called the inverting amplifier circuit.

Figure 8.5 shows an inverting amplifier in the regulator. The feedback resistance $R_f$ can be seen in the feedback wiring.

The input wiring of the regulator consists of the controlled variable channel with resistance $R_i$ and the command variable channel with the resistance $R_s$. The two channels combine at the inverting input terminal $E_-$ of the operational amplifier. The non-inverting input terminal $E_+$ of the operational amplifier is connected with the reference potential $M = 0$ V. This connection is usually made via a resistance $R_M$. The command variable, its reflexion is the reference of voltage $U_s$, can be set with the potentiometer $R$. It is fed to the comparison point in the form of a current $I_s$ via the resistance $R_s$.

The controlled variable, its reflexion is the actual voltage value $-U_i$, can also be adjusted in the case of many rectifiers. It also reaches the comparison point, where it is compared with the command variable, in the form of a current via resistance $R_i$. The reflexions of the command variable and controlled variable in the stationary state must always be of different polarity so that the currents formed from the voltages at resistances $R_s$ and $R_i$ generates a differential current $I_o$, which reflects the error. Comparing the currents at the comparison point has the advantage that the voltage reflecting a quantity can always be applied unbalanced to the reference voltage $M = 0$ V. This enables any number of input quantities to be compared.

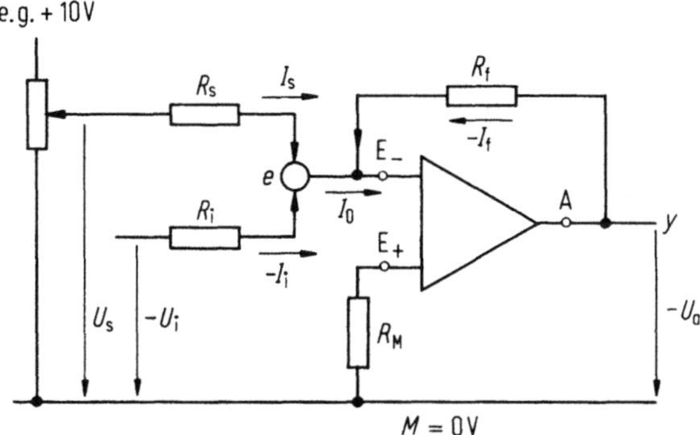


**Fig. 8.5.** Regulator with input and feedback wiring

The constantly adjusted reference voltage $U_s$ for the command variable is positive in the stationary state and the actual value voltage $-U_i$ for the controlled variable changing due to disturbances is negative. The set current $I_s$ flows to the comparison point via resistance $R_s$. Also flowing there is the actual value current $-I_i$ via resistance $R_i$. At the comparison point there is produced from the two currents the current $I_0 = I_s - I_i$ reflecting the error. As the current $I_0$ (at the inverting input $E_-$) flows to the highly resistive input of the operational amplifier, the circuit is closed via the feedback resistance $R_f$.

Only a very small current $I_0$ can flow off to potential M via the highly resistive internal resistance of the operational amplifier and the non-inverting input $E_+$. Thus the output voltage $U_a$ of the regulator becomes negative when the error is reflected by a current $I_0$ flowing towards the inverting input $E_-$:

$$\frac{U_s}{R_s} - \frac{U_i}{R_i} = I_0$$

$$I_0 R_f = -U_a$$

The output voltage $U_a$ of the regulator is the correcting variable $y$ with which the regulator intervenes in the controlled system to bring the controlled variable to the value specified by the command variable. The magnitude of the regulators output voltage must possibly be limited to prevent the controlled system being overridden. In many operational amplifiers two inputs are used for this (not shown in Fig. 8.5), to which a positive and a negative limiting voltage are applied. If necessary, these voltages are set with a potentiometer for each of them.

### 8.4.1 Error and Deviation

If the reference value equals the actual value ($x_s = x_i$), the output is regulated. There is neither a deviation $x_w$ nor an error (control difference) $e$ ($x_w = 0$ and $e = 0$, shown in the centre of Fig. 8.6). In this case the correcting variable

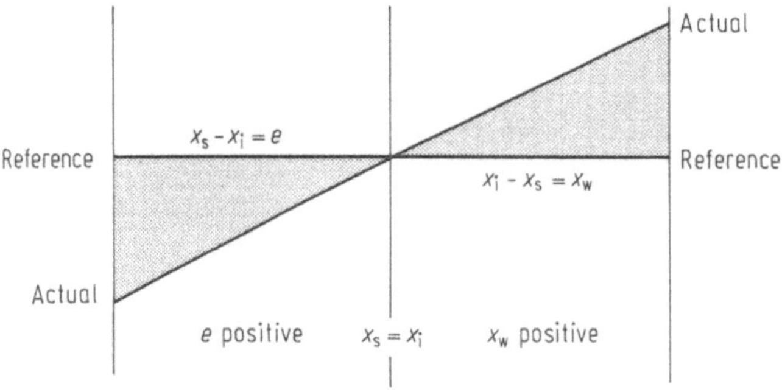


**Fig. 8.6.** Error and deviation

remains constant until a new error, i.e. a new deviation in the controlled variable, makes correction necessary again.

The difference 'command variable minus controlled variable' is called the error. The controlled variable minus command variable, on the other hand, is called the deviation. The error $e$ is positive (or there is a negative deviation) when the reference value is greater than the actual value ($x_s > x_i$). There is a positive deviation $x_w$ (or the error is negative) (right part of figure), when the actual value is greater than the reference value ($x_i > x_s$).

### 8.4.2 Classification of Regulators

A distinction is made between continuous and discontinuous regulators (controllers, Fig. 8.7), e.g. on-off regulators.

There are three basic types of continuous regulators:

1. P regulator (proportional response).
2. I regulator (integral response), also called reset or floating regulator.
3. D regulator (differential response), also called derivative regulator.

Chiefly P regulators, I regulators and PI regulators are used in a telecommunications power supply. Apart from the regulators mentioned above, the combinations of PD regulator and PID regulator are also possible.

*P Regulator.* With the P regulator the output quantity (correcting variable $y$) responds proportionally to the error $e$ on the input side. At equilibrium a certain value for the correcting variable is assigned to each value for the error (Fig. 8.8). For a correcting variable other than zero to occur, the error must be unequal to zero, i.e. the controlled variable does not then have the value specified by the command variable.

The P regulator has an ohmic resistance $R_f$ as its feedback (Fig. 8.8a). There is also a potentiometer $R$ in the feedback wiring, with which the proportional factor $K_p$ (also called the proportional amplification $V_R$) can be set.

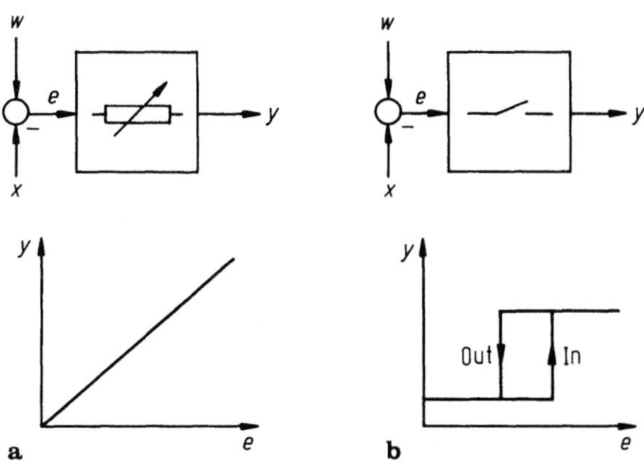


**Fig. 8.7a, b. a** Continuous regulators and **b** discontinuous regulators – here on-off regulator

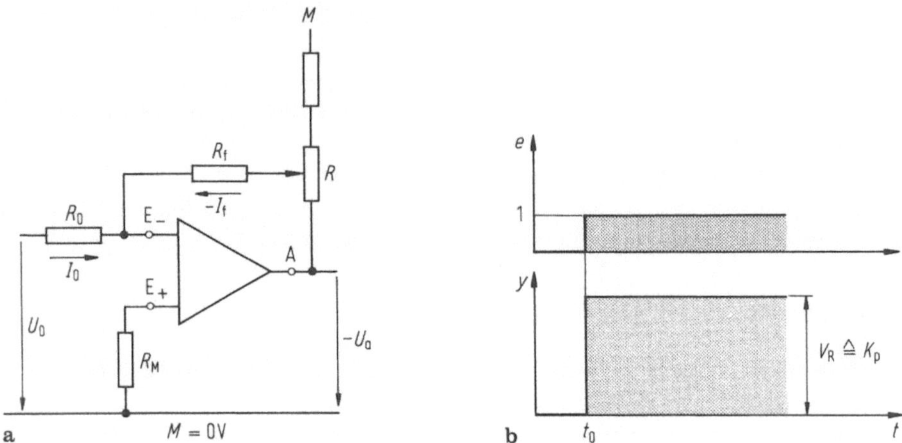


**Fig. 8.8a, b.** P regulator, circuit and transient function

At this point it must be noted that the time response at the output of a transfer element to a step-shaped input quantity is called a step response. When this is related to the height of the input step it is called transient response.

The transient response of the ideal P regulator can be seen in Fig. 8.8b. Here the output voltage of the P regulator follows the voltage jump $\Delta U_0$ at the output without inertia at time $t_0$. The voltage $\Delta U_0$ is to be seen as a reflexion of the error.

The following applies to proportional amplification:

$$V_R = \frac{U_a}{\Delta U_0} = \frac{R_f}{R_0}$$

Thus proportional amplification of, for example $V_R = 3$ is obtained when the feedback resistance $R_f$ is three times greater than the input resistance $R_0$. The output voltage $U_a$ is then three times larger than the input voltage $U_0$. The example in Fig. 8.8b is based on amplification of $V_R = 3$.

P regulators are characterized by rapid correction of the disturbance (good initial response). However, a residual error has to be accepted, for the output quantity of the P regulator is always proportional to the remaining error (poor correction response).

*I Regulator.* With the I regulator a certain rate of change of speed for the correcting variable $y$ is assigned to each value for the error e Fig. 8.9.

The I regulator changes its output value in the direction for correction until the error is practically zero. It has a capacitor $C_f$ as feedback (Fig. 8.9a). In the ideal case the transient function of an I regulator is the same as the diagram in Fig. 8.9b. The feedback capacitor $C_f$ determines, together with the input resistor $R_0$, the time response of the I regulator. If the input quantity, in Fig. 8.9a the

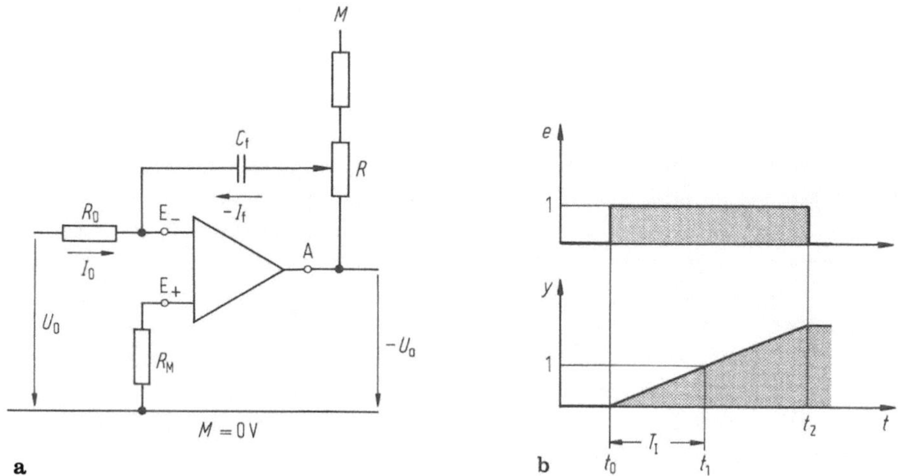


**Fig. 8.9a, b.** I regulator, circuit and transient function

voltage $U_0$, is equal to zero, the feedback capacitor $C_f$ remains in the charging state it has reached.

The output voltage $U_a$ remains at the value it has reached, normally a value other than zero. If the error assumes a value different from zero, the output voltage $U_a$ changes in the opposite direction to the error at a speed determined by the integrating factor:

$$K_I = \frac{1}{R_0 C_f}.$$

This means that the output voltage $U_a$ changes in relation to the initial value $U_{a0}$ as shown in the equation:

$$-U_a(t) = \frac{1}{R_0 C_f} \int_0^t U_0(t)\,dt + U_{a0}$$

If at time $t_0$ the input voltage $U_0$ makes a jump, the integrating time $(T_I = R_0 C_f)$ is that time during which the output voltage $U_a$, starting from the initial voltage $U_{a0}$, undergoes a change; it depends on the height of the voltage jump on the input side (time $t_1$).

The output voltage $U_a$ changes continuously as long as there is a voltage at the input differing from zero. If the input voltage again becomes zero (time $t_2$), the output voltage remains at the final value it has reached.

The I regulator permits very accurate correction of the disturbance (good correction response), as it is only when the error has become zero that the output value no longer changes. A disadvantage is that it intervenes relatively slowly (poor initial response) because the error must first be integrated and a sufficient correcting variable built up.

The integrating time can be set with the potentiometer $R$ when the input resistance $R_0$ is very large compared with the resistance $R$ of the potentiometer.

*PI Regulator.* With the PI regulator the correcting variable $y$ is the same as the sum of the output values of a P and an I regulator (Fig. 8.10).

The PI regulator initially responds like an I regulator, i.e. any error results in integration of the same so that the output quantity moves towards the set point. In addition and simultaneously with the error, a proportional response is superimposed on the integral one until the error has been eliminated. This means that the PI regulator does more than the I regulator. With the aid of the P component it forms a derivative action, the P derivative action.

The PI regulator has in its feedback loop resistor $R_f$ and a capacitor $C_f$ (Fig. 8.10a). This regulator combines the advantages of both regulators, namely the rapid reaction of the P regulator with the accuracy of correction of the I regulator without having the disadvantages of either.

The transient function of the PI regulator in the ideal case is as shown in Fig. 8.10b. If at time $t_0$ the input voltage $U_0$ intended as a reflexion of the error makes a jump from zero, the output voltage $U_a$ also makes a jump from the level $\Delta U_0 V_R$. As the capacitor $C_f$ acts as a short-circuit, only the input voltage $U_0$ and the proportional amplification $V_R$ are decisive for the output voltage $U_a$. The output voltage $U_a$ then alters as a straight line in time from the voltage jump made in accordance with the integration time $T_I = R_0 C_f$. The rise in output voltage results from the magnitude of the charging current $I_f = \Delta U_0/R_0$ and according to the capacitance of the capacitor $C_f$.

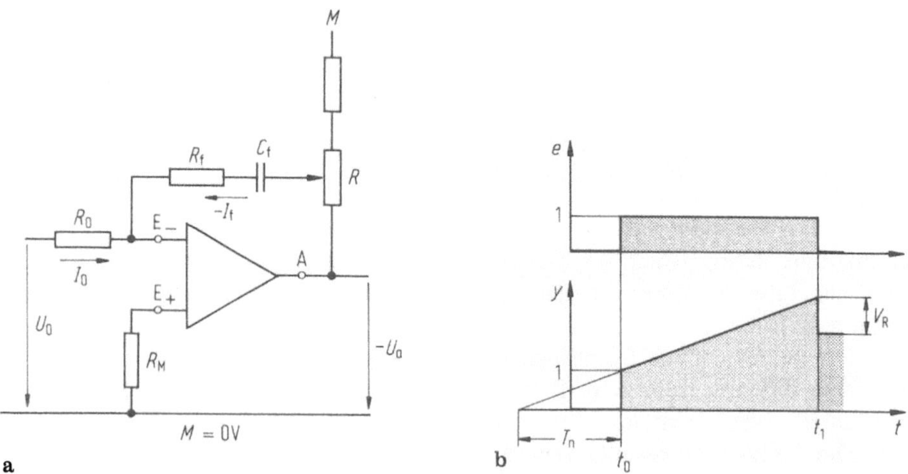


**Fig. 8.10a, b.** PI regulator, circuit and transient function

If the input voltage $U_0$ at time $t_1$ returns to zero, the P derivative action disappears and the PI regulator jumps down on the output side, similarly from the height $\Delta U_0 V_R$. The output voltage of the PI regulator then remains at the value there would have been had only the I component been effective.

As with the P regulator, the proportional factor (the magnitude of proportional amplification $V_R$) is set with the feedback potentiometer $R$. The proportional amplification $V_R$ of a PI regulator indicates, for each jump in the input quantity, the factor for the change in input quantity as referred to the output voltage jump in the first instance.

Integration or reset time $T_n$ is the name given to that time by which the I regulator would have to intervene earlier to achieve the same change in correcting variable as a PI regulator. With the start of the signal jump at the input the PI regulator has requested as its output an input signal $V_R$ times too much, but at the end of the signal jump has reduced the request by the same amount. With the reset time $T_n$ the output signal is adjusted, as if the PI regulator had started integrating earlier by the time $T_n$.

## 8.5 Controlled Rectifiers with Thyristor Power Section

Figure 8.11 shows the construction of a controlled rectifier with a thyristor power section and phase-angle control. Such rectifiers are suitable mainly for the middle and upper power range.

### 8.5.1 Power Section

The alternating supply voltage is transformed by the *main transformer* so that the required maximum d.c. output voltage of the rectifier can be obtained. The main transformer also isolates the mains supply and the load electrically. An auxiliary transformer produces the synchronizing alternating voltage from the supply voltage (see Fig. 8.12). The *thyristor set* together with the *trigger set* act as the final control element. It has the task of providing the required direct voltage.

From the measuring shunt the actual current value reaches the current-limiting regulator as voltage. The direct voltage is smoothed by the *filters*, i.e. filtered to the permissible level of superimposed alternating voltage. The constant direct voltage $U_A$ is tapped at the output terminals of the power section (see Fig. 8.13).

### 8.5.2 Control Section

The Regulation (closed-loop control) assembly (Figs. 8.11 and 8.14) combines all the essential functions used in forming the trigger pulse. There are different versions of this assembly depending on whether it is connected to an alternating current or a three-phase supply.

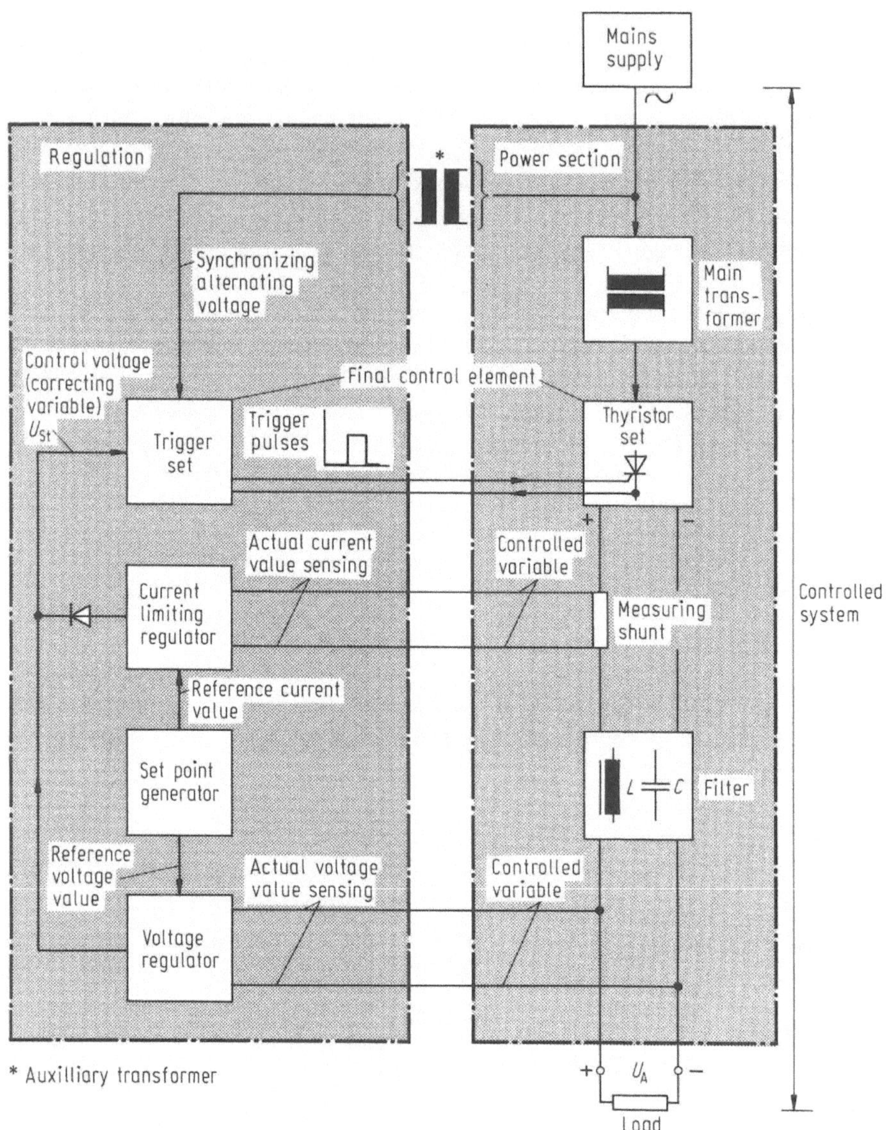


**Fig. 8.11.** Block diagram of a controlled rectifier with thyristor power section and phase-angle control

*Set point generator.* The set point for the direct voltage to be controlled is set in the set point generator with the aid of a potentiometer. It also has a potentiometer for setting the current limit.

*Voltage regulator.* The direct voltage to be controlled is tapped at the output of the power section and fed to the voltage regulator as actual voltage value

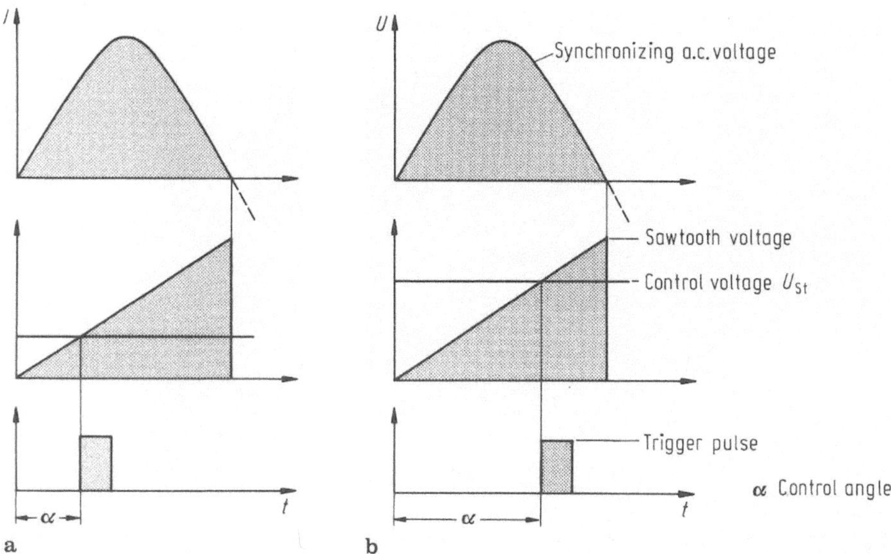



**Fig. 8.12a, b.** Shift of the trigger pulse depending on the coincidence of sawtooth voltage and control voltage $U_{St}$

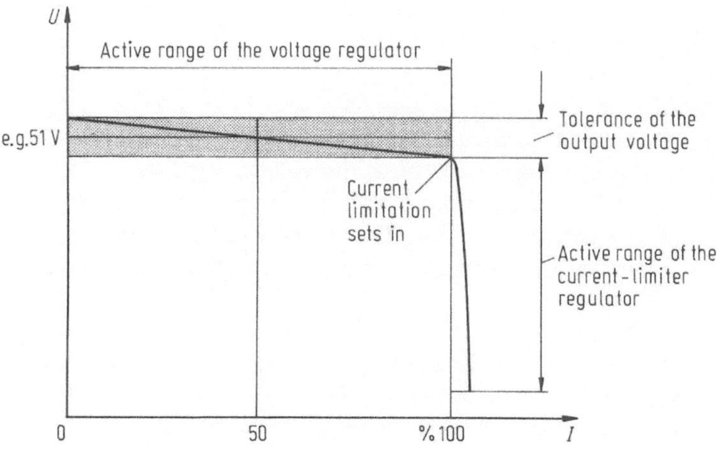



**Fig. 8.13.** Behaviour of the d.c. output voltage

(controlled variable). The reference voltage value passes from the set point generator so that there is a 'reference voltage value/actual voltage value' comparison at the voltage regulator's input. The resultant error is processed by the voltage regulator so that through its proportional and time response a correcting variable (control voltage $U_{St}$) is produced, which with the aid of the trigger set determines a suitable firing time for the thyristor set.

**Fig. 8.14.** Regulation assembly (for a three-phase unit). (Photo by courtesy of Siemens AG)

*Trigger set.* The trigger set supplies the trigger pulses for the thyristors in the thyristor set. It adjusts the output voltage of the thyristor set according to the correcting variable, i.e. the output voltage from the voltage regulator (control voltage $U_{St}$).

For this, depending on the supply voltage, a sawtooth voltage is generated from the synchronizing alternating voltage which starts with each zero passage of the phase (see Fig. 8.12). The output voltage of the voltage regulator, the control voltage $U_{St}$, is compared with the sawtooth voltage. If the values are equal, a trigger pulse is passed to the thyristor set. The direct voltage at the output of the thyristor set is determined by shifting the firing time. In this manner, the d.c. output voltage is held constant.

*Current-limiting regulator.* The current-limiting regulator (referred to below as simply the current regulator) is used to protect the rectifier against overloading. The maximum permissible current is set with the aid of a potentiometer in the set point generator. The current regulator only intervenes in the control circuit if the actual current value starts to rise above the reference maximum value. To limit the current, the current regulator reduces the d.c. output voltage until the rectifier can not deliver more than the reference current (this is usually the rated current 100% ($I_{rated}$), see Fig. 8.13.

*Control processes.* The control processes in four instances are explained below (Figs. 8.11, 8.12, and 8.13):

(1) d.c. output voltage constant,
(2) d.c. output voltage too high,

(3) d.c. output voltage too low,
(4) current limitation.

(1) Let the set point value for the d.c. output voltage $U_A$ of the rectifier be, for example, 51 V. If the d.c. output voltage is now actually 51 V, the voltage regulator detects no difference between the set voltage value and the actual voltage value. The error is zero. Thus the control voltage $U_{St}$ at the output of the regulator remains constant and the trigger set gives, in relation to the zero passage of the supply voltage, the trigger pulse to the associated thyristor at the same time as for the preceding voltage half waves. The d.c. output voltage in the given example therefore remains constant at 51 V (see Fig. 8.13).

(2) With a decreasing, purely resistive load the output voltage of the rectifier might initially become a little larger. In the example $U_A$ becomes $> 51$ V. As soon as the actual voltage value becomes greater than the set voltage there arises a negative error (positive deviation). This results in the voltage regulator delivering a somewhat higher control voltage $U_{St}$. Consequently, the trigger pulses from the trigger set are emitted a little later to the respective thyristor, i.e. shifted in the direction of 180° (see Fig. 8.12b). This means later firing of the associated thyristor so that the d.c. output voltage of the rectifier drops and returns to the set voltage value of 51 V (see Fig. 8.13).

(3) With an increasing, purely resistive load the output voltage of the rectifier initially falls a little below the value of 51 V because a somewhat higher source voltage is now required throughout the whole d.c. circuit to achieve the required set voltage of 51 V for the load. Consequently the source voltage must be adjusted somewhat higher by the control process. As the actual voltage value in the initial moment of stepping up the load drops slightly below the set voltage value, there is a positive error. It follows from this that the voltage regulator delivers at its output a slightly lower control voltage $U_{St}$ to the trigger set. Therefore, the trigger set now generates trigger pulses for the thyristors, which fire the thyristor a little earlier in relation to the voltage zero passage of the half-wave. The trigger pulses are thus shifted slightly in the direction of 0° (see Fig. 8.12a). In this way the d.c. output voltage of the rectifier again rises to the set voltage value of 51 V (see Fig. 8.13).

(4) The control processes considered thus far (examples 1, 2, and 3) are executed by the voltage regulator alone without the current regulator having to intervene, because the current drain remained within the permissible range.

Example (4) now refers to the current-limiting regulator. Let the current limit be set at 100% ($I_{rated}$). As long as the actual current value is less than the reference current value coming from the set point generator, the current limiter does not come into operation. If, however, the actual current value starts to become greater than the reference current value, the current regulator cuts out the voltage regulator by delivering the now higher control voltage $U_{St}$. The diode at the output of the current limiter becomes conducting in the process (see Fig. 8.11).

The trigger set now generates trigger pulses that are so far shifted towards 180° that a d.c. output voltage is produced at the rectifier which drives only the maximum permissible current, thereby protecting the rectifier against overloading (see Fig. 8.13). Figure 8.12 shows the shift of the trigger pulses as a function of the control voltage.

### 8.5.3 Behaviour of the Output Voltage and Dynamic Response

Figure 8.13 shows the *behaviour of the d.c. output voltage* of a controlled rectifier, indicating the static tolerance range and the current limitation. The static tolerance range means the range within which the controlled variable must remain after a control process. The static tolerance for rectifiers with a thyristor power section is, for example, $\pm$ 0.5% (transistor controlled rectifier $\pm$ 1%).

If changes in disturbance (here primarily loading of the rectifier) occur in closed-loop control, the control equipment reacts with a transient of the controlled variable characterizing the *dynamic response* of the control circuit (Fig. 8.15). With stepped changes in load this can result in short-time deviations of the d.c. output voltage. Depending on the extent of the change in load the output voltage will lie more or less far outside the static tolerance range for one or more half-waves of the transient.

There are two comparative values for judging the control response when correcting a stepped disturbance. These are transient overshoot and settling (correction) time.

The transient overshoot denotes the amplitude of the first deflection of the controlled variable when settling the disturbance.

By settling time is understood the time that passes after the occurrence of a stepped change in load, measured from the instant at which the voltage departs

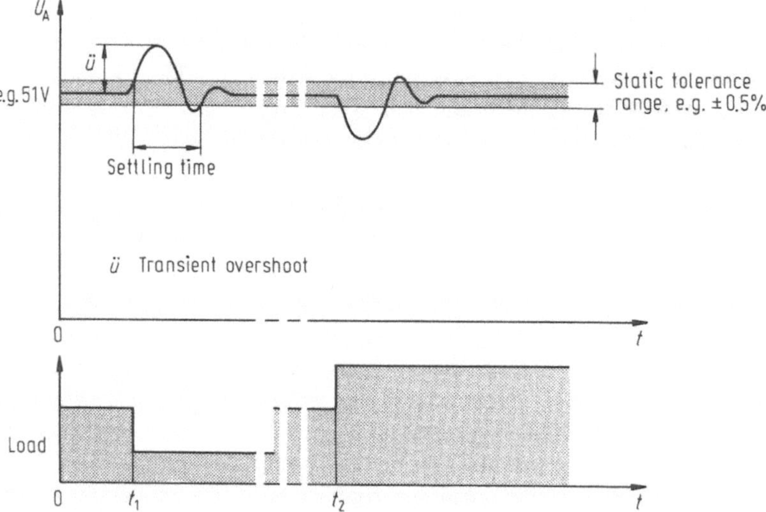


**Fig. 8.15.** Behaviour of d.c. output voltage after a stepped change in load

from the specified tolerance range, until the d.c. output voltage returns to, and remains within, the static tolerance range. At time $t_1$ there occurs a stepped change in the load in the negative direction. The d.c. output voltage will then initially experience a positive deflection. There now is an appreciable error and the regulator returns the voltage to the set value of e.g. 51 V.

At time $t_2$ there occurs a stepped change in load in the positive direction. The first control process now takes place in the opposite direction. The deflection of the voltage to be controlled is here in the negative direction. A deviation of $\pm 4\%$ from the desired d.c. output voltage is considered to be the permitted tolerance range.

### 8.5.4 Type Designation

In this section (Fig. 8.16) the designations of the different types of thyristor-controlled rectifiers are explained. The letter (combinations) mean:

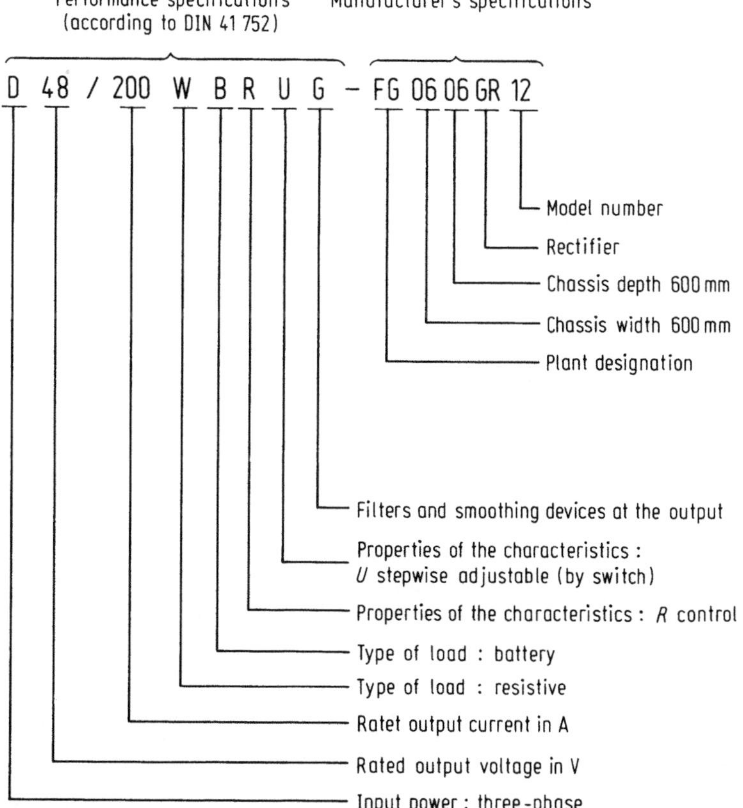


**Fig. 8.16.** Designation of rectifier types

*Type of input power:*
E   single-phase alternating current,
D   three-phase alternating current,
G   direct current (only in designations of inverters or d.c./d.c. converters).

*Type of load:*
W   resistive load.

The rectifier can be directly connected to the load in this type, without connecting a battery in parallel.

B   battery load.

The rectifier can be connected to a battery and supply the necessary charging current/voltage.

*Properties of characteristics:*

R   rectifier with controlled, hence stabilized, characteristic (controlled devices).

Regardless of certain fluctuations in mains voltage, mains frequency or load, the d.c. output voltage is maintained constant to within a tolerance of, e.g., $\pm 0.5$ %.

U   Rectifier with stepwise modifiable (by switch) characteristic.

*Filters and smoothing devices:*

G   Rectifiers with smoothing or filtering devices.

Filters limit the magnitude of the interference voltage to that allowed for the communications system.

*Multiple inputs or outputs:*

No letter: rectifiers with only one input and one output.

D   rectifiers with several d.c. outputs.

Even if several outputs are available, only the values for the principal outputs are included in the type designation. The letter D occurs for thyristor-controlled rectifiers when several outputs are present.
The abbreviation GR (rectifiers) can be taken as an example. If significant changes have been made in the construction or electrical properties of a rectifier, then the *model number* changes as well.


## 8.6 Controlled Power Supply Equipment with Transistor Power Section

A distinction is made between devices with linear (series) regulator (see Sect. 8.6.1) and devices where the power transistor is operated as a switch (switching regulator), (see Sects. 8.6.2 and 8.6.3).

The power transistors used in the switched mode are either bipolar transistors, as shown in the schematic diagrams of Sect. 8.6, or field effect transistors (e.g. power MOS).

### 8.6.1 Rectifiers with Linear Regulators

Figure 8.17 shows a transistor-controlled rectifier with a linear series regulator (longitudinal controller). The power transistor is variably controlled by the control assembly, depending on the level of the measured d.c. output voltage compared with the reference voltage: if the direct voltage at the output of the rectifier V is appreciably higher than the d.c. output voltage $U_A$ to be provided, a voltage drop can be set at this transistor of such a size that a constant output voltage is produced.

The efficiency that can be achieved with this circuit is relatively low. Favourable factors are rapid correction and the particularly simple circuit design. Transistor-controlled rectifiers with a linear regulator are used to advantage in the lower power range, up to about 25 W.

The linear regulator principle is often found in power supply equipment for providing internal supply voltages, for example, integrated circuits in power supply modules.

### 8.6.2 Rectifiers with Switching Regulators

Figure 8.18 shows a controlled rectifier with switching regulator. A supply-side bridge circuit V2 supplies the 'intermediate circuit voltage'. According to the actual voltage value ($\hat{=} U_A$) measured, the control module in each cycle switches

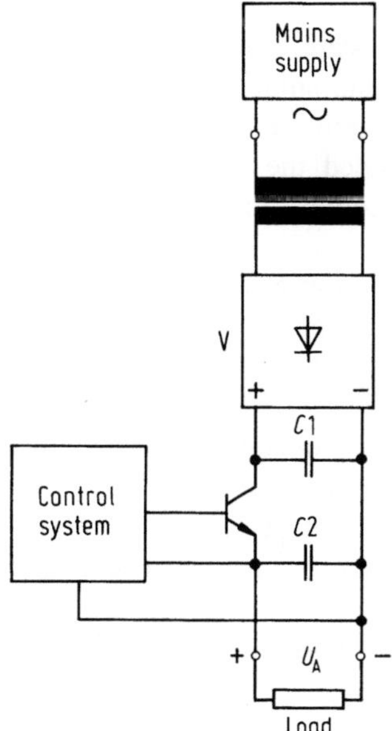


**Fig. 8.17.** Rectifier with transistor power section and linear regulator

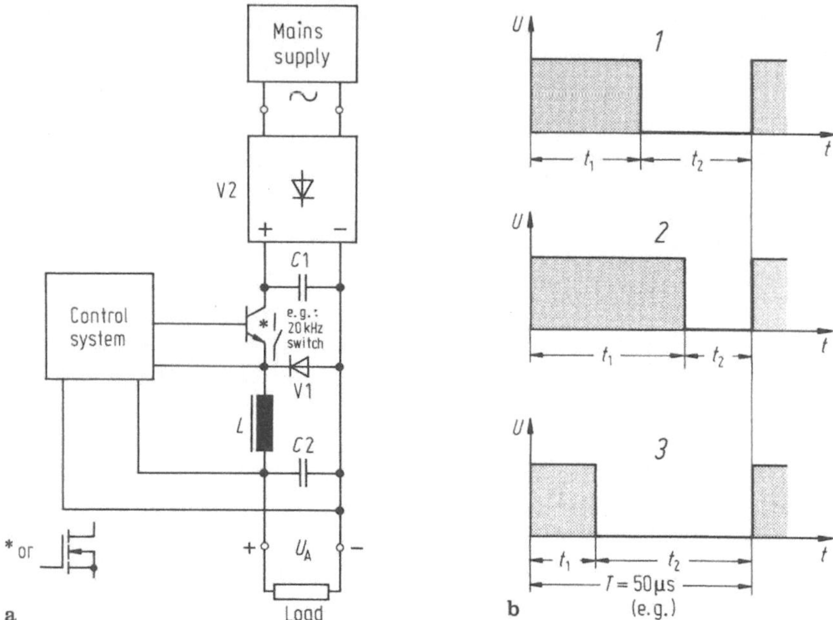


**Fig. 8.18a, b.** Rectifier with transistor power section and switching regulator. $T$ Period (constant), $t_1$ on time, $t_2$ off time

the switching transistor to the conducting state long enough for the mean voltage value of that cycle to match the reference voltage value for the d.c. output voltage $U_A$.

The clock frequency is often 20 kHz or higher. The human ear is insensitive to these frequencies.

With the switching transistor in the conducting state, current flows from the positive pole V2 to the switching transistor, via the smoothing choke $L$ to the charging capacitor $C2$ and via the load to the negative pole V2. When the conducting phase for the switching transistor has ended the current continues flowing via the smoothing choke $L$ and the free-wheeling diode V1, capacitor $C2$ being thereby discharged.

As the clock frequency is relatively high, the filters can be kept small. However, the high frequency does require certain measurements for the suppression of radio interference. Figure 8.18b shows the voltage at diode V1 as a function of time.

In example 1 (Fig. 8.18b) it is assumed that the 'conducting state' $t_1$ of the switching transistor is of the same length as the 'non-conducting state' $t_2$.

With a change in load or in the intermediate circuit voltage either the interval $t_1$ becomes longer and the interval $t_2$ shorter (example 2)[1] or vice-versa (example

---

[1] In practice the interval $t_1$ is always shorter than the interval $t_2$ and the example 2 is not possible.

3). The period $T$ remains constant. This is called *pulse-width control* (see Sect. 7.2.2). The longer the conducting state $t_1$ lasts and thus the larger the voltage-time area, the higher the average value for the d.c. output voltage.

### 8.6.3 Switching-mode Power Supplies

The principle described in connection with the switching regulator (see Sect. 8.6.2) can also be applied to switching-mode power supplies.

As shown in Fig. 8.19, a switching-mode power supply consists of a power section and the control assembly. The power section can be broken down into a supply-side rectifier and a d.c./d.c. converter. The supply-side rectifier supplies the voltage for the d.c./d.c. converter (intermediate circuit voltage). With the a.c. input components can be found for overvoltage protection and radio interference suppression.

The alternating voltage from the supply mains is rectified (V1), filtered (C1) and then 'chopped' by a switching transistor[2]. The resultant alternating voltage (e.g. square wave) is again rectified (V2) after transforming. The transformer is also used for electrical isolation. The rated constant d.c. output voltage $U_A$ is then available after filtering (C2). With the d.c. output components can also be found for radio interference suppression.

The control assembly has the task of keeping the d.c. output constant. Pulse-width control is also used here, as with the switching regulator.

Switching-mode power supplies (Fig. 8.20) are suitable for the lower ($> 25$ W) to medium power range.

There are 'primary switching' and 'secondary switching' switching-mode power supplies. Primary switching ones, which are the type almost exclusively used today, are based on the forward converter principle.

### 8.6.4 D.C./D.C. Converters

**Single-ended Forward Converters**
In the single-ended forward (flow) converter, the consumption of energy from the primary side coincides in time with the delivery of energy from the secondary side. Figure 8.21 shows a single-ended forward converter with *one* switching transistor. Here the capacitor $C1$ and the inductance of the transformer form the 'primary-side resonant circuit'. A second resonant circuit ('secondary-side resonant circuit') is represented by the winding c and capacitor $C2$. Current flows in the primary side resonant circuit during the conducting phase of switching transistor V3 which leads to current flowing to the load in the secondary side

---

[2] Bipolar or field effect transistor (e.g. Power MOS) working frequency is between 20 kHz and 100 kHz. The used voltage wave forms of the energy flow at the collector are in general square, trapezoidal, half sinusoidal or in exceptional cases sinusoidal. The collector currents are triangular or rectangular.

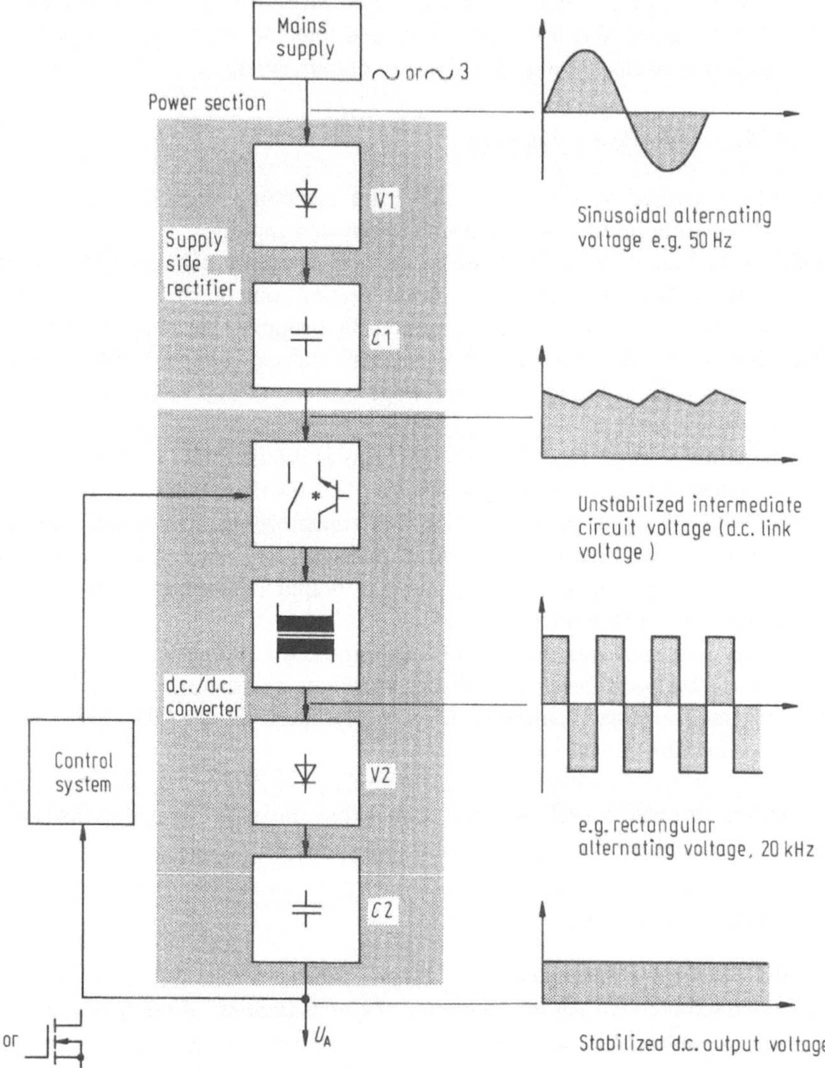


**Fig. 8.19.** Principle of a switching-mode power supply

resonant circuit via the now conducting diode V1. Capacitor $C2$ is at the same time being charged. This is possible because the primary and secondary windings have the same direction of winding (marked by a dot on the same side at both windings a and c (see Fig. 8.21a).

The smoothing choke $L$ stores energy during this process.

During the non-conducting phase of switching transistor V3 the supply of current is interrupted in the secondary circuit. The choke still delivers energy to the load for a short time. The smoothing capacitor C2 is discharged. The

**Fig. 8.20.** Switching-mode power supply GR40 48V/120A (60V/100A). (Photo by courtesy of Siemens AG)

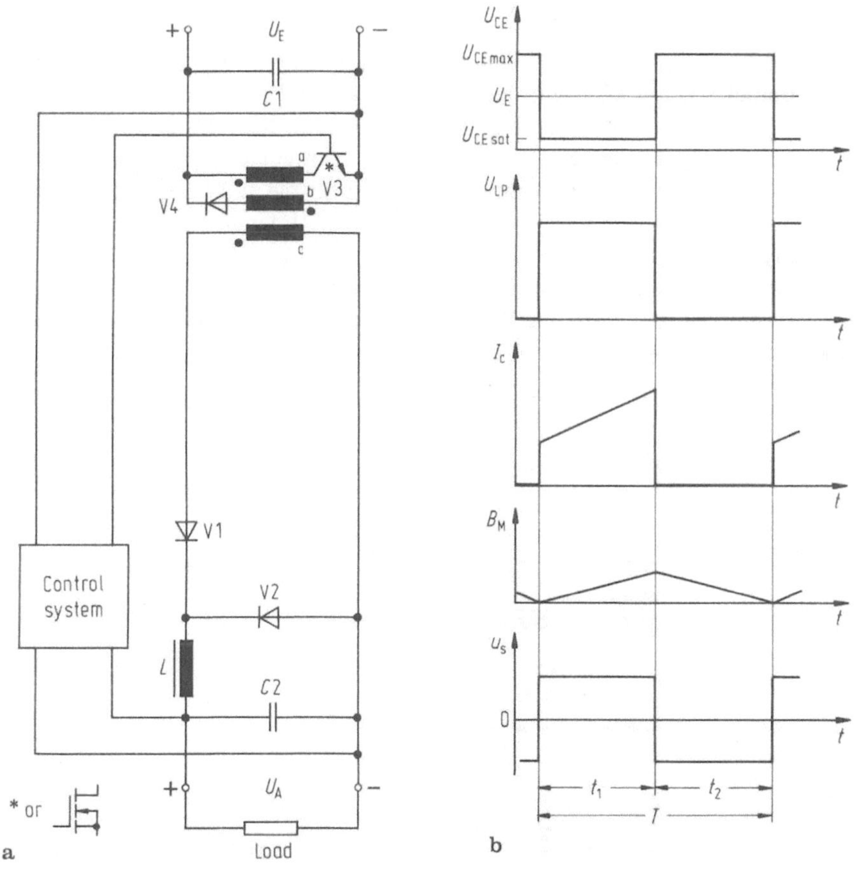

**Fig. 8.21a, b.** Single-ended forward converter with one switching transistor. $U_E$ Input voltage, $U_A$ output voltage, $U_{CE}$ voltage between collector and emitter of transistor V3, $U_{CE\ sat}$ saturation voltage, $U_{CE\ max}$ max. voltage between collector and emitter, $U_{LP}$ voltage at primary winding (**a**) of the transformer, $I_c$ collector current (primary current), $u_S$ secondary voltage, $B_M$ Magnetic flux density, $T$ period, $t_1$ on time of transistor V3, $t_2$ off time of transistor V3, pulse duty factor $V_T = 0.5$

free-wheeling diode V2, now in the conducting state, lies in parallel so that the current still continues to flow.

The control system switches transistor V3 to the conducting state long enough for the d.c. output voltage to be kept constant with a fluctuating input voltage or changing load.

There is a demagnetizing winding b (see Fig. 8.21a) so that the transformer does not become saturated. The primary side oscillation circuit turns over during the non-conducting phase of switching transistor V3. Current flows in the reverse direction via diode V4 and the demagnetizing winding, thereby preparing the transformer for the next conducting phase. Figure 8.21b illustrates the ideal behaviour of the individual variables.

Using the single-ended forward converter, voltages with narrow tolerance (e.g. component voltages of 5 to 24 V ±4 %) and large output power can be produced.

Figure 8.22 shows a single-ended forward converter which supplies a *fully regulated* output voltage $U_{A1}$ with narrow tolerance and wide power range. The actual voltage value for the control system is tapped at this output voltage. The *partly regulated* output voltage $U_{A2}$ benefits from the regulation to a minor extent.

The level of this voltage depends on the load on the fully regulated output. For this reason it is only possible to achieve a wide tolerance of, e.g., ± 10% for the partly regulated output.

If a narrow tolerance is required for the partly regulated output voltage, too (for low power), a (continuous) linear regulator (series regulator, adjustment controller) must also be provided (Fig. 8.23).

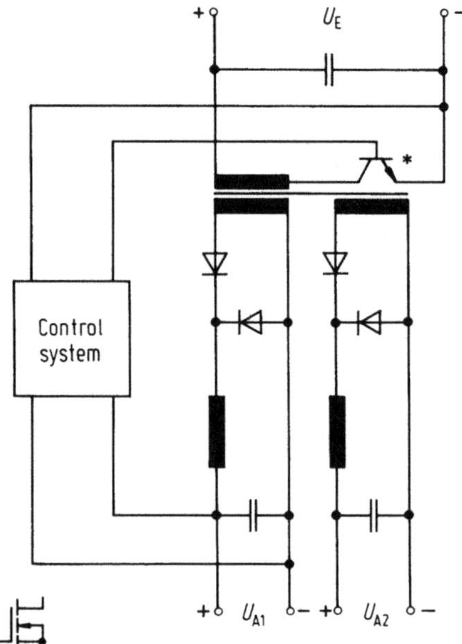


**Fig. 8.22.** Single-ended forward converter with one fully regulated and one partly regulated output

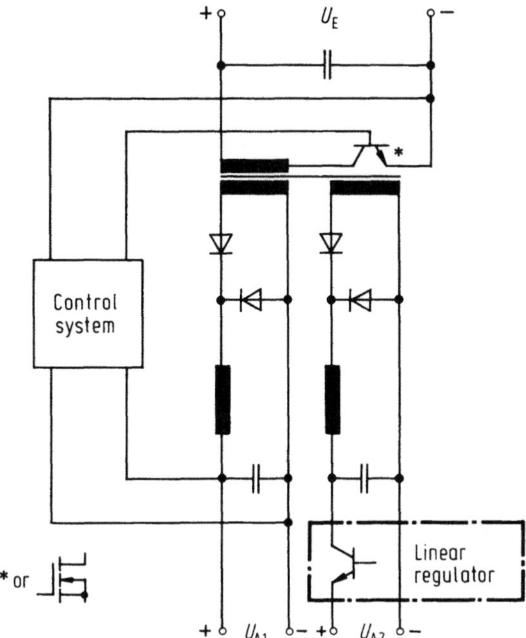

**Fig. 8.23.** Single-ended forward converter with one fully regulated and one partly regulated output with linear regulator

## Single-ended Flyback Converter

In contrast with the single-ended forward converter (as in Fig. 8.21), in the case of the single-ended flyback (blocking) converter (Fig. 8.24) energy consumption and energy delivery are shifted by 180° in time. The inductance of the transformer acts together with the capacitance of the capacitor $C1$ as a parallel resonant circuit. During the conducting phase of the switching transistor V3 current flows through the transformer (Fig. 8.24a). Primary and secondary windings are wound in opposite directions. Thus, in the conducting phase of the switching transistor V3 the diode V1 is polarized in the reverse direction and no current can therefore flow through the secondary winding. The load is supplied exclusively from smoothing capacitor $C2$.

In the non-conducting phase of the switching transistor the polarity of the voltage at the transformer reverses. Diode V1 now becomes conducting. The energy stored in the transformer during the conducting phase is delivered to the load. At the same time capacitor $C2$ is recharged. Diode V2 attenuates inductive voltage peaks.

Figure 8.24b illustrates the ideal behaviour of the individual variables.

## Push-pull Forward Converter

The push-pull forward (flow) converter (Fig. 8.25) is a combination of two single-ended forward converters which within a period are operated by the control system with a phase shift of 180°. When transistor V3 is conducting, current

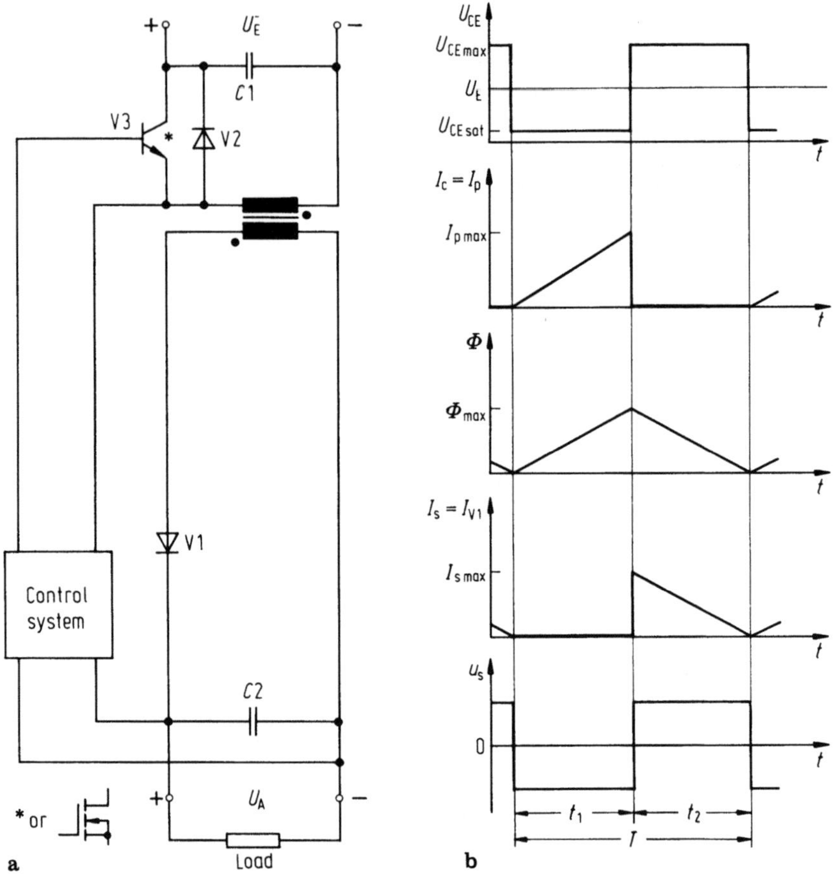



**Fig. 8.24a, b.** Single-ended flyback converter. $U_{CE}$ voltage between collector and emitter, $U_{CE\ sat}$ saturation voltage, $U_E$ d.c. input voltage, $U_A$ d.c. output voltage, $I_c$ collector current, $I_p$ primary current, $I_{p\ max}$ maximum primary current, $\Phi$ magnetic flux, $\Phi_{max}$ maximum magnetic flux, $I_s$ secondary current, $I_{s\ max}$ maximum secondary current, $I_{V1}$ current through diode V1, $u_s$ secondary voltage, $T$ period, $t_1$ on time of transistor V3, $t_2$ off time of transistor V3

flows through the primary winding P1 of the transformer. This produces a voltage in both secondary windings S1 and S2. As these windings are wound in the opposite direction to primary windings P1 and P2, a voltage of positive polarity now lies at the outside of winding S1 so that diode V1 becomes conducting. This causes current to flow in the circuit: winding S1, diode V1, choke $L$, load, and then back to the winding S1.

The smoothing capacitor $C2$ is charged.

At the secondary winding S2, which is wound in the opposite direction to the primary winding P1, there is now a voltage of negative polarity at the outside so that diode V2 becomes non-conducting.

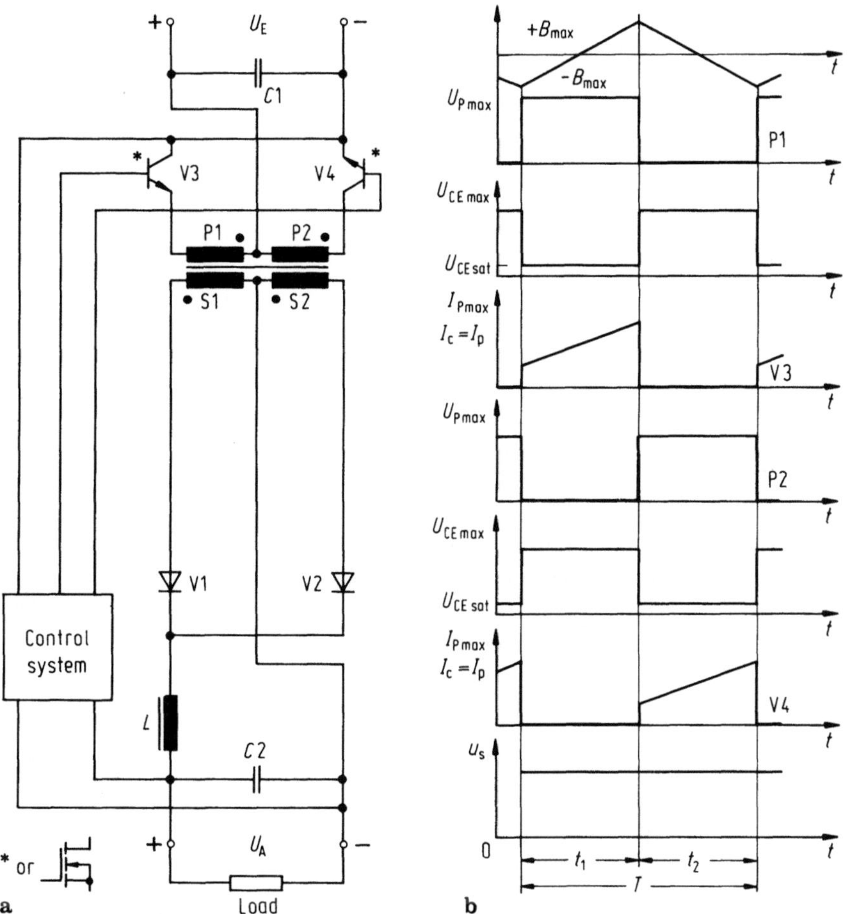



**Fig. 8.25a, b.** Push-pull forward converter. $U_E$ Input voltage, $U_A$ output voltage, $U_{CE\ max}$ maximum voltage between collector and emitter, $U_{CE\ sat}$ saturation voltage, $U_{p\ max}$ maximum primary voltage, $I_c$ collector current, $I_p$ primary current, $I_{p\ max}$ maximum primary current, $+B_{max}$ positive maximum flux density, $-B_{max}$ negative maximum flux density, $u_s$ secondary voltage, $T$ period, $t_1$ on time of transistor V3, off time of transistor V4), $t_2$ on time of transistor V4, off time of transistor V3), P1 primary winding 1, P2 primary winding 2, S1 secondary winding 1, S2 secondary winding 2

The duration of the conducting phase of the switching transistor V3 is determined by the control system. Once the conducting phase has ended there follows a state in which both switching transistors V3 and V4 are non-conducting: V3 is already off and V4 is not yet conducting. During this time the choke $L$ acts as 'current source'; together with capacitor $C2$ it continues supplying the load with current (at a constant voltage). Half of the current (load current) flows through each of the two parts of the transformer's secondary winding. Diodes V1 and V2 act here as free-wheeling diodes.

The second half-period runs as a mirror image of the first half-period. First the switching transistor V4 becomes conducting, and with it also diode V2. Diode V1, on the other hand, now becomes non-conducting. With the end of the transition time of transistor V4 both switching transistors V3 and V4 are again non-conducting. A new period starts. Pulse-width control is used for the control system as in the other power converter types.

The ideal behaviour of the individual variables can be seen from Fig. 8.25b.

### 8.6.5 Pulse Inverters

*Devices with transistor bridge input circuits.* The d.c. input voltage $U_E$ (battery voltage) is applied to the fully controlled transistor bridge V1, V2, V3, V4 through the input filter $C1$, $L1$, $L2$ (Fig. 8.26).

Depending on the control system, these pulse-width controlled power transistors connect the transformer T1 alternately to the input supply with either transistors V1 and V4 or V2 and V3 being turned on. The bridge circuit is triggered by $2 \times 800$-Hz pulses, whose pulse width is modulated with a 50-Hz fundamental. Since the phase displacement between the pulses to the two branches of the bridge is 180°, the result is a 50-Hz square wave voltage pulsating at 1600 Hz at the primary winding of transformer T1.

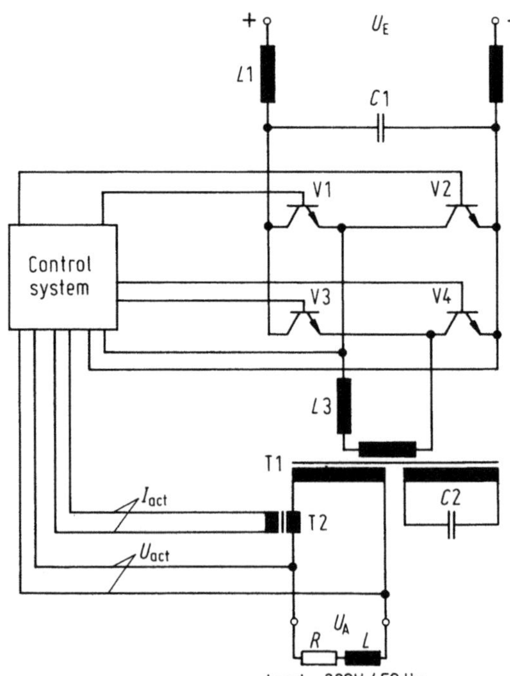


Fig. 8.26. Pulse inverter (e.g. 7.5 kVA). $U_A$ a.c. output voltage, $U_E$ d.c. input voltage, $L1, L2, C1$ input filter, V1, V2, V3, V4 transistors in fully controlled bridge circuit, $L3, C2$ output filter, T1 output transformer, T2 current converter, $I_{act}$ actual value of current, $U_{act}$ actual value of voltage

The pulse frequency of 1600 Hz is filtered out by the choke $L3$ and the capacitor $C2$. Thus, a sinusoidal alternating voltage of typically 220 V, 50 Hz is produced at the secondary winding of T1.

Figure 8.27 illustrates in a simplified manner the formation of the output voltage waveform by pulse-width modulation. In practice the modulation frequency is higher.

*Devices with transistor bridge output circuit.* In the pulsed modulated (e.g. 2.5 kVA) inverter (Fig. 8.28) the principle of the forward converter (20 kHz, pulse-width control) is employed. From the 48 or 60 V input voltage (40 to 75 V), a single-phase a.c. voltage of 220 V at 50 Hz (60 Hz) is produced.

The input direct voltage $U_E$ (as shown in Fig. 8.28) passes through a filter (1) – which serves to reduce the interference – to the d.c./d.c. converter (2); the latter is triggered by the regulator 1 (7) at a constant frequency (20 kHz). The sinusoidal reference voltage (4) is rectified by a rectifier circuit (5) and input to the regulator 1 (7) as reference value $U_{RM}$. The modulated converter voltage $U_M$ is fairly similar to a rectified mains voltage. Using the polarity-inverting bridge circuit (3) the output alternating voltage $U_A$ is formed from it, here the bridge control (6) switches the bridge branches S1/S4 and S2/S3 (transistors as switches) alternately at every zero crossing of the sine-wave generator (4). In general a complex load must be expected. The energy stored in the reactive load distorts the waveform of the output voltage and increases the distortion factor. For this reason a compensation is provided in the form of a reactive-load converter (10), which feeds the superfluous energy back into the energy source.

The oscillogram in Fig. 8.29 displays the voltage distortion at the output of the inverter for a cos $\varphi$ of about 0.7 (inductive).

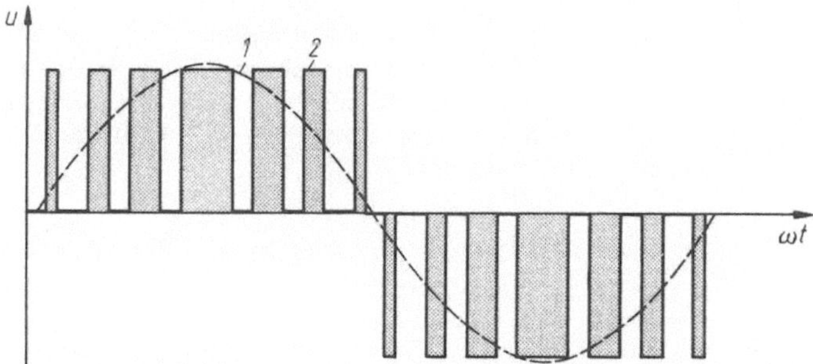


*1* Sinusoidal a.c. output voltage
*2* Pulse-width controlled square wave a.c. voltage

**Fig. 8.27.** Formation of the inverter output voltage waveform by pulse-width control (modulation, PWM) as in schematic diagram. In reality the pulse frequency is much higher

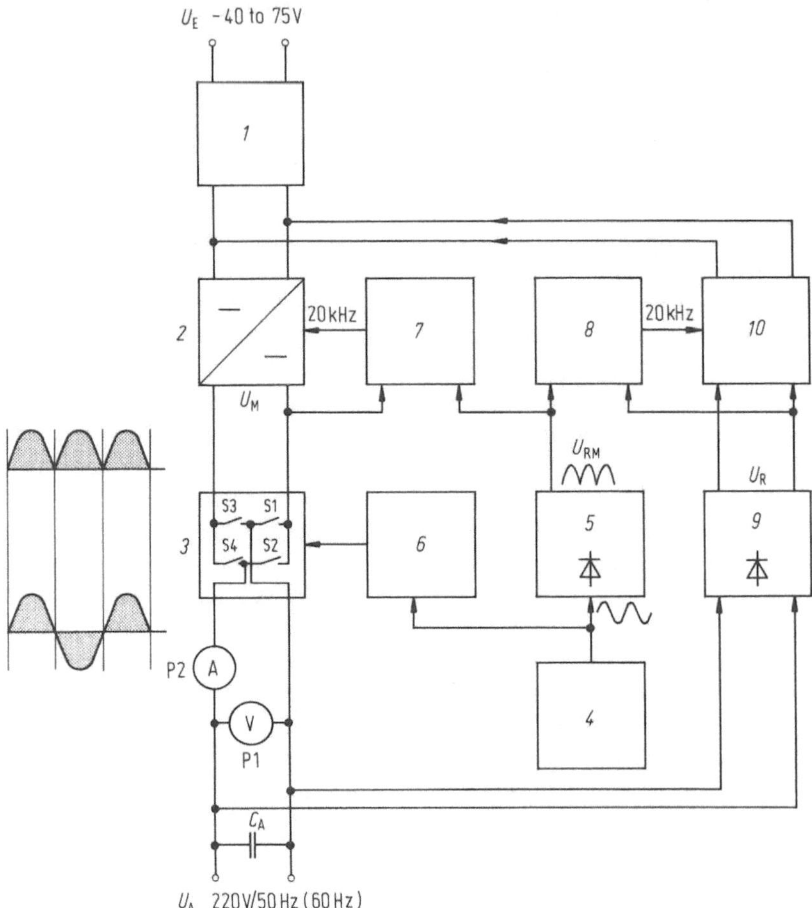



**Fig. 8.28.** Block diagram of a pulsed, modulated d.c./a.c. inverter (e.g. 2.5 kVA WR 20, source: Siemens AG). $C_A$ output capacity, S1 to S4 controlled bridge circuit (transistor switch), $U_A$ a.c. output voltage, $U_E$ d.c. input voltage, $U_M$ modulated converter voltage, $U_R$ rectified output voltage, $U_{RM}$ reference voltage of the modulated converter, P1 voltmeter, P2 ammeter, 1 filter (low/high frequency), 2 pulsed modulated d.c./d.c. converter, 3 polarity-inverting bridge circuit, 4 50-Hz sine generator (reference generator), 5 rectification of sinusoidal reference voltage, 6 bridge control, 7 regulator 1, 8 regulator 2, 9 reactive current rectification, 10 reactive load converter

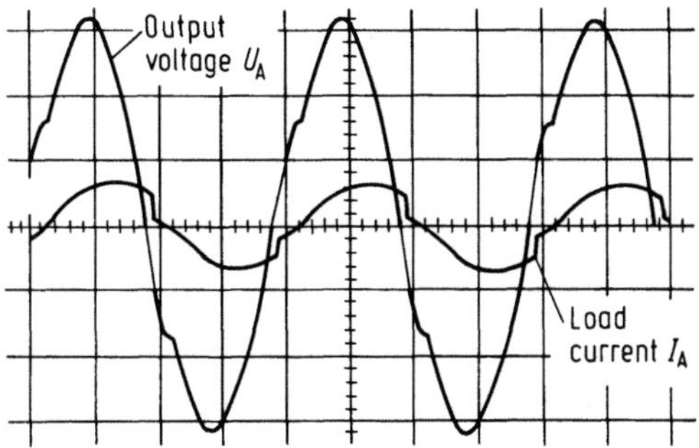


**Fig. 8.29.** Output voltage and load current of the inverter (scale: per notch, $U_A = 100$ V; $I_A = 10$ A; $t = 12.5$ ms)

# 9 Switching Mode Power Supplies

At this point a selection of equipment will be described; they all employ the principle of pulse-width control.

## 9.1 Type 48 V/12 A (60 V/10 A)

### 9.1.1 General and Application

The transistor-controlled switched-mode power supply units 48V/12A (60V/10A), GR40[1] (universal slide-in unit), (see Fig. 9.1), serve for example as a rectifier module as used in power supply unit SVE40, different types of telecommunications systems (with lower power consumption) such as transmission systems or RDLU's (Remote digital line unit) of EWSD (Digital electronic switching system) systems.

The rectifier module is designed to draw a sinusoidal current with no significant conducted EMI. On the output side, it delivers a regulated d.c. voltage of

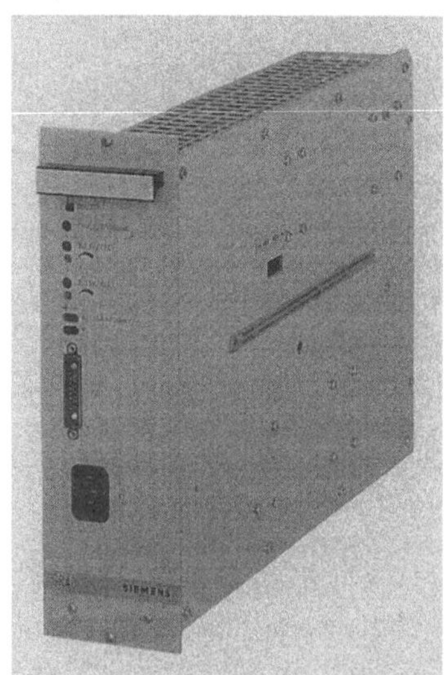

**Fig. 9.1.** Switched-mode power supply (rectifier module) 48 V/12 A (60 V/10 A), GR40. (Photo by courtesy of Siemens AG)

---

[1] *Source*: Siemens AG.

48 V or 60 V. The switch-mode technology used has a high efficiency, which gives it very accurate control and response dynamics, and keeps it small, light and quiet. The low conducted EMI also means that standby power sources can be fully utilized. Six modules are generally used in the SVE40 power supply unit. In special cases, up to twelve can be connected in parallel to provide a larger system. When overloading occurs, current limiting is enforced. This reduces the voltage until the current falls back to the rated value.

### 9.1.2 Modes of Operation

The rectifier modules 48 V/12 A (60 V/10 A) can be operated in the rectifier mode (see Sect. 3.1) or in the standby parallel mode (see Sect. 3.3).

### 9.1.3 Survey Diagram of the Power Supply System

The series SVE40 power supply unit consists of a frame accomodating up to 6 series GR40 rectifier modules, protective devices and a monitoring and control board (A40) (see Figs. 9.2 and 9.3).

These equipment units, along with a storage battery, form a no-break d.c. power supply system operating in the standby parallel mode for the supply of telecommunications systems with 48 V or 60 V rated power. It is possible to ground the positive (normal) or the negative d.c. conductor. The SVE40 can be connected to single- or three-phase a.c. supply systems. When the SVE40 is connected to a three-phase system and the rectifier module load is evenly distributed between the phases, the neutral conductor carries no current. The remote signals cause a relay to drop out in the event of a fault.

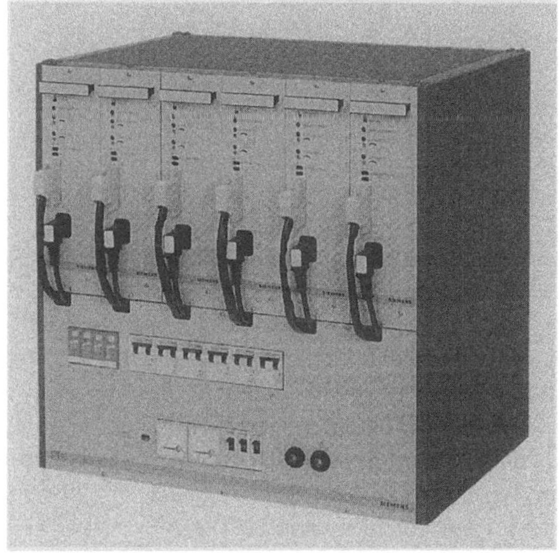

**Fig. 9.2.** Power supply unit SVE40. (Photo by courtesy of Siemens AG)

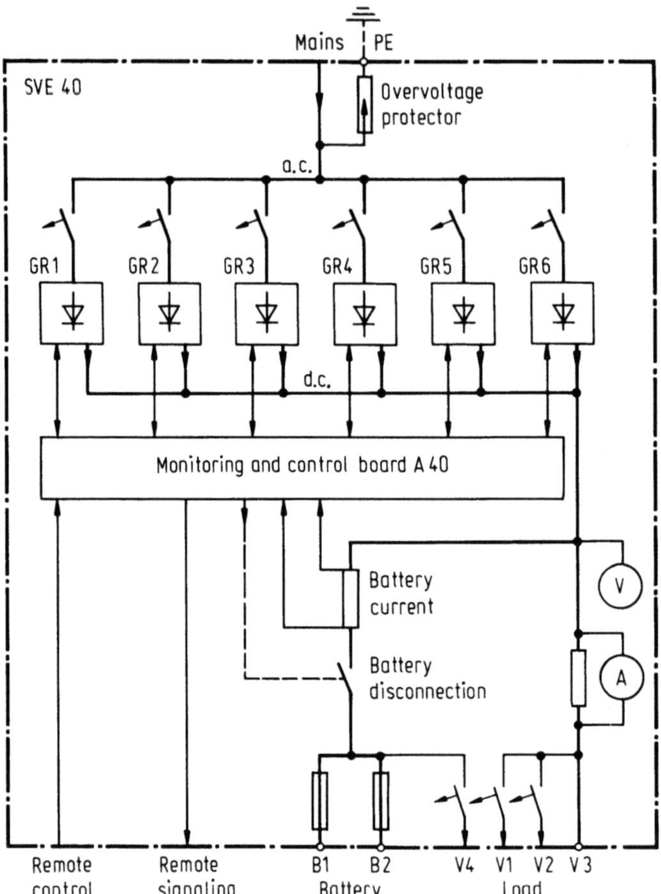



**Fig. 9.3.** Survey diagram of the power supply unit SVE40

### 9.1.4  Survey Diagram and Functioning Principle of the Rectifier Module 48 V/12 A (60 V/10 A)

Figure 9.4 shows the survey diagram of the switched mode power supply unit (rectifier module) 48 V/12 A (60 V/10 A), GR40.

The GR40 converts the mains voltage into a controlled d.c. voltage available at its output, namely plug X2. This plug also provides information about the input mains and about faults in the module. In addition, it can be used to control the module remotely that is, to switch it on and off, to select charge or trickle charge mode and to alter the output voltage, when, for example, the temperature in the battery room changes.

To minimize RF interference, the input voltage is fed through an *LC* input filter. The input rectifier bridge, which converts the a.c. voltage into a pulsating d.c. voltage is connected to the output side of this filter.

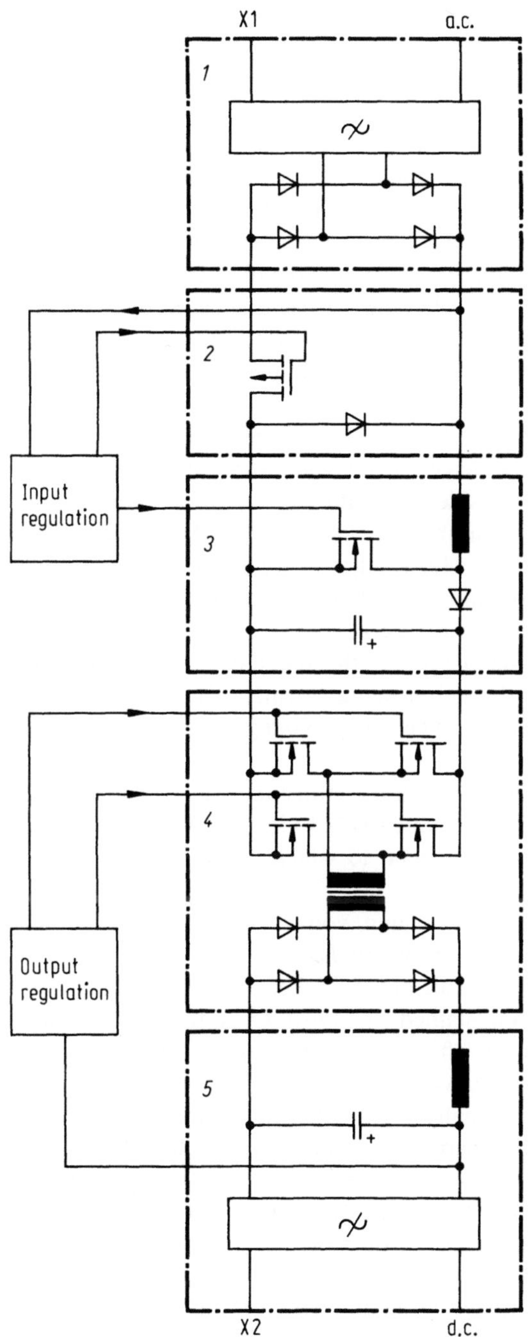

**Fig. 9.4.** Survey diagram of switched-mode power supply unit (rectifier module) 48 V/12 A (60 V/ 10 A), GR40 (see Section 3.1). *1* Input filter, input rectifier, *2* input protection circuit, *3* boost converter, *4* d.c./d.c. converter, *5* output filter

Switch-on current limiting (through the input protective circuit) is a feature of the GR40. During the switch-on phase, an IGBT (Insulated Gate Bipolar Transistor) ensures that the current increases slowly. When the unit is switched on, the IGBT protects the downstream components from overvoltages.

The function of the boost converter is to draw a sinusoidal current from the mains with low conducted EMI. The semi-conductor circuit is activated in such a way that the charging current drawn by the input capacitor becomes sinusoidal. The advantage of the principle behind the boost capacitor is that most of the current drawn flows past the controlling element to the d.c./d.c. converter, making this part of the circuit's high efficiency. The type of converter used operates at a circuit frequency of 38 kHz and makes it possible to transform the rectified a.c. voltage into a link d.c. voltage that is infinitely variable over a large range. This d.c. voltage is used for the d.c./d.c. converter.

The d.c./d.c. converter converts the d.c. link voltage produced by the boost converter into the lower output d.c. voltage. At the same time, the mains side is separated electrically from the output side. A control IC takes over the control and activation of the circuit transistors in the fully controlled bridge circuit. These are switched at a frequency of $2\times40$ kHz, so that a regulated rectangular a.c. voltage is produced at the secondary side of the transformer. This a.c. voltage is converted to the output d.c. voltage in a bridge rectifier.

There is an *NC-LC* filter at the output of the d.c./d.c. converter to limit surge and also an *HF-LC* filter at the unit's output to maintain the permissible RF interference values (output filtering). The output voltage is regulated by the d.c./d.c. converter.

Current sharing for parallel operation of the rectifier module is regulated via a bus, i.e. the current bus. From this bus, which is led via plug X2, the regulator is fed with a mean value of the output current, to which each GR40 can match the current it has to deliver.

Providing the difference between the output voltages is less than 1%, the difference between the GR40 currents is $< 5\%$ of $I_N$.

The output of the GR40 is short circuit proof. If there is overvoltage on the d.c. voltage side, the GR40 goes to maintained shutdown. Once the fault has been rectified, the module must be switched off and then on again in order to unlock it. If there is undervoltage on the d.c. voltage side, only the signal GR fault is given. The GR40 is protected internally from overheating by a temperature sensor. This is situated at the hottest part of the unit and switches it off when the temperature becomes too high. It also causes the LED fault to light up, and sends the signal GR fault to plug X2. The signal GR mains fault is given when the input voltage of a GR40 is outside the permissible voltage limits.

A GR mains fault is passed on only as a remote signal with floating changeover contacts of a releasing relay. No additional LED lights up.

### 9.1.5 Technical Data

The principal technical data for the rectifier modules types 48 V/12 A (60 V/10 A), GR40 are listed in Table 9.1.

**Table 9.1.** Technical data for rectifier modules 48 V/12 A (60 V/10 A), GR40

| Mains input supply | | | | |
|---|---|---|---|---|
| Voltage (V) | 230 (175 to 264) | | | |
| Frequency (Hz) | 50 or 60 (47.5 to 63) | | | |
| Current, sinusoidal (A) | max. 3.7 (at 230 V) | | | |
| Degree of radio interference | Limit class B (VDE 0878) | | | |
| **D.C. output** | | | | |
| Operating mode or condition | Rated direct voltages (V)<br>(tolerances for 1.5 to 100% rated curent) | | | |
| | Equipment voltage $\hat{=}$ load voltage | | | |
| | 48 V Systems | | 60 V Systems | |
| | Lead-acid battery with | | | |
| | 24 cells | 25 cells | 30 cells | 31 cells |
| Rectifier mode or parallel operation/ float (trickle) charging (2.23V/C) | 53.5 ± 1% | 56 ± 1% | 67 ± 1% | 69 ± 1% |
| Parallel operation/charging (2.33 V/C) | 56 ± 1% | 58.5 ± 1% | 70 ± 1% | 72 ± 1% |
| Rated direct current (A) | 12 | | 10 | |
| Interference voltage (mV) | $\leq$ 0.7 (frequency-weighted with CCITT A-filter) | | | |
| Dimensions (H×W×D) (mm) | 234 × 84 × 374 | | | |

## 9.2 Type 48 V (60 V)/30 A and 100 A

### 9.2.1 General and Application

The microprocessor-controlled switched-mode power supply units 48 V (60 V)/30 A and 100 A, GR20[2] (built-in construction (Fig. 9.5) serve for example as rectifier modules in power supply panels (power supply-series 200) for the supply of mobile radio telephone networks. The rectifier modules GR20 100 A are applied in the power supply system 200 for the supply of the digital switching system EWSD. The units GR20 have various special features such as high efficiency, high control precision and good dynamic response, small volume and weight. Due to an operating frequency of 2×20 kHz the rectifier modules do not develop any noise, so that the power supply system is suitable for installation in office rooms.

Up to thirty modules can be connected in parallel. The present layout of the compact power supply panels (as in series 200) allows the provision of four

---

[2] *Source*: Siemens AG

**Fig. 9.5.** Switched-mode power supply (rectifier module), 48 V(60V)/100 A, GR20. (Photo by courtesy of Siemens AG)

rectifier modules per panel, each module with 30 A. With regard to the power supply system 200 up to four rectifier modules (100 A) can also be connected per panel (in the rectifier cabinet).

### 9.2.2 Modes of Operation

The rectifier modules 48 V (60 V)/30 A and 100 A, GR20 can be operated in the rectifier mode (see Sect. 3.1) or in the standby parallel mode (see Sect. 3.3).

### 9.2.3 Survey Diagram of the Power Supply System

*Application in compact power supply systems.* The compact power supply, series 200 or 201, consists of two panels, a power supply panel with rectifier modules, d.c. and a.c. distribution equipment and boards, as well as a battery panel with an integrated maintenance-free lead-acid battery which provides a complete power supply system for the supply of telecommunication facilities (e.g. mobile radio telephone network). Due to its compact modular design and execution it is suitable for various applications. For a power demand of up to max. 240 A (equipped in each case with 4 rectifier modules GR20 48 V (60 V)/30 A) two power supply panels can be switched in parallel.

For systems with greater power demands (up to 600 A) maximum 2 power supply panels, each of which is equipped with 3 rectifier modules GR20 48 V (60 V)/100 A, can be applied (see Figs. 9.6 and 9.7). The operating reserve in cases of mains failure may be augmented even further by larger, separately installed batteries.

The *power supply panel* (①see Fig. 9.7) serves to house a maximum of 3(4) rectifier modules GR20 (G1 to G3 or G1 to G4). Both consumer and battery are switched in parallel and operate in the standby parallel mode. The power supply panel type 200 is designed without battery and thermo disconnection, as opposed to the power supply panel type 201 (with A16 and K10).

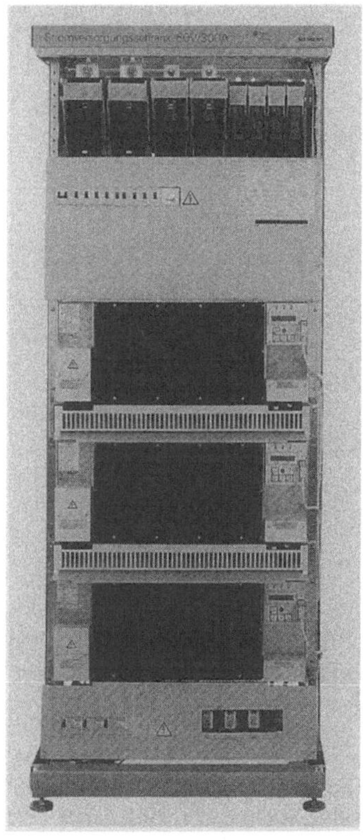

**Fig. 9.6.** Compact power supply panel AS201, equipped with three rectifier modules, 48 V (60 V)/100 A, GR20. (Photo by courtesy of Siemens AG)

The *battery panel* (②see Fig. 9.7) serves to house a 24-, 25-, 30- or 31-cell maintenance-free battery (e.g. dryfit). The capacity of the battery depends on the type of panel and can be 75 Ah, 200 Ah or 250 Ah.

The *battery-* and *thermo disconnection* module A16 in the power supply panel consists of a battery undervoltage monitor and a relay control for the thermo disconnection.

On the *signal module* A80 in the power supply panel the signal outputs of the rectifier modules are interconnected, the load fuses and battery fuses are monitored and the fault signals of the power supply panel are recorded.

In the case of the rectifier modules release fault signals for mains failure and equipment failure these signals are forwarded separately by remote signalling, but they are indicated as one signal by the yellow LED 'FAULT' at the top location.

As long as the rectifier modules are in operation, the green LED 'OPERA-TION', likewise located at the top, lights up.

For operation with maintenance-free batteries it is necessary to block the charging characteristic (2.33 V/C) and the initial charging characteristic (charging up to 2.7 V/C). These batteries may only be charged with a trickle charging voltage of 2.23 V/C. The blocking of charging characteristics is achieved by

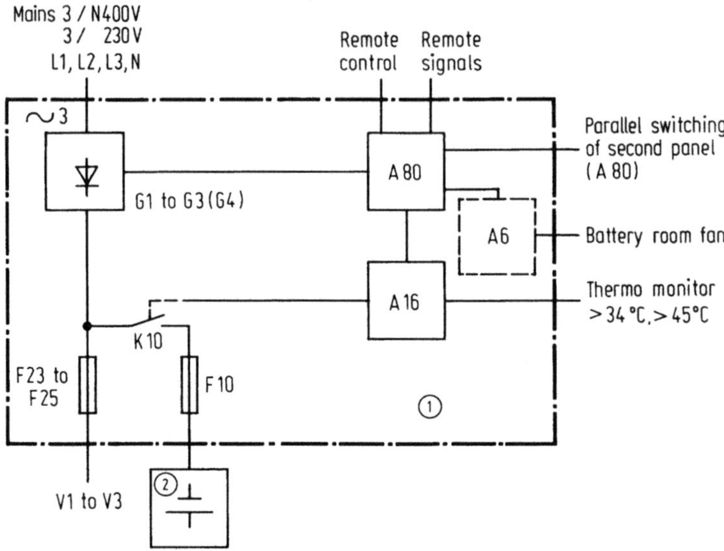



**Fig. 9.7.** Survey diagram compact power supply series 200. (basic cabinet AS201) A6 Voltage monitor (control of battery room fan and undervoltage monitor signalling), only used in power panel AS201, 300 A, A16 Battery- and thermo disconnection module, A 80, signalling board F10 Battery fuse, F23 to F25 Load fuses, G1 to G4 Rectifier modules (GR20, 30 A) or G1 to G3 Rectifier modules (GR20, 100 A), K10 Battery disconnecting contactor, V1 to V3 Main consumer connections, ① Power supply panel 48 V (60 V) 120 A (or 300 A), ② Battery panel with valve regulated maintenance-free lead-acid battery 24, 25, 30 or 31 cells

means of a bridge (X5) on module A80 (not represented in the figure). Through this bridge and via a signal plug connector (X2) positive potential is laid to the input of the microcontroller in the rectifier module. Thus, the controller blocks the switch-over to the charging characteristics.

Table 9.2 shows the application of the rectifier module GR20, 30 A and 100 A in the compact power supply and the system configuration at different current consumptions (also applicable for the rectifier module GR40/12 A see Sect. 9.1).

*Application in power supply system 200.* The power supply system 200 (see Figs. 9.8 and 9.9) is designed for the range of 100 A to 2500 A current consumptions. The following basic components may be used in that system:

– rectifier cabinet 48 V (60 V)/400 A,
– battery and d.c. distribution cabinets 48 V (60 V) 630 A, 1250 A or 2500 A and
– a.c. power boards 250A or 630 A.

**Table 9.2.** Planning information for 48 V power supply systems series 40 and 200 with switched-mode rectifier units (rectifier modules) types GR40/12 A and GR20/30 A or 100 A; max. 600 A systems

| Current consumption (A) | Rectifier module (n+1) 12 A GR40 | Power supply unit SVE 72 A (2 Batt.) SVE40 | Rectifier module (n+1) 30 A GR20 | 100 A GR20 | Compact power supply 120 A (2 Batt.) AS201 B | 120 A AS200 E | 300 A (2 Batt.) AS201 B | 300 A AS200 E | Mains distribution 250 A NV200 |
|---|---|---|---|---|---|---|---|---|---|
| 12 | 1+1 | 1 | | | | | | | |
| 24 | 2+1 | 1 | | | | | | | |
| 60 | 5+1 | 1 | [2+1] | | [1] | | | | [1] |
| 90 | | | 3+1 | [1+1] | 1 | | [1] | | 1 |
| 120 | | | 4+1 | [2+1] | 1 | 1 | [1] | | 1 |
| 150 | | | 5+1 | [2+1] | 1 | 1 | [1] | | 1 |
| 210 | | | 7+1 | [3+1] | 1 | 1 | [1] | [1] | 1 |
| 300 | | | | 3+1 | | | 1 | 1 | 1 |
| 400 | | | | 4+1 | | | 1 | 1 | 1 |
| 500 | | | | 5+1 | | | 1 | 1 | 1 |
| 600 | | | | 6+2 | | | 1 | 2 | 1 |

(n+1) = supply of load + battery charging (max. 20%) standby rectifier, E extension, B basic, value in the bracket [ ] = alternative

**Fig. 9.8.** Power supply system series 200. (Photo by courtesy of Siemens AG)

Each rectifier cabinet is designed to accommodate four 100-A rectifier modules GR20. Terminals for connecting the modules are provided.

Table 9.3 shows the application of the rectifier module GR20, 100 A in the power supply system 200 and the system configuration at different current consumptions.

### 9.2.4 Survey Diagram, Block Diagram and Functioning Principle of the Rectifier Module 48 V (60 V)/30 A and 100 A

Figure 9.10 shows the survey diagram of the switched-mode power supply unit (rectifier module) 48 V (60 V)/30 A respectively 100 A, GR20. Here the principle of the double single-ended forward converter (pulse width control, $2 \times 20$ kHz = 40 kHz) is applied. The two converters operate in push-pull operation.

*Power section.* The three phase a.c. mains voltage $U_E$ is fed via a *RC* radio interference suppression filter (①)(see Fig. 9.10) to the power rectifier (three-phase bridge configuration V1) and is thereby rectified. The second part of the interference suppression filter (①) is located at the output-side of the device.

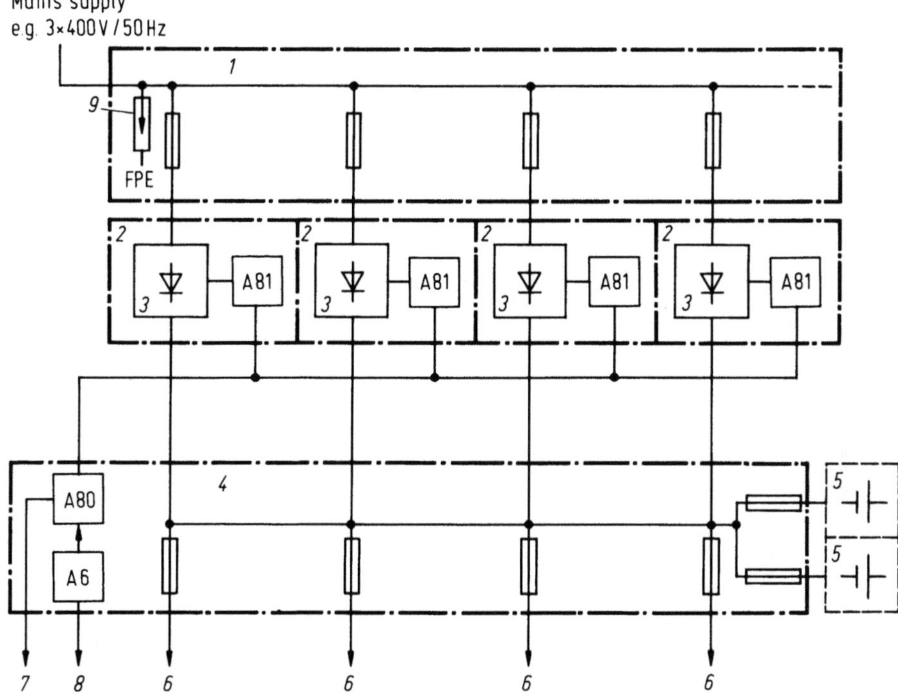


**Fig. 9.9.** Survey diagram of the power supply system series 200. *1* A.C. power board 400 V/250 A NV400 or 630 A NV200, *2* rectifier cabinet 48 V(60 V)/400 A GS200, *3* rectifier module 48 V(60 V)/100 A GR20, *4* battery and d.c. distribution cabinet 48 V (60 V)/ 630 A, 1250 A or 2500 A VS200 (VS201), *5* battery cabinet or battery groups (each e.g. 24 cells or 25 cells), *6* communications system (load), *7* signalling and remote signalling, *8* battery room fan, *9* overvoltage protection, A6 voltage monitor, A80 signalling board, A81 signal bus

Thus the interference voltage is limited to permissible values (that is limited to class B as in VDE 0871). The power rectifier is followed by an input filter (②) (see Fig. 9.10), which consists of reactors and capacitors and serves to store energy. Through link variants the unit can be matched to differing a.c. supply voltages (terminal 1 to 4).

The rectified mains voltage $U_R$, varying along with the mains voltage, is fed to the power converter modules A4 and A5 as converter input voltage $U_1$ and $U_2$. Every converter operates with a frequency of 20 kHz. Through interconnection on the secondary side of the two converters with their staggered voltage blocks a voltage of 40 kHz frequency is produced.

The bipolar power switching transistors V1 and V2 in Darlington-configuration 'chop' and regulate the converter input voltage (final control element). The square voltage blocks are transformed (matching and electrical isolation) by means of power transformers T1 and T2. Then follow the rectifier diodes V3, V4, free-wheeling diodes V5/V6 and output filters $L_7, C_9$.

**Table 9.3.** Planning information : for 48 V power supply systems series 200 with switched mode rectifier unit (rectifier module) type GR20/100 A; max. 2500 A systems

| Current consumption (A) | Rectifier module $(n+1)$ | Rectifier connecting panel | Battery and distribution panel | | | Mains distribution switch board | Mains switch panel |
|---|---|---|---|---|---|---|---|
| | 100 A | 400 A | 630 A (2 Batt.) | 1250 A (2 Batt.) | 2500 A (2 Batt.) | 250 A | 630 A |
| | GR20 | GS200 | VS200 | VS200 | VS200 | NV200 | NV200 |
| 100 | 1+1 | 1 | 1 | | | 1 | |
| 200 | 2+1 | 1 | 1 | | | 1 | |
| 300 | 3+1 | 1 | 1 | | | 1 | |
| 400 | 4+1 | 2 | 1 | | | 1 | |
| 600 | 6+2 | 2 | 1 | | | 1 | |
| 800 | 8+2 | 3 | | 1 | | 1 | |
| 1000 | 10+2 | 3 | | 1 | | 1 | |
| 1200 | 12+3 | 4 | | 1 | | 1 | |
| 1400 | 14+3 | 5 | | | 2 | | 1 |
| 1600 | 16+3 | 5 | | | 2 | | 1 |
| 1800 | 18+4 | 6 | | | 2 | | 1 |
| 2000 | 20+4 | 6 | | | 2 | | 1 |
| 2200 | 22+4 | 7 | | | 2 | | 1 |
| 2500 | 25+5 | 8 | | | 2 | | 1 |

$(n+1)$ = supply of load + battery charging (max. 20%) standby rectifier

Due to its good dynamics the voltage and current regulator A6 dampens the input ripple content, and for this reason the passive filters in the power circuit are used exclusively for the filtering of high-frequency interferences.

The switching transistors are protected by a *RCV*-circuit (⑥) and (⑦) (see Fig. 9.10).

The actual current value is recorded by the shunt resistor $R_9$ and then evaluated on module A6 for current limitation. The block diagram of the rectifier module GR20, 48 V (60 V)/30 A or 100 A in Fig. 9.11 shows the layout of the modules and their interaction.

*Input module A1.* On the input module are located radio interference suppression capacitors. For protection against transient mains overvoltages (e.g. lightning and switching processes) varistors and surge diverters are provided.

*Auxiliary converter module A2.* For the electrically safely isolated functional groups on the input and output side of the rectifier module auxiliary voltages are required which are generated with an auxiliary converter from the input voltage. The converter supplies the auxiliary voltages of $\pm$ 5 V and $\pm$ 12 V for the trigger circuit of the main converters, as well as for the logic and regulation circuits.

An optocoupler, which transmits a signal to the microcontroller on the regulation module A6 in cases of mains undervoltage, is integrated into this module. The controller then displays an appropriate message via the LCD displays on the operating panel module A7. The protection against mains overvoltage is ensured by a varistor.

*Regulation, control and monitors*

*Logic module A3.* The central unit of the logic module is a switched-mode power supply IC regulator component and performs the following functions:

- protection of the power transformers against saturation,
- current limitation of instantaneous value,
- pulse width control,
- generation of clock frequency,
- voltage monitor,
- softstart,
- dynamic current limitation and
- symmetrical control of the two push pull channels.

During the *symmetrical control* the two converter voltages $U_1$ and $U_2$ are recorded, a voltage drop $\Delta U$ is formed and then compared to a reference voltage. If $\Delta U$ exceeds a certain tolerance range, i.e. if one of the converter voltages is too high, blocking pulses are generated depending on the polarity of $\Delta U$. These pulses pass to certain inputs of the regulator. The regulator then reduces the trigger pulses of the converter with the lower voltage until $\Delta U$ has fallen below its tolerance value. The *trigger pulses* for the power switching transistors of both

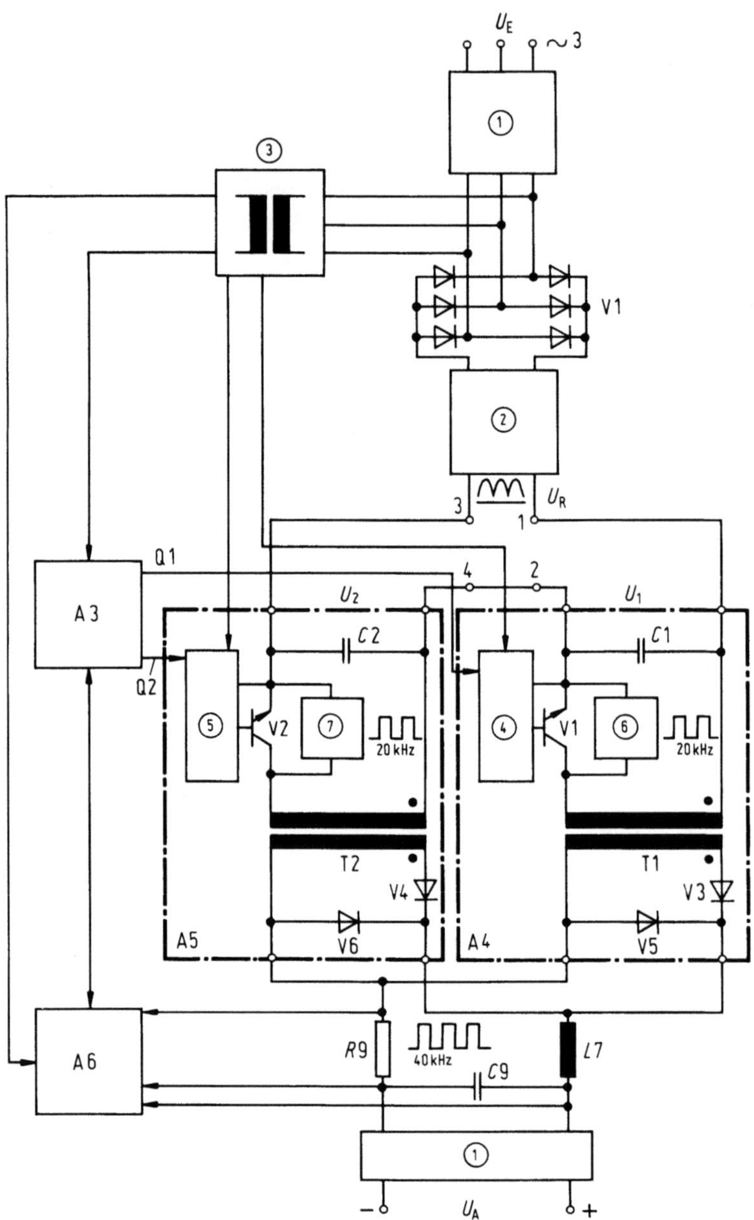


**Fig. 9.10.** Survey diagram of switched-mode power supply unit (rectifier module) 48 V (60 V)/30 A respectively 100 A, GR20. $U_E \sim 3$ Three-phase input a.c. voltage: terminal connection for 220 V ..... 230 V: 1–4, 2–3 (parallel switching of the two converters), terminal connection for 380 V ..... 415 V: 2–4 (series switching of the two converters), V1 Supply side rectifier (uncontrolled), $U_A$ Output d.c. voltage, $U_R$ Rectified supply voltage (intermediate circuit d.c. link voltage), $U_1$, $U_2$ Converter input voltage, $C_1$, $C_2$ Input capacity of the power modules, $C_9$ Output capacity, $V_3/V_4$ Secondary rectifier, $V_5/V_6$ Free-

converters are released via the regulator outputs Q1 and Q2 as in Fig. 9.10. The regulator also performs the evaluation of the under- and overvoltage monitor of the intermediate circuit voltage $U_R$. For this, the voltages $U_1$ and $U_2$ are recorded and evaluated. The signals generated are forwarded to the regulator and compared to the reference voltages. If there is an input overvoltage or an input undervoltage, the regulator blocks the trigger pulses of the power switching transistors. The rectifier module switches off and after a delay time of approximately 5s it tries a 'softstart' which only succeeds if the input voltage has returned to within its normal limits.

During *softstart* the pulse duty factor of the power transistor trigger pulses is continually increased by the regulator. Furthermore, a *duty factor current limitation* and a rapid *intermittent current limitation* are also located on this module. Both current limitations serve to protect the power switching transistors.

*Regulation module A6.* The main components of this module are the microcontroller component, a EPROM, and a secondary-side auxiliary converter. This converter is fed from the associated battery and supplies the regulation module with the auxiliary voltages needed even when there is a mains voltage failure. The converter thus ensures the operation of the microcontroller in cases of mains failure.

On the primary side the microcontroller continually monitors the mains voltage. If there is a mains failure, the microcontroller blocks the triggering of the optocoupler. Thereby, trigger pulses for the power transistors are not generated anymore and the rectifier module switches off. If the mains returns to its normal state the rectifier module comes into operation with a 'softstart'.

On the secondary side the microcontroller receives informations on the magnitude of the output current and output voltage. If the latter is too high, the processor switches the device off (locked) via the regulator of the logic module A3. Afterwards, the rectifier module can only start-up again if the device has been unlocked by operating the ON/OFF switch on the operating panel.

The regulation module A6 also has the function of secondary current limitation which comes into operation after a short period of delay. The signal for current limitation is provided by the shunt resistor $R_9$ (see Fig. 9.10). Via a measuring circuit on the module it is ensured that for excessively high output currents which are not permitted, the reference voltage at the regulator (reference input variable) is reduced. Due to this measure the output voltage can be reduced

---

**Fig. 9.10.** (*Continued*)
wheeling diodes, $L_7$ Output reactor, $R_9$ Shunt resistor ($I_{act}$ sensing),$V_1$, $V_2$ Power switching transistor, $T_1$, $T_2$ Power transformer, ① Interference suppression filter, ② Input filter, ③ Primary auxiliary converter module A2, ④ Trigger module 1, ⑤ Trigger module 2, ⑥ *RCV* protective circuit for transistor 1, ⑦ *RCV* protective circuit for transistor 2, A3 Logic module, A4 Power module 1 (d.c./d.c. converter module), A5 Power module 2 (d.c./d.c. converter module), A6 Regulator module (voltage regulator, current regulator and secondary auxiliary converter), Q1, Q2 Trigger pulses for power transistors

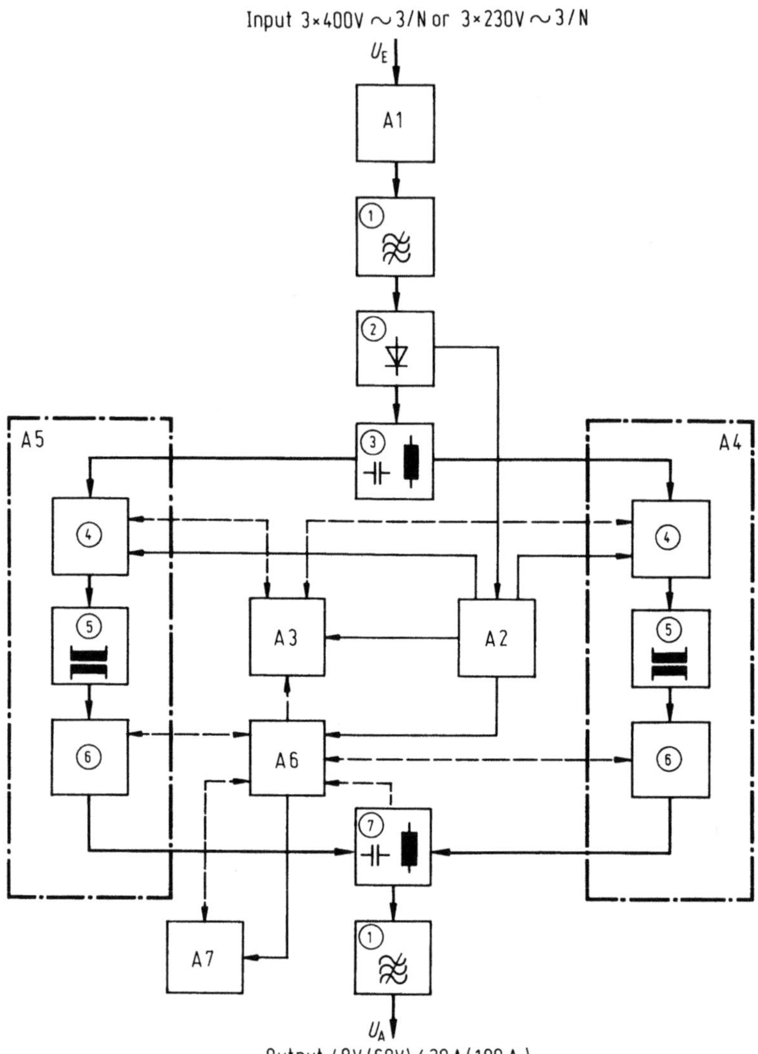



**Fig. 9.11.** Block diagram switched-mode power supply unit (rectifier module) 48 V (60 V)/30 A or 100 A, GR20. A1 Input module, A2 Auxiliary converter module, A3 Logic module, A4 Power module 1, A5 Power module 2, A6 Regulation module, A7 Operating panel module, ① Interference suppression filter, ② Supply side rectifier, ③ Input filter, ④ Primary power section, ⑤ Power transformer module, ⑥ Secondary power section, ⑦ Output filter, ⟶ Energy flow, – – ⇒ Signal flow, ⟶ Auxiliary energy (internal power supply)

to '0 V', whereas the output current remains approximately the nominal current value.

Furthermore, the microcontroller receives the state of the voltages (e.m.k.) present immediately after the converters. If the e.m.k. is below a certain value (output voltage becomes too low), the controller releases a corresponding message via the display (operating panel module A7) and switches on the relay for trouble indication K2 (also on module A7). The rectifier module is not switched off.

Via the changeover contact of the relay which rests potential-free on the signal plug connector X2 (A7) it is possible to start a remote signalling process. The microcontroller switches this relay on even if there is an overvoltage at the output, or if the secondary auxiliary converter is faulty.

A relay to indicate trouble K1 (which is also on operating panel module A7) is triggered by the processor as soon as there is a mains failure or phase cut. Here, too, a potential-free changeover contact lies on the signal plug connector X2 and can be used for remote signalling.

In addition, the microcontroller records the duration of a mains failure. On return of the mains voltage and depending on the adjusted charging timer settings the processor performs a switchover of the output voltage from trickle charging to charging in order to recharge the connected battery (this is only possible with operation of the switch 'AUTO' on the operating panel module A7, not with manual operation). If the adjusted charging time of the battery has elapsed, the microcontroller switches the rectifier module from charging voltage back to trickle charging voltage.

In order to allow simultaneous switchover to the new characteristics for several rectifier modules switched in parallel, a synchronizing of the characteristics is realized via the microcontroller.

If a locked, maintenance-free battery (e.g. dryfit), which must not receive a charging voltage of 2.33 V/C, is connected in parallel, this situation can be signalled (from outside) to the microcontroller via the signal plug connector X2 so that the microcontroller blocks the charging characteristics. Now only the switch ON/OFF is still effective. The initial charging characteristics (special charging up to 2.7 V/C) are also blocked.

The processor controls the display of the individual devices' status which is indicated via the display on the operating panel module A7.

If several rectifier modules are switched in parallel the regulator receives, via the signal plug connector (X2), a total average current value to which it can match its own device current value (current balancing). Furthermore, the microcontroller switches the devices on and off as soon as it receives a control signal via the plug connector X2. For the various control and regulation tasks it performs, the microcontroller requires a program which is stored in an EPROM memory.


*Operating panel module A7.* The elements needed for the operation of the rectifier module listed in the following are located on the operating panel module A7, see Fig. 9.12:

– 16-digit LCD-display,
– potentiometer $I_A$ 2.7 V/C; current limitation for initial charging of the battery,

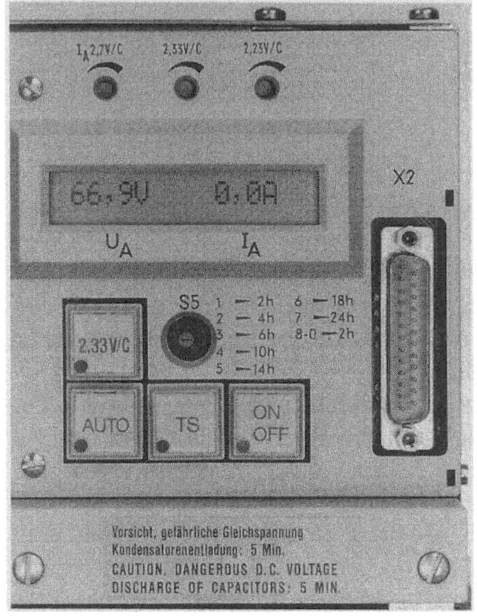



**Fig. 9.12.** Operating module A7. (Photo by courtesy of Siemens AG)

– potentiometer 2.33 V/C; fine adjustment of the charging voltage,
– potentiometer 2.23 V/C; fine adjustment of the float (trickle) charging voltage,
– key 2.33 V/C[3]; operation mode charging,
– switch S5; adjustment of charging timer 2 to 24 h,
– key AUTO[3]; automatic operation mode,
– key TS[3]; time reduction for testing operation and changing of the language English-German-English at the display,
– key ON/OFF[3]; output operation switch and
– signal plug connector X2; remote signals and remote control, serial interface.

Also accommodated on the A7 board is the fault-signalling relay K1. Its contacts as well as those of relay K2 (on the A6 board) are employed for connection via the signalling connector X2,

– relay K1 is operated by the microcontroller when undervoltage, overvoltage, power failure or phase failure occurs on the a.c. side and
– relay K2 is operated by the microcontroller when overvoltage occurs at the rectifier output, or the auxiliary voltage supply on the regulator board A6 is faulty.

Each of the relays has a changeover contact connected to the signalling connector X2.

---

[3] Each key has a LED which lights up when the key is pressed.

The operating panel has two further controls:

- potentiometer for adjusting the LCD brightness and
- dip-fix switch bank S6 (setting and matching 48 V or 60 V system, 24, 25, 30 or 31 battery cells and special charging).

The connector X2 is in the compact power supply to be connected to the signal board A80 or in the power supply system 200 in the rectifier cabinet to be connected with the signal bus A81.

A remote control unit (e.g. power controller) can also be connected via the signalling connector X2 to operate the rectifier modules through a serial interface and/or to obtain their operational condition (current-voltage-characteristic).

**Table 9.4.** Technical data for rectifier modules 48 V (60 V)/30 A and 100 A, GR20

| Type | 30 A | | 100 A | |
|---|---|---|---|---|
| **Mains input supply** | | | | |
| Voltage (selectable) (V) | 3a.c.  175 to 235<br>3/N a.c. 323 to 456 | | | |
| Frequency (Hz) | 47.5 to 63 | | | |
| Equipment fuse protection (A) | 3×16 (at 230 V) | | 3×32 Agl | |
| | 3×10 (at 400 V) | | 3×20 Agl | |
| Degree of radio interference | Limit class B (VDE 0871/0878) | | | |
| **D.C. output** | | | | |
| Operating mode or condition | Rated direct voltages (V)<br>(tolerances for 1.5 to 100% rated<br>current) | | | |
| | Equipment voltage $\hat{=}$ load voltage<br>(except for initial charging with<br>communications system disconnected) | | | |
| | 48 V systems | | 60 V systems | |
| | Lead-acid battery with | | | |
| | 24 cells | 25 cells | 30 cells | 31 cells |
| Rectifier mode or parallel operation/<br>float (trickle) charging (2.23V/Cell) | 53.5 ±<br>1% | 56 ±<br>1% | 67 ±<br>1% | 69 ±<br>1% |
| Parallel operation/charging<br>(2.33 V/Cell) | 56 ±<br>1% | 58.5 ±<br>1% | 70 ±<br>1% | 72 ±<br>1% |
| Initial charging (at rated mains<br>supply voltage with communication<br>system disconnected) | 65 | 67.5 | 84 | 84 |
| Rated direct current (A) | 30 A | | 100 A | |
| Interference voltage (mV) | ≤ 1.8 (frequency-weighted with<br>CCITT A-filter) | | | |
| Dimensions (H×W×D) (mm) | 145×654×340 | | 265×654×340 | |

### 9.2.5 Technical Data

The principal technical data for the rectifier modules types 48 V (60 V) / 30 A and 100 A, GR20 are listed in Table 9.4.


## 9.3 Type 48 V/120 A (60 V/100 A)

### 9.3.1 General and Application

The transistor-controlled switched-mode power supply units (rectifier modules) 48 V/120 A (60 V/100 A) GR40[4] (for the slide-in unit see Fig. 8.20) are applied in the power supply system 400 for the supply e.g. of the digital switching system EWSD. The rectifier modules convert the a.c. supply voltage into a regulated, smoothed d.c. voltage. This d.c. voltage meets all the requirements for supply needed for telecommunications systems.

The GR40 draws a sine wave current. This means that there is nearly no interaction with the a.c. supply system. This non-interactive load allows also 100% utilization of standby generating equipment.

The design of GR40 is modular. If more rectifier modules are connected in parallel (n+1 operation) greater systems can be built up. Up to five rectifier modules can be connected per one rectifier cabinet.

The GR40 can be used in the compact power supply 400 or in the power supply system 400.

The GR40 includes as a standard a microcontroller interface for the communication with a higher-order control board A42.

The function of the rectifier module GR40 120A (100 A) is in general very similar to the rectifier module GR20 (see Sect. 9.2) and GR40 12 A (10 A) (see Sect. 9.1).


### 9.3.2 Modes of Operation

The rectifier modules GR40 48 V/120 A (60 V/100 A), can be operated in the rectifier mode (see Sect. 3.1) or in the standby parallel mode (see Sect. 3.3).


### 9.3.3 Survey Diagrams of the Power Supply System 400

*Application in compact power supply system 400.* In their basic configuration, series 400 compact power supplies (CPSs) consist of two different power supply cabinets, a basic cabinet KS400/KS401 and an extension cabinet KS400E (see Figs. 9.13 and 9.14). Thanks to their compact dimensions, they are ideal for installation in confined spaces, such as containers, for supplying power to digital

---

[4] *Source*: Siemens AG

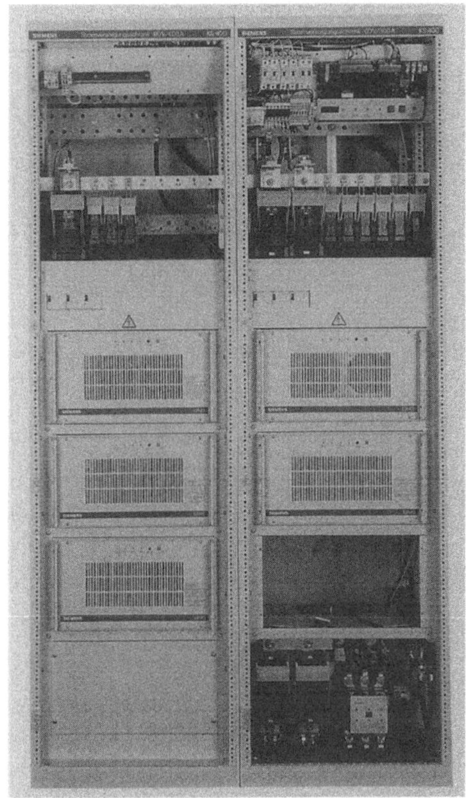

**Fig. 9.13.** Compact power supply system-power supply cabinets series 400, basic (right) KS401 and extension (left) KS400E. (Photo by courtesy of Siemens AG)

switching equipment, radio relay stations, mobile radio and the like. Depending on the amount of current required, the systems consist of one or two of the above-mentioned cabinets. One KS400/KS401 cabinet and one KS400E cabinet can handle up to 840 A for 48 V systems and up to 700 A for 60 V systems. In view of the size of the battery fuse, a maximum of 630 A is therefore available for supplying the loads; the remaining rectifier capacity is used for charging the batteries or as a safety margin.

Compact power supplies operate in parallel standby mode. In other words, the rectifier equipment, batteries and loads are connected in parallel in all operating states. To protect the battery from excessive discharge it can be isolated from the system by means of an optional battery disconnection unit (KS401). With the aid of a Siemens Power Supply Controller (SPSC40) all the operating data can be monitored and called down by the power (supply) control center PCC via a modem. Conversion of the power from the public a.c. mains supply is performed by clocked 19″ rack-mounted rectifier modules of type GR40 with rated output currents of 120 A and 100 A (for rated voltages of 48 V and 60 V respectively). These modules are characterized by high efficiency, sine-wave current input, high

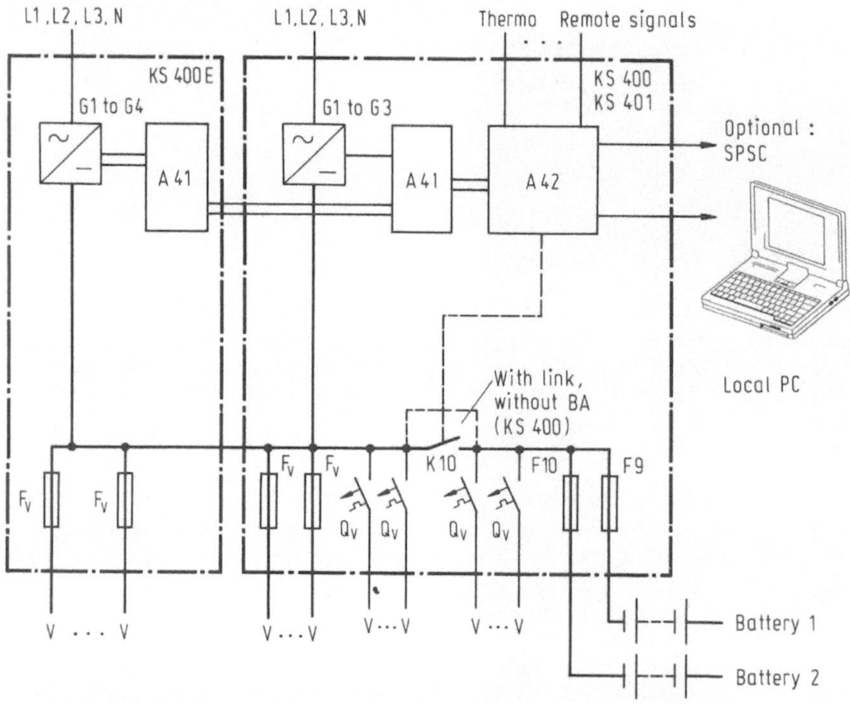


**Fig. 9.14a.** Survey diagram compact power supply series 400 (basic cabinet KS401 (KS400) and extension cabinet KS400E). A41 Signal board, A42 control board, Fv, Qv load fuse/automatic circuit breaker, F9, F10 battery fuse, V load connection, BA battery deep discharging protection, battery disconnecting unit, G1...G4 rectifier modules GR40 120 A, (100 A), L1, L2, L3, N mains 3/N a.c. 400 V, 3 a.c. 230 V, K10 battery disconnection contactor, Thermo temperature monitor $> 34°C$, $> 45°C$, KS400 compact power supply cabinet, basic, without battery disconnecting unit (360 A), KS401 compact power supply cabinet, basic, with battery disconnecting unit (360 A), KS400E compact power supply cabinet, extension (480 A), SPSC Siemens power supply controller

**Fig. 9.14b.** Operation, adjustments and monitoring with notebook and software of the monitor program series 400. (Photo by courtesy of Siemens AG)

control accuracy, high dynamic range, low volume and low weight. To ensure the safety of operating personnel, all live parts in the compact power supply are enclosed and all fuses employ shock hazard protected disconnectors.

In addition there is a a.c. power distributor 250 A necessary.

Table 9.5 shows the application of the rectifier module GR40, 120 A in the compact power supply system 400 and in the power supply system 400 with the system configuration at different current consumptions (also applicable for the rectifier module GR40, 12 A, see Sect. 9.1)

*Signal board A41* is designed as a pc board and is located at the top of the compact power supply cabinets. It contains seven 25-pin and 9-pin subminiature connector strips for connecting the control and signalling lines (SGST) and the serial interface (SRST). The rectifiers are connected via X31 to X35 (for SGST) and X41 to X45 (for SRST). The adjacent expansion cabinets are connected via X50, X51 and X20, X21. If there are no further cabinets with an A41 module on one side, the terminating resistors for the SRST must be activated by closing dip fix switches S1 and S2. Signal board A41 forwards the signals and also contains connections for fuse monitoring and recording standardized battery and load currents (0 to 20 mA).

There is in general only one *control board A42* (see Figs. 9.15 and 9.16) in each power supply system. This module, which acts as a central unit, performs control, signalling and monitoring functions and also enables the rectifier modules to exchange signalling traffic via the signal interface (SGST). It is connected via the SGST and SRST to board A41 and via terminal strips to the signal lines of the power supply system. As an option, the Siemens Power Supply Controller (SPSC40) can be controlled via the second connecting strip for the SRST. Rotary switch S3 enables 12 standard configurations to be selected (see Fig. 9.17). The settings are shown in Table 9.6. In addition, the system settings or rectifier module settings can be individually modified via the service interface (SEST, RS232) by means of a PC and a suitable software (see Fig. 9.14 b).

Once the desired configuration has been set up, the control module is reset with the aid to the reset switch S4 so that the configuration comes into force. It is also possible to cycle through the value for the shunts with the aid of pushbutton mode S1 (see Fig. 9.17). The value displayed is then automatically used.

*Application in power supply system 400.* The power supply system 400 (see Figs. 1.6 and 9.21) designed for the range of 120 A (100 A) up to 5000 A (2500 A) current-consumptions, and is very similar to the power supply system 200 (see Sect. 9.2.3). Only the rectifier cabinet (see Fig. 9.18) in the power supply system 400 is designed to accommodate five 120 A (100 A) rectifier modules GR40. The following basic components are used in this system:

- rectifier connecting cabinet GS400 48 V/600 A (60 V/500 A),
- battery switching and d.c. distribution cabinets: BV400 48 V (60 V)/630 A, 1250 A and 2500 A (see Fig. 9.19) and
- a.c. power distribution equipment; power distributor NV400, 250 A or power distribution cabinet NV400, 630 A (see Fig. 9.20).

**Table 9.5.** Planning information: for 48 V power supply systems series 40 and 400 with 5000 A systems

| | Rectifier module (n+1) | SVE Power supply unit | Rectifier module (n+1) | Control board | Compact power supply cabinet | |
|---|---|---|---|---|---|---|
| Load current requirement (A) | 12 A GR40 | 72 A (2 batt.) SVE40 | 120 A GR40 | A42 (only in KS400/401 (S)* or BV400) | 360 A (2 batt.) KS400/401 B | 480 A KS400E E |
| 12 | 1+1 | 1 | | | | |
| 24 | 2+1 | 1 | | | | |
| 60 | 5+1 | 1 | | | | |
| 80 | | | 1+1 | 1 | 1 | |
| 120 | | | 1+1 | 1 | 1 | |
| 240 | | | 2+1 | 1 | 1 | |
| 360 | | | 3+1 | 1 | 1 | 1 |
| 480 | | | 4+1 | 1 | 1 | 1 |
| 600 | | | 5+1 | 1 | 1 | 1 |
| 720 | | | 6+1 | 1 | | |
| 840 | | | 7+2 | 1 | | |
| 1000 | | | 9+2 | 1 | | |
| 1200 | | | 10+2 | 1 | | |
| 1500 | | | 13+3 | 1 | | |
| 1800 | | | 15+3 | 1 | | |
| 2000 | | | 17+4 | 1 | | |
| 2400 | | | 20+4 | 1 | | |
| 2500 | | | 21+5 | 1 | | |
| 3000 | | | 25+5 | 1 | | |
| 3600 | | | 30+6 | 2 | | |
| 4200 | | | 35+7 | 2 | | |
| 5000 | | | 42+8 | 2 | | |

B Basic
E Extension
(n+1) = load feeding+battery charging (20% max.)/standby unit, values in brackets [ ] = alternative

The modular series 400 power supply is used as a secure supply for communication equipment operating at rated voltages of 48 V d.c. or 60 V d.c. The lead-acid batteries which are needed to provide power if the public supply network should fail are supplied with the necessary trickle (float) charge or recharge. Depending on the configuration selected, it is possible to control the 2.33 V/cell characteristic on a time or capacity basis.

The power supply operates in parallel standby mode. In other words, the rectifier modules (RFs), batteries and loads are connected in parallel in all operating

switched-mode rectifier units (rectifier modules) types GR40, 12 A and GR40 120 A; max

| EWSD power supply cabinet | Rectifier cabinet | Battery and d.c. distribution cabinet | | | A.C. power distribution | A.C. power distribution cabinet |
|---|---|---|---|---|---|---|
| 480 A (2 batt.) | 600 A | 630 A (2 batt.) | 1250 A (4 batt.) | 2500 A (4 batt.) | 250 A | 630 A |
| KS400S* KS400SE* | GS400 | BV400 | BV400 | BV400 | NV400 | NV400 |
| 1 | 1 | 1 | [1] | | 1 | |
| 1 | 1 | 1 | [1] | | 1 | |
| 1 | 1 | 1 | [1] | | 1 | |
| 1 | 1 | 1 | [1] | | 1 | |
| 2 | 1 | 1 | [1] | | 1 | |
| 2 | 2 | [1] | [1] | | 1 | |
| 2 | 2 | [2] | 1 | [1] | 1 | [1] |
| 3 | 2 | [2] | 1 | [1] | 1 | [1] |
| 3 | 3 | [2] | 1 | [1] | 1 | [1] |
| 3 | 3 | [2] | 1 | [1] | 1 | [1] |
| 4 | 4 | | [2] | 1 | [2] | 1 |
| 5 | 4 | | [2] | 1 | [2] | 1 |
| 6 | 5 | | [2] | 1 | [2] | 1 |
| 6 | 5 | | [2] | 1 | [2] | 1 |
| | 6 | | [2] | 1 | [2] | 1 |
| | 6 | | [3] | 2 | | 1 |
| | 8 | | [3] | 2 | | 2 |
| | 9 | | [4] | 2 | | 2 |
| | 10 | | [4] | 2 | | 2 |

* *Remark to the S-version*: When integrating the power supply into the EWSD switching system, only a cabinet version of series 400 – namely power supply cabinet 480 A KS400S (KS400SE) of SIVAPAC design – is used. EWSD switching systems with a current requirement of 120 A to approx. 3000 A can be supplied by connecting several cabinets in parallel.

states. These power supplies are particularly suitable for digital exchanges such as EWSD, mobile radio, radio relay stations and the like.

Series 400 power supplies can be linked to a PC and/or a Siemens Power Supply Controller (SPSC40) so that all operating data can be monitored. This data can then be interrogated via a modem.

Conversion of the three-phase or single-phase a.c. voltage from the mains supply to the desired d.c. voltage is performed by clocked 19″ rack-mounted rectifier modules with rated output currents of 120 A and 100 A (for rated voltages

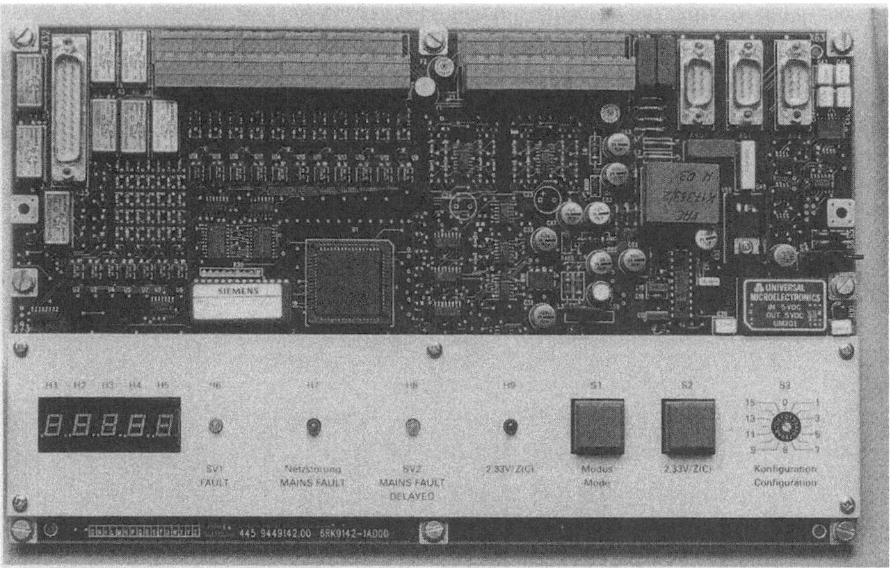

**Fig. 9.15.** Control board A42 (Photo by courtesy of Siemens AG)

of 48 V and 60 V). These rectifier modules are characterized by high efficiency, sine-wave current input, high control accuracy, high dynamic range, low volume and low weight.

With the available components (rectifier modules, rectifier cabinets, battery switching and distribution cabinets and power distribution equipment) it is possible to construct power supplies with ratings between 100 A and 5000 A.

To ensure the safety of operating personnel all live parts of the system are enclosed and all fuses employ shock-hazard protected disconnectors.

Since the standby parallel operating mode is being used, all the Rectifier modules (RFs) are connected in parallel with the batteries and the loads in all operating states. When mains voltage is present, the RFs supply load current and the necessary current for the batteries.

If mains voltage fails, the batteries supply uninterrupted power to the loads. When mains voltage returns, the RFs continue to supply the loads and recharge the batteries at increased voltage (2.33 V/C) for a preselected duration or according to the level of charge. If the batteries are valve regulated (maintenance-free, sealed) batteries which may not be operated at this high charge voltage, the 2.33 V/C mode must be disabled on configuration of control board A42. These batteries will then only ever be operated at a voltage of 2.23 V/C.

In standby parallel mode the load voltage depends on the battery voltage and varies between the following values:

- 42 V and 56 V for –48 V systems and 24-cell lead batteries
- 44 V and 58.5 V for –48 V systems and 25-cell lead batteries
- 54 V and 70 V for –60 V systems and 30-cell lead batteries.

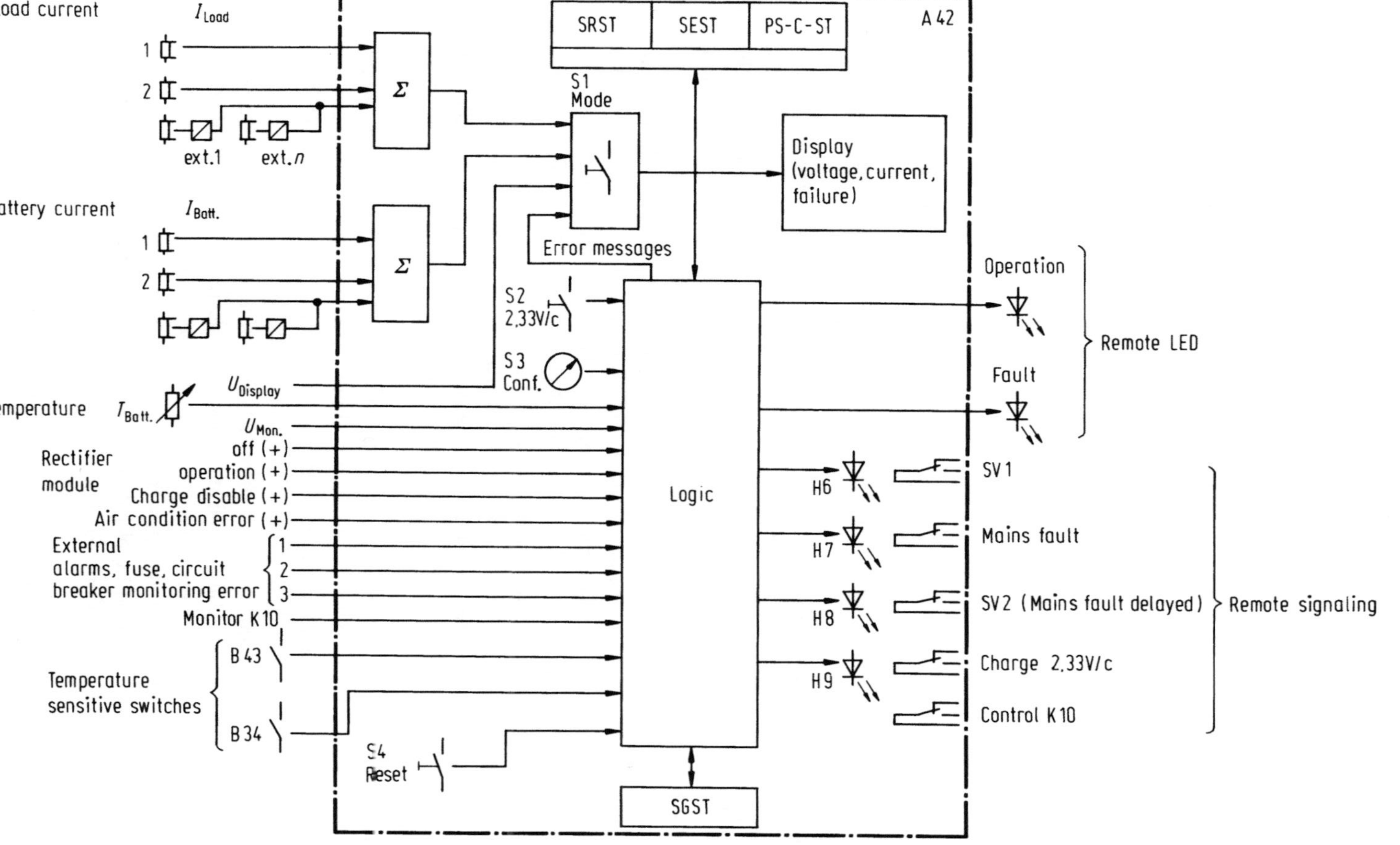



**Fig. 9.16.** Block diagram control board A42. SGST signal interface for normal operation–connection via signal bus module A41 to rectifier modules GR40, 120 A, SRST serial, data interface (RS485) for commissioning or maintenance–connection via signal bus module A41 to rectifier modules GR40, 120 A, SEST service, data interface (RS232)–connection for local PC–commissioning, maintenance and monitoring, PS-C-ST ($\hat{=}$ SRST), optional: Siemens power supply controller (SPSC), PC personal computer (IBM–compatible)

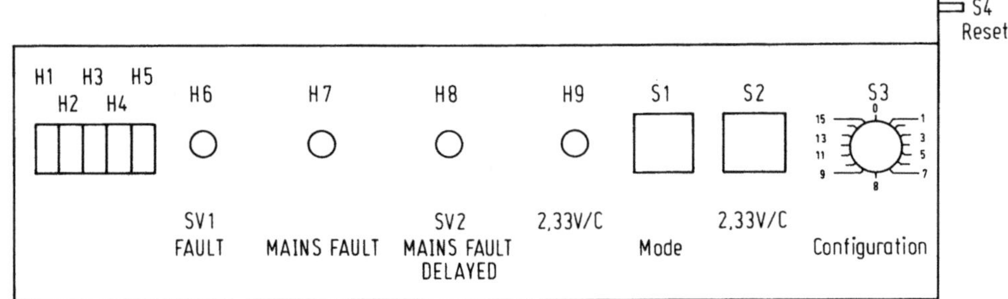

**Fig. 9.17.** Control board A42-arrangement of controls and indicators

It is possible to reduce the input power of the power supply system with NEA (standby) operation. To do this, the part of the RF which is specified for battery charging can be switched off and the load characteristics of the other RFs disabled.

The *rectifier cabinet* GS400 is designed to accommodate five RFs which are installed on site. The mounting locations for the RFs are arranged one above the other; the necessary connecting lines are already in place. If fewer than five RFs are used the lower or lowest mounting locations remain unoccupied. Mounting locations which are not used are covered by a blanking plate.

The RFs can be inserted in the cabinet from the front and are held in place by screws. The electrical connections for the RFs are made on the a.c. mains side via 25 A power switches and on the d.c. side via NH (Low-voltage-high-breaking capacity) circuit breaker assemblies (size 00) fitted with 125 A NH fuses. These components are located at the top of the cabinet.

The control and signalling terminals of the RFs are connected to the signal board A41 via plug-in lines with 25-pin subminiature connectors for the signal interface (SGST) and 9-pin subminiature connectors for the serial interface (SRST).

The signal board A41 (see Section 'application in compact power supply system 400') is connected to the adjacent cabinets (the rectifier cabinet and the battery switching and distribution cabinet) at connector strips X20, X21 (SRST) and X50, X51 (SGST) with the control and signalling lines.

The *battery switching and d.c. distribution cabinets* for rated currents of 630 A, 1250 A and 2500 A are of virtually identical design. The main difference lies in the power ratings for the battery fuses and battery connections. In addition to the central equipment of the power supply system, control board A42 and signal board A41 (see Section 'application in compact power supply system 400'), they contain all the connections for batteries, for loads and for a mobile power supply unit.

Switch S1 can be used to select the battery or load voltage for the digital display on the A42. Next to the switch there are jacks for connecting external instruments for measuring the battery and load currents. The remote signalling lines for SV1 and SV2 are connected directly to the terminals of the A42 board.

At the bottom of the cabinet are the connections for batteries and mobile power supply units.

In the middle and at the top of the cabinet are the load-switchable circuit breaker assemblies of different sizes for the outgoing load circuits installed on three $100 \times 10$ mm Cu clips. As an option, each outgoing load circuit can

**Table 9.6.** Configurations of the control board A42. (See also next page)

<table>
<tr><td></td><td colspan="4">48 V Systems</td><td colspan="8">60 V Systems</td><td></td></tr>
<tr><td>Conf. switch<br>S3 position:<br>(Standard conf.)</td><td>1</td><td>2</td><td>3</td><td>4</td><td>5</td><td>6</td><td>7</td><td>8</td><td>9</td><td>10</td><td>11</td><td>12</td><td>15</td></tr>
<tr><td>Conf. switch<br>S3 position:<br>(can be modified<br>via a personal<br>computer PC)</td><td>0</td><td></td><td></td><td></td><td>13</td><td>14</td><td></td><td></td><td></td><td></td><td></td><td></td><td></td></tr>
<tr><td>Battery disconnection</td><td colspan="2">no</td><td colspan="2">yes</td><td colspan="2">no</td><td colspan="4">yes</td><td colspan="3">no</td></tr>
<tr><td>Disable charge characteristic<br>(y: yes, n: no)</td><td>y</td><td>n</td><td>y</td><td>n</td><td>y</td><td>n</td><td>y</td><td>n</td><td colspan="5">yes</td></tr>
<tr><td>Control charge characteristic</td><td></td><td>*</td><td></td><td>*</td><td></td><td>*</td><td></td><td>*</td><td></td><td></td><td></td><td></td><td></td></tr>
<tr><td>Signal for 'RF mains fault'</td><td colspan="4">SV2 and mains fault signal (Nst)</td><td colspan="5">SV2</td><td>SV1</td><td>SV2</td><td>SV1</td><td></td></tr>
<tr><td>Signal for 'all' RF-<br>'RF mains fault'</td><td colspan="4">SV2</td><td colspan="9">SV2 and mains fault signal (Nst)</td></tr>
<tr><td>Signal for 'RF off'</td><td colspan="4">SV1</td><td colspan="5"></td><td colspan="4">SV1</td></tr>
<tr><td>Signal for 'last' RF-'RF off'</td><td colspan="4">SV1</td><td colspan="5">SV2</td><td colspan="4">SV1</td></tr>
<tr><td>Signal for 'RF fault'</td><td colspan="13">SV1</td></tr>
<tr><td>Signal for 'fault' at fuse</td><td colspan="13">SV1</td></tr>
<tr><td>Remote control</td><td colspan="13">not active</td></tr>
<tr><td>(V)</td><td colspan="4">24C 62.0</td><td colspan="9">30C 73.0</td></tr>
<tr><td>Overvoltage threshold  (V)</td><td colspan="4">25C 64:0</td><td colspan="9">31C 75.0</td></tr>
<tr><td>(V)</td><td colspan="4">24C 44.0</td><td colspan="9">30C 57.0</td></tr>
<tr><td>Undervoltage threshold (V)</td><td colspan="4">25C 46.0</td><td colspan="9">31C 58.0</td></tr>
<tr><td>Battery emerg. shut off (V)</td><td colspan="4">24C 42.0</td><td colspan="9">30C 53.0</td></tr>
<tr><td>(V)</td><td colspan="4">25C 44.0</td><td colspan="9">31C 55.0</td></tr>
</table>

**Table 9.6.** (*Continued*)

Remarks and explanations to Table 9.6:

| Conf. switch<br>S3 position: | | |
|---|---|---|
| 3 and 4<br>    7, 8 and 9 | (48 V  Systems)<br>(60 V  systems) | settings for compact power<br>supply systems (eg. KS 401) with<br>battery disconnection |
| 1 and 2<br>    5, 6, 11, 12 | (48 V  systems)<br>(60 V  systems) | settings for power supply systems<br>without battery disconnection |
| 5 | as delivered condition | |
| 13 (progr.) | in general for valve-regulated lead-acid batteries<br>(factory progr.: disable charging characteristic) | |
| 14 (progr.) | in general for liquid-electrolyte lead-acid<br>batteries (factory progr.: enabling charging<br>characteristic) | |
| 15 | spare and factory set up | |

RF Rectifier module
* The charging characteristic is activated for 120 minutes in the event of
a power failure lasting at least 15 minutes

be equipped with a shunt resistor with test jacks. The outgoing load circuits are
installed on site. Permanently installed on the bottom bar are three size 00 circuit
breaker assemblies without shunt resistors for the low-current loads such as the
safety lights.

### 9.3.4 Survey Diagram, Block Diagram and Functioning Principle of the Rectifier Module 48 V/120 A (60 V/100 A)

Figure 9.22 shows the block diagram of the switched-mode power supply unit
(rectifier module) 48 V/120 A (60 V/100 A), GR40 (see also Fig. 9.23).

For the three-phases of the mains there is a separated functional unit of
conversion from a.c. in a controlled d.c. supply. Such a functional unit contains
every input including the supply side rectifier, surge voltage protective circuitry,
boost converter and d.c/d.c. converter. After the output filter, there is a parallel
connection of those circuits. Therefore there is a redundancy because in the case
of only one of three units it is faulty or there is a phase cut of the rectifier
module which is still in operation (with reduced power).

The rectifier module GR40 can be controlled from a remote point via the
floating control inputs of signalling interface X3 (SGST) or serial interface X4
(SRST), (RS485).

Setting the GR40 (e.g. trickle (float)-charge voltage as a function of number
of battery cells) is carried out at start-up with configuration (mode selector)
switch S2, which is designed as a rotary BCD (binary coded decimal) switch

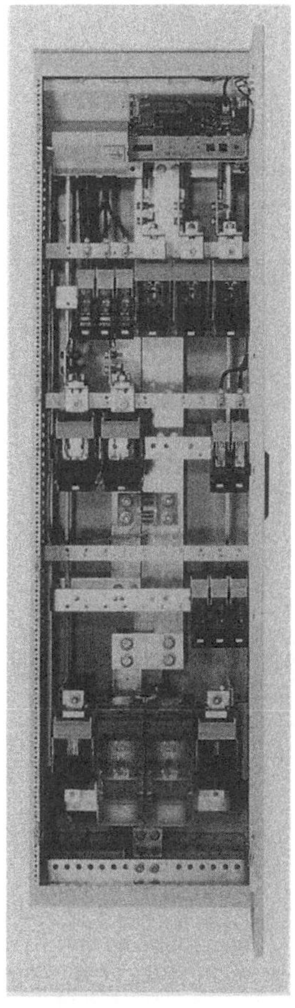
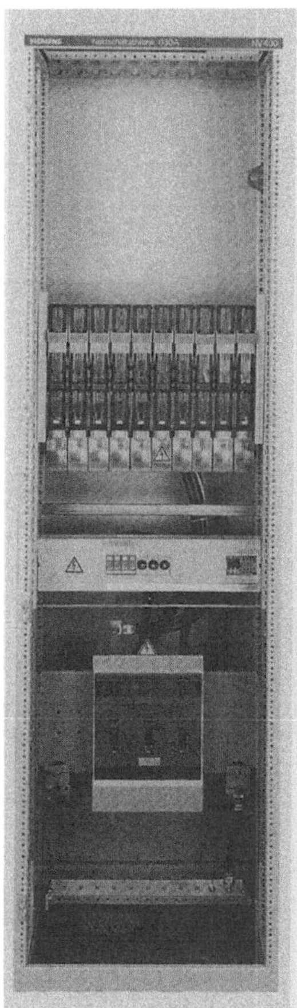

**Fig. 9.18**                **Fig. 9.19**                **Fig. 9.20**

**Fig. 9.18.** Rectifier connecting cabinet GS400 48 V/600 A (60 V/500 A) in the power system 400 (Photo by courtesy of Siemens AG)

**Fig. 9.19.** Battery switching and d.c. distribution cabinet BV400 48 V (60 V)/1250 A for the power system 400 (Photo by courtesy of Siemens AG)

**Fig. 9.20.** A.C. power distribution cabinet NV400/630 A for the power system 400 (Photo by courtesy of Siemens AG)

and is located behind the front panel on the board A5 (see also Fig. 9.24). The operating modes (trickle charge, recharge) are set via signalling interface X3.

*Power section.* The input voltage is passed through a *multi-stage input filter* to minimize line-side EMC. Connected to the output of this filter is a rectifier for converting the a.c. voltage into a pulsating d.c. voltage.

Each functional unit has a protective circuit current limiting circuit consisting of an IGBT (insulated gate bipolar transistor) to prevent the inrush current from

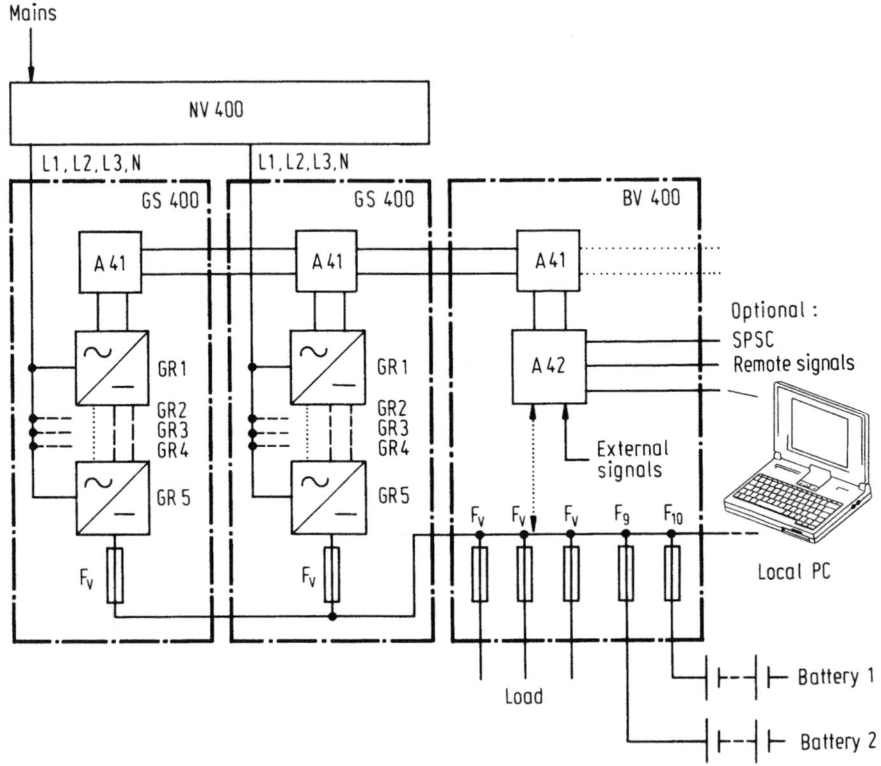



**Fig. 9.21.** Survey diagram power supply system 400. GR1...GR5 rectifier modules GR40/ 120 A (100 A), GS400 rectifier connecting cabinet, BV400 battery switching and d.c. distribution cabinet, NV400 power distributor or power distribution cabinet, A41 signal board, A42 control board, SPSC Siemens power supply controller

rising above the rated current. While the rectifier module is in operation, the IGBT protects the downstream components against overvoltage. The circuit meets the requirements of VDE 0160, eg. a surge withstand capability of $2.3 \times \hat{U}_{rated}$.

The function of the *boost converter* is to match the waveform of the a.c. input current to that of the a.c. system with the least possible harmonic content. The semiconductor switch is turned on and off so as to make the charging current drawn by the link circuit capacitor sinusoidal. It is used to provide optimum efficiency for the circuit because most of the input current flows past the controlling element to the d.c./d.c. converter. Operating at a switching frequency of 50 kHz, the booster converts the rectified a.c. voltage into the continuous d.c. link voltage required for feeding the downstream d.c./d.c. converter. It operates as an autonomous unit which is suitably monitored and controlled.

The function of the *d.c./d.c. converter* is to step down the link voltage generated by the boost converter to the d.c. output voltage level. At the same time it provides safe electrical isolation between a.c. and d.c. circuits.

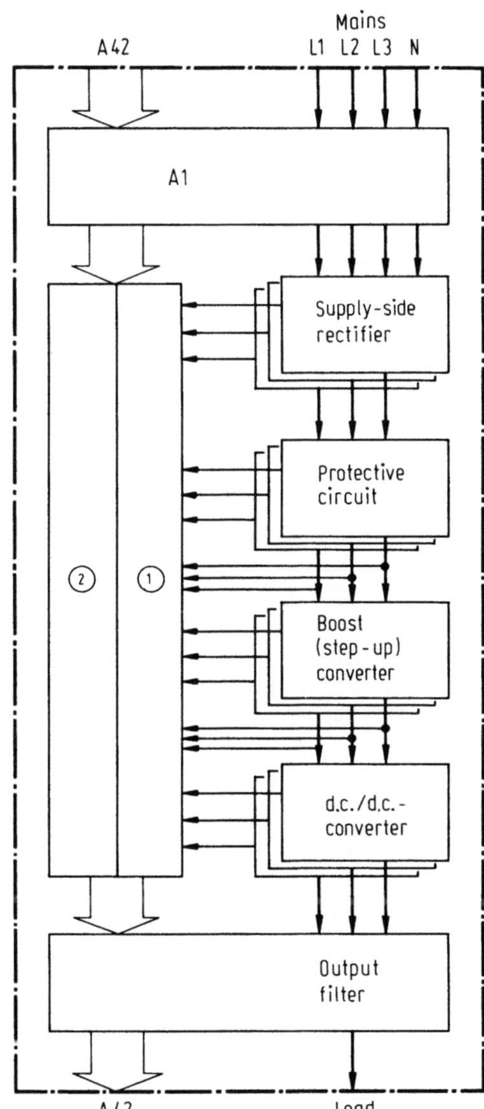



**Fig. 9.22** Block diagram of switched-mode power supply unit (rectifier module) GR40 48 V/120 A (60 V/100 A). (1) Regulation and monitoring, (2) display and control panel, A1 input filter LF/HF and overvoltage protection, A42 control board (external)

An IC assumes the task of regulating and driving the switching transistors connected in a fully controllable bridge configuration. The transistors are switched at a frequency of 50 kHz to produce a regulated 100-kHz rectangular a.c. voltage on the secondary side of the transformer. A bridge rectifier then converts this a.c. voltage into the d.c. output voltage. The module's rated output power is obtained by connecting the outputs of the three functional units in parallel.

The module has an *LF/LC-filter* connected in the d.c./d.c. converter's output circuit to limit ripple as well as an RF/*LC*-filter across its output to meet EMI requirements.

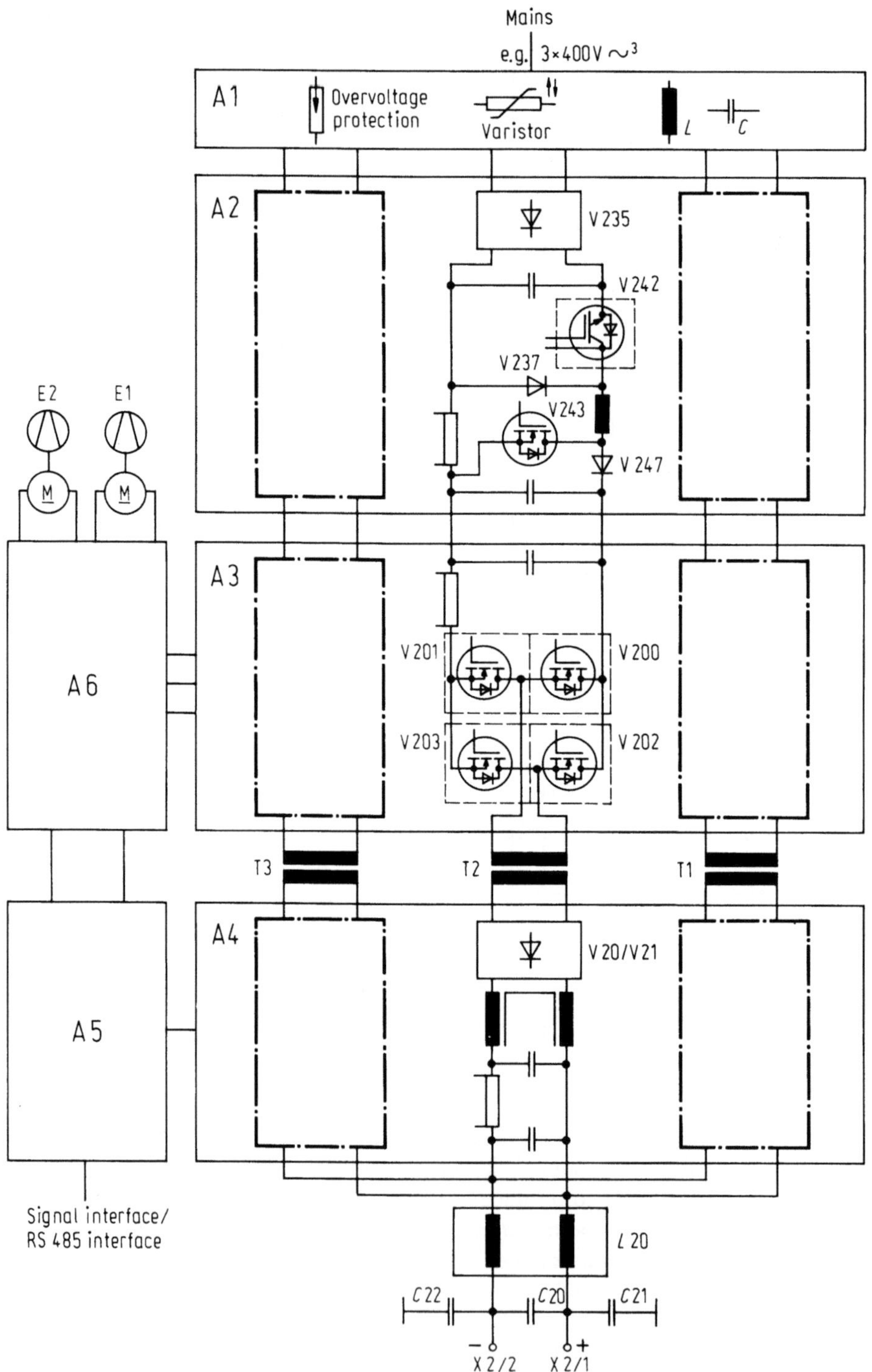
Mains
e.g. 3×400V ∼³
A1
Overvoltage
protection
Varistor
L
C
A2
V 235
V 242
V 237
V 243
V 247
E2
E1
M
M
A6
A3
V 201
V 200
V 203
V 202
T3
T2
T1
A5
A4
V 20/V 21
Signal interface/
RS 485 interface
L 20
C 22
C 20
C 21
−
+
X 2/2
X 2/1

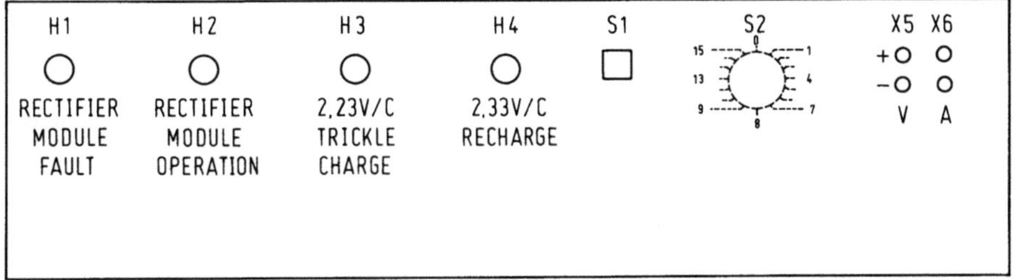


**Fig. 9.24.** Controls, indicators and operating elements (on A5) of the rectifier module 48 V/120 A (60 V/100 A), GR40. H1, H2, H3, H4: LED light-emitting diodes, X5/X6: measuring points voltage (V)/current (A), (6 V $\triangleq$ 120 A, 5 V $\triangleq$ 100 A), S1: ON/OFF switch, S2: mode selector switch (behind the front panel)

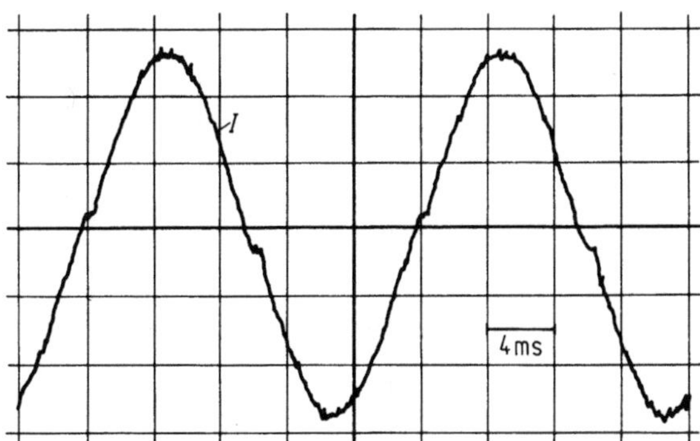


**Fig. 9.25.** Rectifier module GR 40/120 A. Input current wave form at a.c. input voltage: 400 V; d.c. output voltage: 53.5 V; d.c. output current: 120 A

The module's *output voltage* is *regulated* by the individual *d.c./d.c. converters*.

Rectifier modules *operating in parallel* share current via a power bus. Routed via signalling interface X3, the bus feeds an average current rating to the regulator to enable each GR40 to provide its share of the total output current. The deviation between the separate, individual branches is therefore < 5% of $I_{rated}$.

The *output* of the rectifier module GR40 is *short-circuit-proof*. The current limiting mechanism ensures that in the event of overcurrent the voltage is reduced

---

**Fig. 9.23.** Survey diagram of rectifier modules 48 V/120 A (60 V/100 A), GR40. A1 LF/HF filter and overvoltage protection, T1, T2, T3 main transformers, A2[5] supply side rectifier, IGBT (insulated gate bipolar transistor) for the surge voltage protection, MOS-FET boost sinusoidal controller, A3[5] d.c./d.c. converter (fully controlled transistor bridge circuit), A4[5] secondary-rectifier, HF filter, A5 regulation and control board, A6 internal power supply, L20 output filter, C20, C21, C22 HF filter, E1/E2 fan

---

[5] Remark: with regard to A2, A3, A4: All three phases may be used for the same components which are then shown for one phase.

until the needed power equals the rated output power. In extreme cases (eg. short circuit across the terminals) the output voltage is 0 V.

*Overvoltage* on the d.c. side will inevitably lead to shutdown of the GR40. If the unit carries more than 95% of rated current and the output voltage reaches

**Table 9.7.** Configurations of the rectifier module GR40 120 A (100 A)

| | 48 V Systems | | | | | 60 V Systems | | | | |
| --- | --- | --- | --- | --- | --- | --- | --- | --- | --- | --- |
| Conf. switch S2 position: (Standard conf.) | 0 | 1 | 2 | 3 | | 5 | 6 | | 8 | 9 |
| Conf. switch S2 position: Values as per default setting (can be changed via a personal computer PC) | | | | | 4 | | | 7 | ↑ | ↑ |
| Number of lead-acid battery cells | 24 | | 25 | | 24 | 30 | | 30 | | |
| | | | | | 25 | | | 31 | | |
| Float (trickle) charging voltage 2.23 V/C (V) | 53.5 | | 56.0 | | 53.5 | 67.0 | | 67.0 | | |
| | | | | | 56.0 | | | 69.0 | | |
| Charging voltage 2.33 V/C (V) | 56.0 | | 58.5 | | 56.0 | 70.0 | | 70.0 | | |
| | | | | | 58.5 | | | 72.0 | | |
| Rectifier mode (basic characteristic) (V) | | 51.0 | | 51.0 | 53.5 | | 62.0 | 67.0 | | |
| | | | | | 56.0 | | | 69.0 | | |
| Overvoltage (fast) (V) | 67.0 | | | | 67.0 | 79.0 | | 79.0 | | |
| Overvoltage (slow, single operation) (V) | 64.0 | | | | 64.0 | 75.0 | | 75.0 | ↓ | ↓ |
| | | | | | | | | 77.0 | | |
| Overvoltage (slow) with 95% $I_N$ (V) | 62.0 | | | | 62.0 | 73.0 | | 73.0 | | |
| | | | | | | | | 75.0 | | |
| Undervoltage (V) | 44.0 | | | | 44.0 | 57.0 | | 57.0 | | |
| Output current (A) | 120 | | | | | 100 | | | ↓ | ↓ |

Remarks:
In all of rectifier modules must set the conf. switch S2 to the same position!
Conf. switch S2 position 0 as delivered condition
S2 Position 8 : Spare (not used)
S2 Position 9 : Address setting

62 V/73 V (at 48 V or 60 V $U_{\text{rated}}$) the time-delay monitor effects maintained shutdown of the module ($t_{\text{del}} = 500$ ms). If the module delivers less than 95% of rated current, it is not turned off until the voltages reaches 64 V/75 V.

A second, fast-acting monitor turns off the module at 67 V/79 V ($t_{\text{del}} = 1.2$ ms). After a short pause the latter resumes operation. Should overvoltage of 67 V/79 V occur once again within the next 10 seconds, the module responds with a maintained shutdown.

In the event of undervoltage on the d.c. side, 'rectifier fault' is signalled, with the rectifier module remaining in operation.

Temperature sensors protect the module components that are subjected to maximum thermal stress against overheating by reducing the output power or turning off one of the functional units ('rectifier fault' signal).

Undervoltage or overvoltage on the a.c. side lead to instantaneous shutdown of the rectifier module GR40. Operation is automatically resumed after the fault has been removed.

Failure of a phase will not lead to complete rectifier module failure but merely to a reduction in output power of 33% per failed phase and the issue of a 'rectifier a.c. fault' signal.

The rectifier module (see also Fig. 9.24, Table 9.7) is *configured* by means of rotary switch (configuration switch) S2 which is located behind the front cover. Any other settings can only be made via a PC.

Fine adjustment of the output voltages between the devices is not necessary since the factory settings have an accuracy of $+1\%$.

Figure 9.23 shows a survey diagram with pc boards of the rectifier module and Fig. 9.24 the elements of operation. For additional cooling two fans are inserted.

### 9.3.5 Technical Data

The principal technical data for the rectifier module types 48 V/120 A (60 V/100 A), GR40 are listed in Table 9.8.

## 9.4 Type 48 V (60 V, 67 V)/50 A and 100 A

### 9.4.1 General and Application

The transistor-controlled switched-mode power supply units 48 V (60 V, 67 V)/ 50 A and 100 A, WGS-U[6] (slide-in unit), as shown in Fig. 9.26, serve for as rectifier modules for different types of telecommunications systems. Those

---

[6] Source: Gustav Klein GmbH & Co. KG.

**Table 9.8.** Technical data for rectifier modules 48 V/120 A (60 V/100 A) GR40

| Mains input supply | | | |
|---|---|---|---|
| Voltage (V) | 3 N a.c.<br>400<br>(300 to 460) | 3 a.c.<br>230<br>(173 to 266) | 1 N<br>230 |
| Frequency (Hz) | 50 or 60 (47.5 to 63) | | |
| Current<br>sinusoidal (A) | $11.2^c$ | $19.4^c$ | $33.6^c$ |
| Degree of radio<br>interference | Limit class B (EN[a] 55022) | | |

| D.C. output | |
|---|---|
| Operating mode<br>or condition | Rated direct voltage (V)<br>(tolerances for 1.5 to 100% rated current) |
| | Equipment voltage $\stackrel{\wedge}{=}$ load voltage |

| | 48 V Systems | 60 V Systems |
|---|---|---|
| Rectifier mode[b] | $51^b < \pm 1\%$ | $62^b < \pm 1\%$ |
| Parallel operation | Setting[b] ranges: | |
| Float (trickle)<br>charging[b]<br>2.23 V/cell | 42 to 64 $< \pm 1\%$<br>(56 V, 25 cells) | 52 to 75 $< \pm 1\%$<br>(67 V, 30 cells) |
| Charging[b]<br>2.33 V/cell | 42 to 64 $< \pm 1\%$<br>(58, 5 V, 25 cells) | 52 to 75 $< \pm 1\%$<br>(69, 5 V, 30 cells) |
| Rated direct<br>current (A) | 120 | 100 |
| Interference<br>voltage (mV) | $\leq$ 1.0 (frequency-weighted with CCITT A-filter) | |

| Dimensions<br>(H×W×D) (mm) | 267 × 483 × 462 ($\stackrel{\wedge}{=}$ 6 height units × 19 pitches × 462 mm) |
|---|---|

[a] Normal Europeans standards.
[b] For settings and for the different number of lead-acid battery cells and for the rectifier mode see Table 9.4.
[c] At rated a.c. line voltage, trickle charging voltage (25 or 30 cells), rated output current.

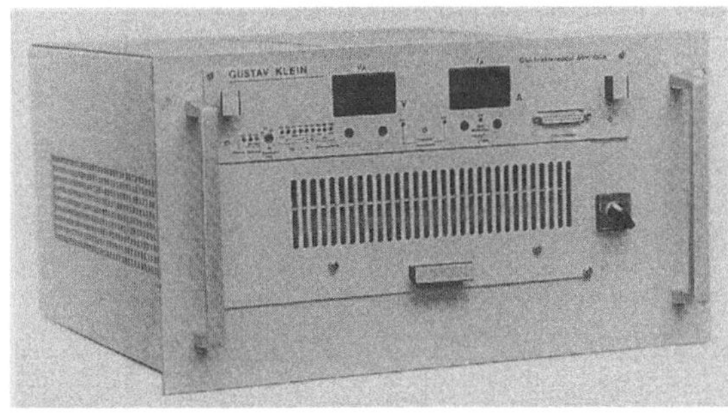

**Fig. 9.26.** Switched-mode power supply (rectifier module) 48 V (60 V, 67 V)/100 A, WGS-U (Photo by courtesy of Gustav Klein GmbH & Co. KG)

rectifier modules are used in compact power supply systems ranging from 150 A or 300 A[7] or in power supply systems ranging from 50 A to 2500 A[7].

In general the systems which are described below are very similar to those in Sects. 9.2 and 9.3.

### 9.4.2 Modes of Operation

The rectifier modules 48 V (60 V, 67 V)/50 A and 100 A, WGS-U can be operated in the rectifier mode (see Sect. 3.1) or in the standby parallel mode (see Sect. 3.3).

### 9.4.3 Survey Diagram of the Power Supply System

The power supply system (see Figs. 9.27 and 9.28) contains:

- mains switch panel or mains distribution switchboard,
- battery and distribution panel and
- rectifier connecting panel (see Fig. 9.29) each with a maximum of four rectifier modules.

In addition the following are also available:

- distribution panels and
- compensators (only for 60 V-systems).

In compact power supply systems the rectifier modules $3 \times 50$ A or $3 \times 100$ A may be found all with fuses, a.c. and d.c.-distributions, a battery and load connections accommodated within one cabinet (see Fig. 9.30).

### 9.4.4  Survey Diagram and Functioning Principle of the Rectifier Module 48 V (60 V, 67 V)/50 A and 100 A

Figure 9.31 shows the block diagram of the switched-mode power supply unit (rectifier module) 48 V (60 V, 67 V)/50 A and 100 A, WGS-U (see also

---

[7]As in the German Postal Administration Telekom Network (since January 1995 Telekom AG).

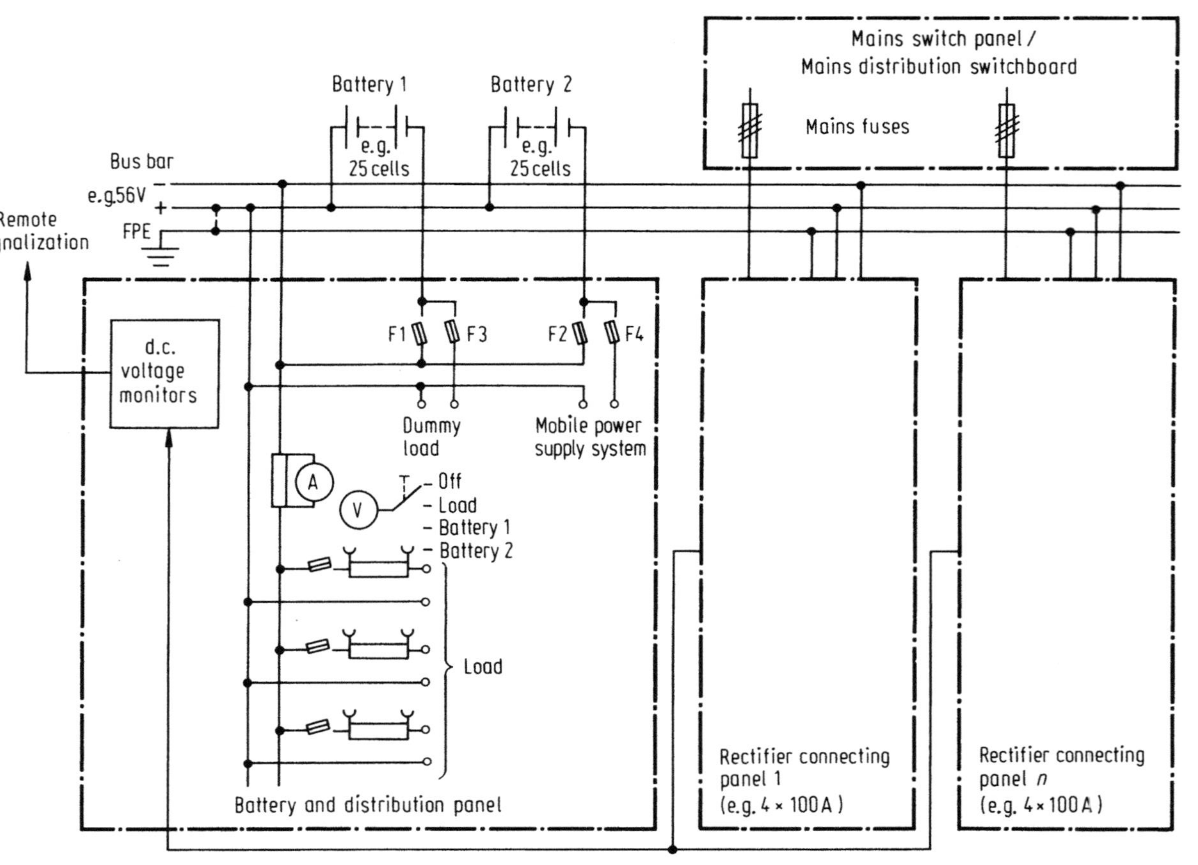


**Fig. 9.27.** Survey diagram of the power supply system

**Fig. 9.28.** Power supply system 48 V (60 V)/1600 A with battery and distribution panel 2500 A (Photo by courtesy of Gustav Klein GmbH & Co. KG)

Fig. 9.29). For additional cooling a fan is connected which increases the operational reliability and life time of the components. This fan is designed for a long service life.

The switched-mode power supply unit (rectifier module) uses the principle of pulse-width control. The output voltage ($IU$ characteristic) is controlled at the primary side of the main transformer.

At the start-up phase there is a current limited charging of the d.c. link circuit filter and so the current at the supply side is reduced to maximum of $1.2 \times I_{rated}$.

All rectifier modules can be part of the current balancing independent from the number of modules used for the mode of operation. Each rectifier module can have its own control, regulation and monitoring system with two digital displays for voltage/current measurement and three LED: operation, mains failure (not urgent), fault (urgent) are the operation states involved.

### 9.4.5 Technical Data

The principal technical data for the rectifier modules types 48 V (60 V, 67 V)/ 50 A and 100 A, WGS-U are listed in Table 9.9.

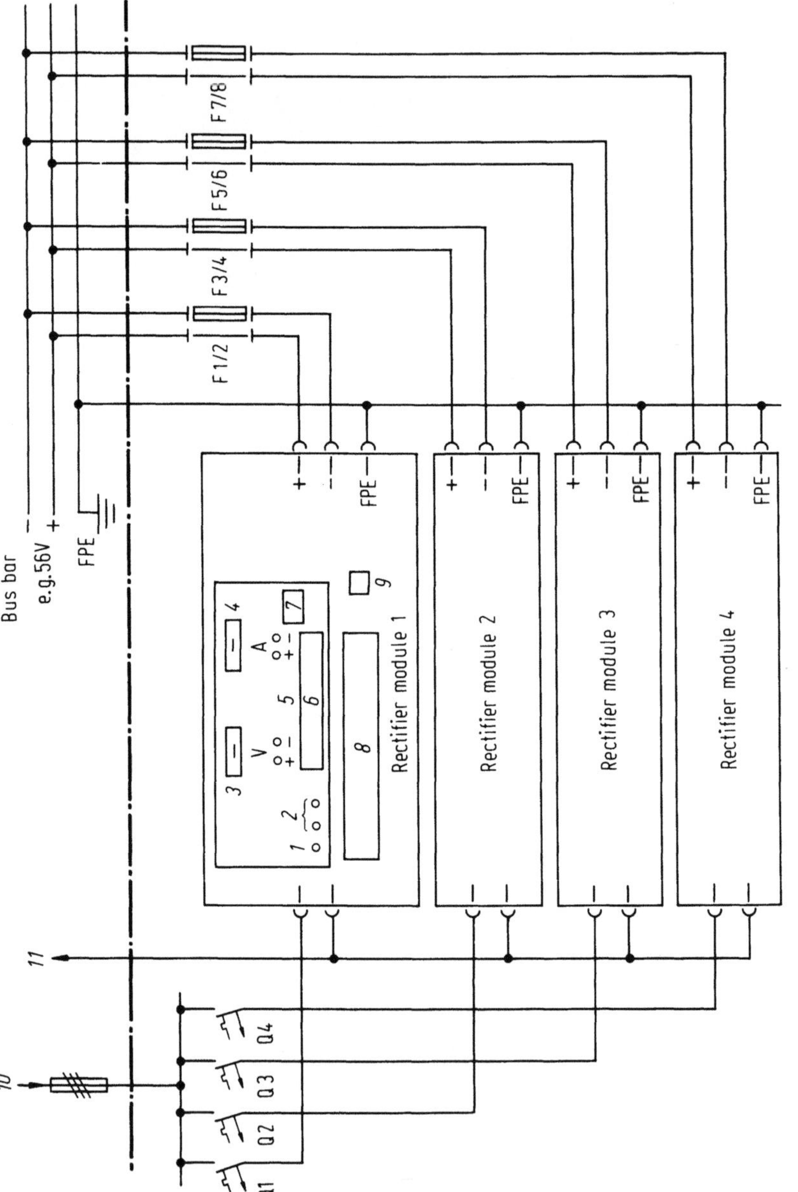


**Fig. 9.29.** Survey diagram of the rectifier connecting panel. *1* LED Operation, *2* LED's fault signalization, *3* digital display – voltage, *4* digital display – current, *5* measuring points, *6* monitoring, *7* test and remote diagnosis connector, *8* fan, *9* mains switch, *10* supply from the mains switch panel or mains distribution switchboard, *11* remote signalization to the battery and distribution panel

**Fig. 9.30.** Compact power supply system 48 V (60 V)/300 A (Photo by courtesy of Gustav Klein GmbH & Co. KG)

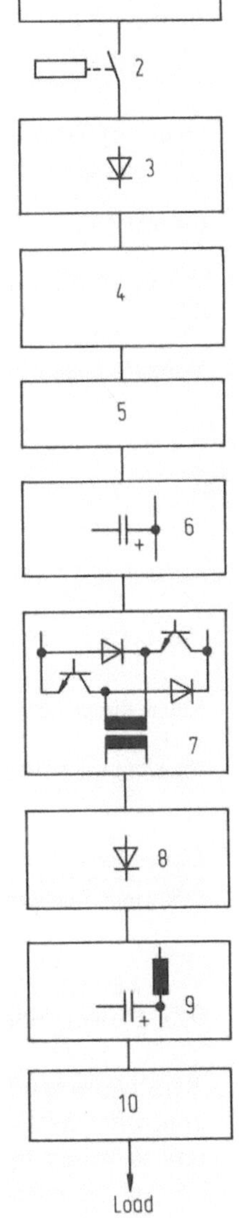


**Fig. 9.31.** Survey diagram of switched-mode power supply unit (rectifier module) 48 V (60 V, 67 V)/50 A and 100 A, WGS-U. *1* HF-filter, *2* mains contactor, *3* supply-side rectifier, *4* filter-charging current limitation, *5* LF-filter, *6* filter-d.c. link circuit, *7* power switching transistors with main transformer, *8* output rectifier, *9* output filter, *10* HF-filter

**Table 9.9.** Technical data for rectifier modules 48 V (60 V) /50 A and 100 A, WGS-U

| Type | 50 A module | 100 A module |
| --- | --- | --- |
| Mains input supply | | |
| Voltage (V) | $\sim$230/400 + 10%<br>$\qquad$ − 15% | |
| Frequency (Hz) | 50 ± 3% | |
| Current (A) | 3 × 5.9 | 3 × 11.5 |
| Degree of radio interference | Limit class B (EN[a] 55011) | |
| D.C. output | | |
| Voltages characteristics (V) | 50A and 100A: | |
| | 48 V Systems: | 60 V Systems: |
| | 49.6 ± 1% | 62 ± 1% |
| | 53.6 ± 1% | 67 ± 1% |
| | 56  ± 1% | 70 ± 1% |
| | 67.2 | 84 |
| Rated direct current (A) | 50 | 100 |
| Interference voltage (mV) | ≤ 2.0 (frequency-weighted<br>with CCITT A-filter) | |
| Dimensions (H×W×D)(mm) | 221.4 × 483 × 440 | 265.9 × 483 × 440 |

[a] Normal European Standards

## 9.5 Power Supply Controller

*Reliable measurements, supervision and operation.* The *Siemens power supply controller SPSC40*[8] (Fig. 9.32) enables remote supervision and maintenance centers to be set up for power supply systems (network management systems).

Its *future-oriented features* allow the SPSC40 not only to supervise such systems but also to measure, record and indicate their operating data.

The SPSC40 can be integrated into both new and existing power supply systems. Operation, maintenance and personnel costs are kept to a minimum through its *ability to communicate fully with the power supply system*; the efficiency and

---

[8] *Source*: Siemens AG.

**Fig. 9.32.** Siemens power supply controller SPSC40 with display (Photo by courtesy of Siemens AG)

speed of service is greatly enhanced through the detailed information transferred between controller and power supply system.

Information about a variety of operating data (such as voltages, currents and operating temperatures) and about the operating conditions of the power supply systems and their infrastructure facilitate *statistical evaluation* and *cost-optimized planning*.

*Siemens power supply controller-overview, interfaces and service features:*

- Connection to control center for supervising several power supply systems,
- connection for local PC for direct operation and interrogation,
- load supervision – current and voltage recording,
- battery supervision – current and voltage recording,
- signal outputs – relays – control outputs,
- operating panel connection (optional),
- signal inputs for individual assignments – air conditioning, generating set etc.,
- fuse supervision,
- temperature supervision,
- interface for connection to equipment series 400,
- connection to rectifiers series 10/12 and
- connection to systems series 200.

*Function.* The SPSC measures voltages, currents and temperatures. The measured values can be displayed on a LCD display (option) or interrogated locally with a PC or remotely via a modem. It is possible to specify which measured values are to be displayed as well as the text displayed with the measured value. Measured values can also be evaluated, i.e. it is tested whether the values exceed or is below a configurable limiting value.

The following are possible: Displays on the LCD display, output with relay or semiconductor relay contacts. The results of the evaluation of measured values

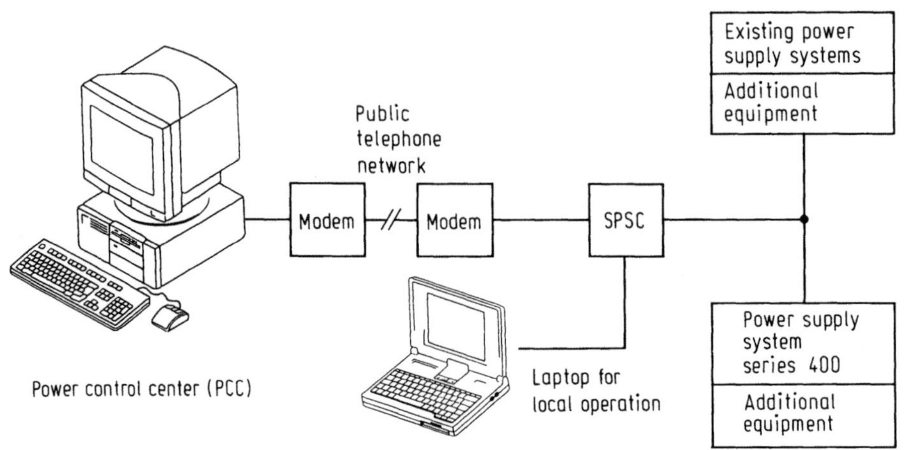



**Fig. 9.33.** Operational area of the Siemens power supply controller SPSC40

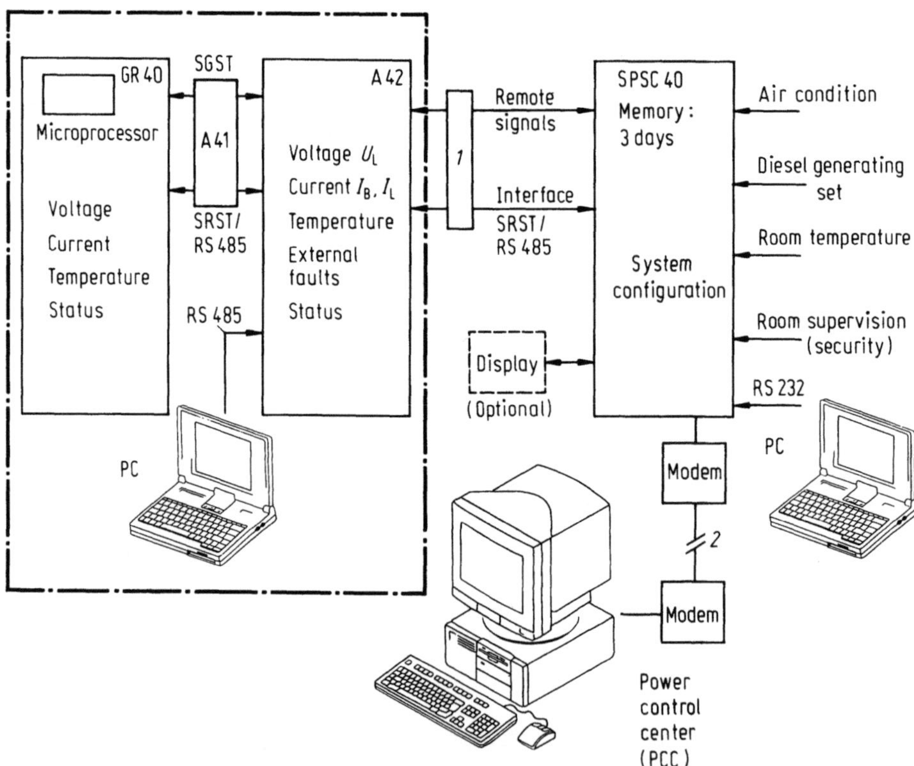



**Fig. 9.34.** Supervision system of Siemens power supply controller SPSC40, control module A42 and rectifier modules GR40/120 A, Interfaces: SGST Signal interface, RS232 (SEST) Service data interface, RS485 (SRST) Serial data interface; PC Personal Computer (e.g. Notebook) for local operation; GR40 Rectifier modules; A41 Signal board; A42 Control board; SPSC40 Siemens power supply controller; *1* Special connecting board; *2* Public switching system

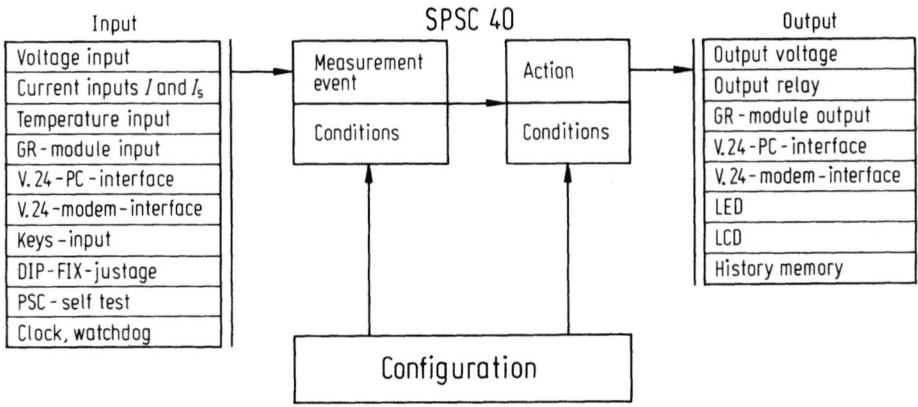



**Fig. 9.35.** Input- and output conditions of the Siemens power supply controller SPSC40

can, for example, be used to operate undervoltage monitoring, overtemperature disconnection, etc. The controller also processes functions such as fuse testing, total current and maximum power value. The inputs and outputs can also be configured for these functions. A further task of the controller is to process a defined protocol which serves remote interrogation and fault indications via a modem.

Control, supervision and remote operation is possible with the Siemens power supply controller SPSC40 (Fig. 9.33).

Figure 9.34 shows the supervision system and connections between the Siemens power supply controller, control module A42 and the rectifier modules GR40/120 A.

In Fig. 9.35 are shown the input- and output conditions of the Siemens power supply controller SPSC40.

*Features:*

Monitoring
– voltage,
– current,
– temperature,
– additional equipment (generating set/aircondition),
– mains supervision,
– rectifier supervision,
– fuses, and
– self test.

Processing
- calculation:
- output power/total current,
- control function,
    temperature dependent voltage regulation,
    overvoltage cut off,
    batterytest,
- reset function, and
- test function.

Alarming
- according to the configuration:
    Local alarms on a display or/and relays and
    Remote with modem.

Configuration
- password,
- voltage inputs,
- current inputs,
- temperature inputs,
- alarms,
- relay outputs,
- functions, and
- optional display.

Operation with local PC
- password,
- actual system parameter,
- history memory, and
- configuration parameter.

Operation with remote PC in the power control center (PCC)
- password,
- actual system parameter,
- history memory,
- configuration parameter, and
- statistics.

*Technical Data:*

| | |
|---|---|
| Feeding voltage | 40...70 V d.c. |
| Power consumption | $\leq$ 750 mA |
| No-load power consumption | 150 mA (typical) |
| Modem feeding | 9 V/250 mA |
| Noise suppression | Limit value class B EN55022 |

| Ambient temperature | $0\ldots45\,°C$ |
| Dimensions (W × H × D) (mm) | $240 \times 242 \times 80$ (without display D: 48) |
| Weight | 2 kg |
| Wall mounting or cabinet | |

*Inputs and Outputs:*

| 40 voltage measurement inputs | $0\ldots\pm75$ V |
| 2 temperature measurement inputs | $-15\,°C\ldots+60°C$ |
| 4 current measurement inputs | with 60 mV shunt |
| 4 current measurement inputs | $0\ldots\pm20$ mA |
| 5 relay outputs (floating contacts) | 80 V/$\leq$ 2 A; 60 W |
| 8 outputs (electronic switches) | $\leq$ 100 V/100 mA |
| 2 analog control voltages | $0\ldots\pm10$ V/5 mA |

# 10 Thyristor Controlled Rectifiers

At this point a selection of equipment will be described; they all employ the principle of phase-angle control.


## 10.1 Type 48 V (60 V)/200 A

### 10.1.1 General and Application

The rectifier units (cabinet type assembly) serve to supply various communications systems such as EWSD, EWSP, EDX, EDS, KNS, ETS and EMS. Four rectifier units may be switched in parallel which thus cover the power consumption of the communications system and the battery up to 800 A.

The following rectifier units are essential:

- 48 V (60 V)/100 A GR12[1] and
- 48 V (60 V)/200 A GR12[1] (GR121).

Compared to the rectifier units GR12, the rectifier units GR121 have, additionally, a device for battery disconnection (battery undervoltage monitor A16 and battery disconnection contactor K10).

For the units GR12N and GR121N the interference voltage can be reduced even further (from 2.0 mV to 0.5 mV) by increased output filtering. This can be achieved by mounting supplementary filter capacitors (PABX using).

### 10.1.2 Modes of Operation

The rectifier unit can be used in the rectifier mode (see Sect. 3.1) or in standby parallel mode (see Sect. 3.3).

### 10.1.3 Survey Diagrams of the Power Supply System

Figure 10.1 shows a basic survey plan of a power supply system for a digital switching system EWSD. The example in the survey plan on Fig. 10.2 represents a power supply system with three rectifier units.

---

[1] *Source*: Siemens AG.

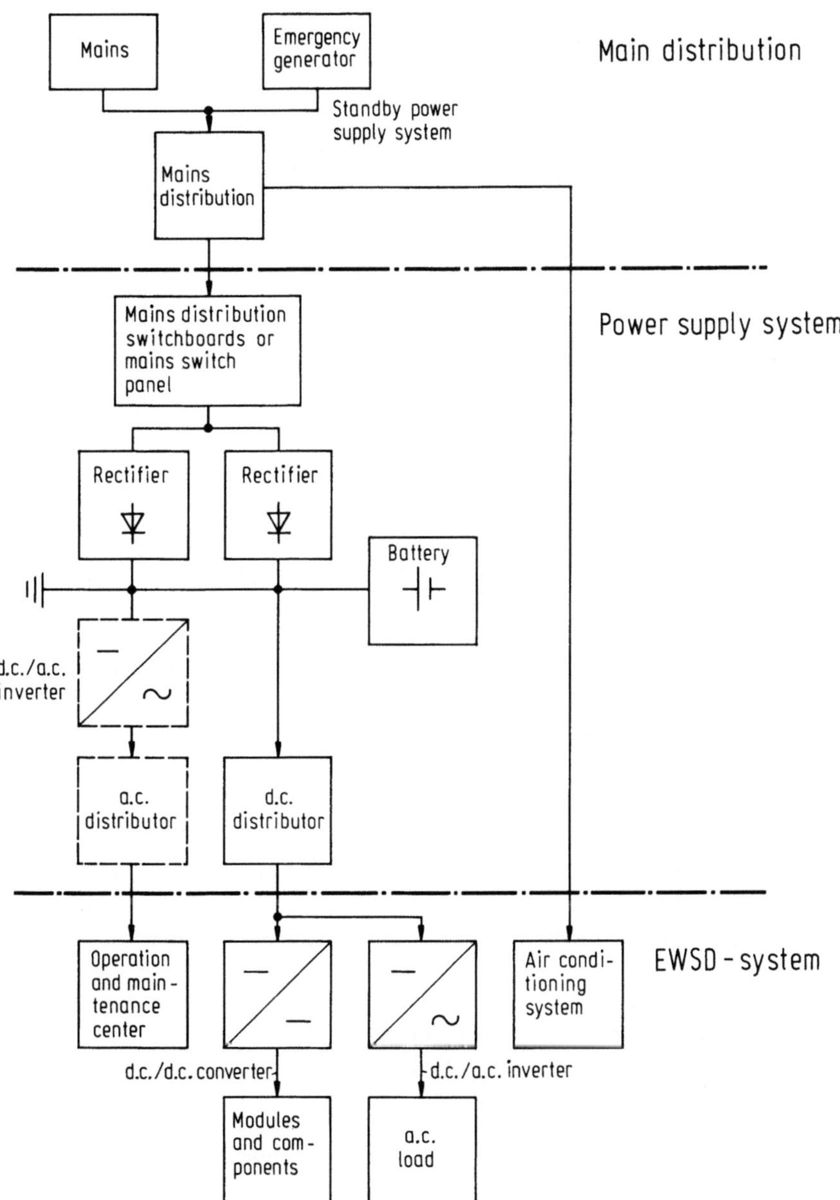


**Fig. 10.1.** Survey diagram for digital switching system EWSD

Table 10.1 shows the allocation of the rectifier units and switching panels depending on the nominal current of the power supply system. The methods indicated of combining the devices are suggestions requiring the precondition that in each case one rectifier unit is intended for battery charging and, at the

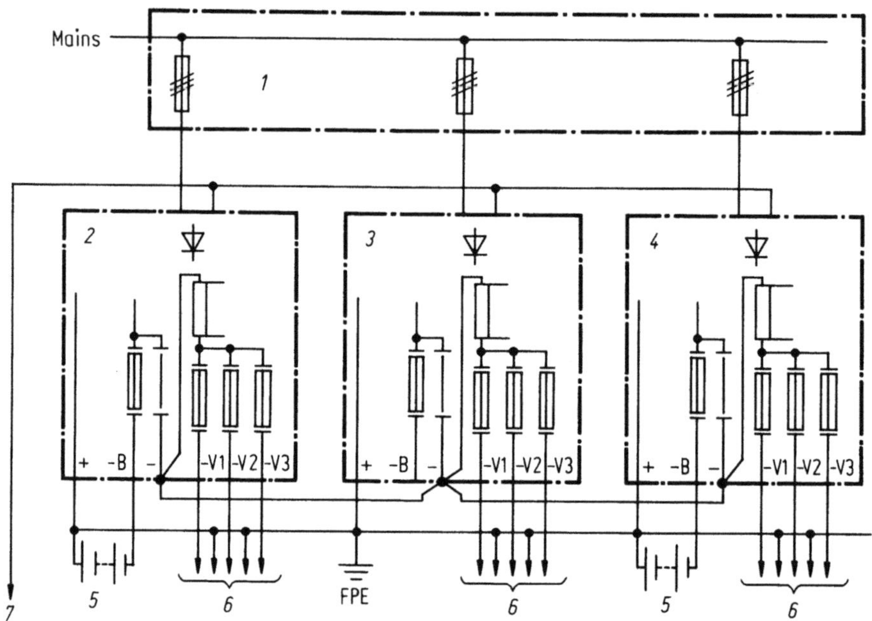

**Fig. 10.2.** Survey diagram for power supply system 48 V/400 A – standby parallel mode. *1* Mains distribution switchboard, *2* to *4* rectifier units 48 V/200 A GR12 with regulation, control, monitoring and measuring, *5* battery with e.g. 25 cells, *6* communications system (load), *7* remote signalling, FPE functional earthing and protective earthing conductor

same time, functions as redundancy. For reasons of safety, the total capacity of the batteries is divided into at least two groups; thus the number of needed battery panels is determined (at least two hours back-up time). Apart from the examples illustrated, further combinations (with devices for different current ratings) can also be effected.

### 10.1.4 Survey Diagram and Functioning Principle of the Rectifier Unit 48 V (60 V)/200 A

Figure 10.3 shows the survey diagram of the rectifier units GR12 (GR121). Module A16, together with the battery disconnecting contactor K10, is integrated only into the units GR121.

*Power section.* The a.c. supply voltage is fed via the input terminals U,V,W as well as via the mains contactor K1 to the main transformer T1 which performs the electrical separation of the communications system from the supply system and matches the level of d.c. voltage at the output of the thyristor set A1 (fully controlled three-phase bridge configuration). Here, at this output, a constant, regulated d.c. voltage is available. The thyristor set A1 performs the following tasks: rectifying and regulating the device output voltage.

**Table 10.1.** Application of the rectifier units and switching panels, depending on the nominal current of the power supply system

| Current consumption (A) | Rectifier units Consumer feeding and battery charging (max. 20%)/ spare unit (n+1) | | | | Battery switching panel | | Mains distribution switch-board | Mains switch panel | |
|---|---|---|---|---|---|---|---|---|---|
| | 200 A (BA 630 A) GR121 | 200 A GR12 | 500 A GR12 | 1000 A GR12 | 1500 A (1 Batt.) BF12 | 2000 A (1 Batt.) BF12 | 160 A NV12 | 600 A NF12 | 1000 A NF12 |
| 200 | 1 | 1 | | | | | 1 | | |
| 400 | 1 | 2 | | | | | 1 | | |
| 600 | 1 | 3 | | | | | 1 | | |
| 800 | | 4+1 | [2+1] | | 2 | | 1 | [1] | |
| 1000 | | 5+1 | [2+1] | | 2 | | 1 | [1] | |
| 1500 | | | 3+1 | | 2 | | | 1 | |
| 2000 | | | 4+1 | | 3 | | | 1 | |
| 3000 | | | 6+1 | | 3 | | | 1 | |
| 1500 | | | | 2+1 | | 2 | | | 1 |
| 2000 | | | | 2+1 | | 2 | | | 1 |
| 4000 | | | | 4+1 | | 3 | | | 1 |
| 6000 | | | | 6+1 | | 4 | | | 2 |
| 8000 | | | | 8+2 | | 5 | | | 2 |
| 10000 | | | | 10+2 | | 6 | | | 2 |

BA Battery disconnection unit, Values in the bracket [ ]=alternative

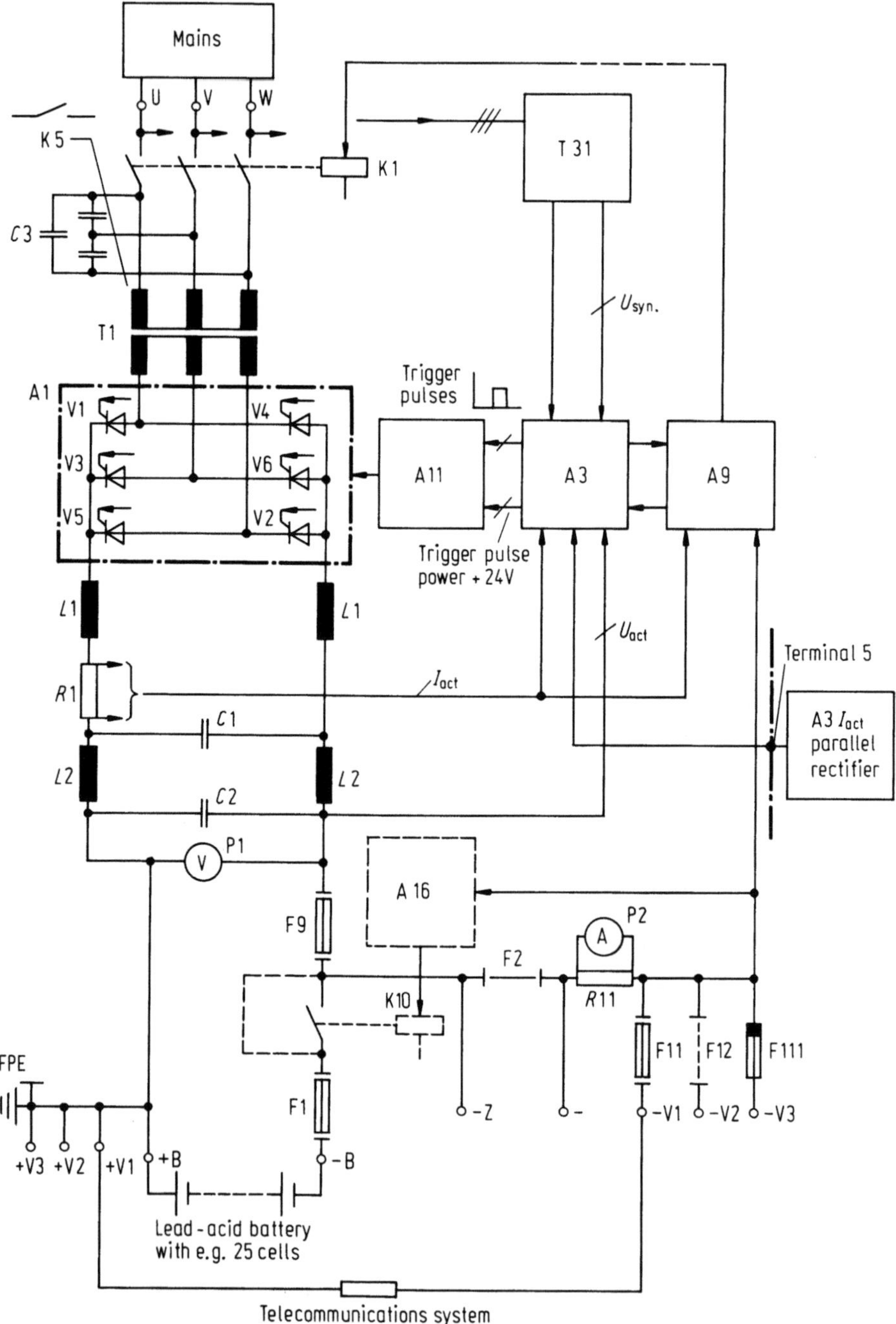


**Fig. 10.3.** Survey diagram for rectifier units 48 V (60 V)/200 A GR12 (GR121). A1 Thyristor set, A3 regulation module, A9 control module, A11 trigger pulse transformer module, A16 battery and thermo disconnection module, FPE function and protective earth-

By means of the filters $L1$, $L2$, $C1$ and $C2$ connected on the d.c. supply side the voltage is smoothed and the frequency weighted interference voltage is reduced below the maximum permissible level of 2 mV. Due to their ability of storing and giving off energy, these filters keep voltage distortions relatively low in cases of load surges. The measuring shunt $R1$ records the actual current value $I_{act}$ for current-limitation and overcurrent-cutout.

The battery is connected to -B via the battery fuse F1 with the equipment for float (trickle) charging of the 2.23 V/cell or charge of the 2.33 V/cell.

For the purpose of testing a changeable load resistor can be connected between $+V$ and $-Z$. F2 is a splitting strip (connector/disconnector). All rectifier units of the system are connected via the $-$ (Minus)-connection (compare this with Fig. 10.2). The terminals V (consumer) serve to connect the communications system (load) for the small consumers. The output voltage is measured at the voltmeter P1, whereas the consumer current is recorded at the amperemeter P2.

*Regulation, control and monitoring.* The auxiliary transformer T31 receives the a.c. line voltage and steps it down. T31 provides the auxiliary a.c. supply voltages and the synchronizing voltages for the regulation module A3. From here the trigger pulses are forwarded via the trigger pulse transformer module A11 to the thyristor set A1. In addition to the control units, such as the setpoint generator, voltage regulator, current limiter regulator and trigger set, the module A3 contains various supplementary and monitoring functions, for example:

- light-emitting diodes (LED) for the indication of operation and malfunction,
- current balancing,
- mains monitor,
- ripple contents monitor.

The enabling of the module regulation A3 is performed by the control module A9. During normal operation A3 supplies the trigger pulse transformer module A11 with $+24$ V (trigger pulse power supply).

The control module A9 is equipped with a display and operation panel (LEDs for load voltage and device current, as well as momentary-contact control switches). The following control commands are forwarded from module A9 to regulation A3:

- float (trickle) charging 2.23 V/cell,
- charging 2.33 V/cell,
- initial charging up to 2.7 V/cell.

---

**Fig. 10.3.** (*Continued*)
ing conductor, K1 mains contactor, K5 power factor correcting network contactor, three contacts (on/off), K10 battery disconnecting contactor, T1 main transformer, T31 auxiliary transformer, $C3$ power factor correcting network, $L1, L2, C1, C2$ output filter

The control module A9 also comprises various monitoring units, as for instance:

– undervoltage monitor,
– overvoltage monitor,
– current monitor,
– overcurrent cutout and
– short circuit cutout.

## 10.2  Type 48 V(60 V)/1000 A

Figures 10.4 and 10.5 illustrate the thyristor-controlled rectifier type 48 V/1000 A GR12.

### 10.2.1  General and Application

As to application, compare the following with the corresponding explanations in Sect. 10.1. Depending on the requirements it is possible to switch in parallel any number of rectifier units (usually up to 10 + 1 units). Besides the unit 48 V/1000 A GR12 there are also other variants such as 48 V/500 A GR12; 60 V/1000 A and 60 V/500 A GR12. Basically, these rectifier units are all built up in the same way (cabinet-type assembly) and they are set up together with battery switching

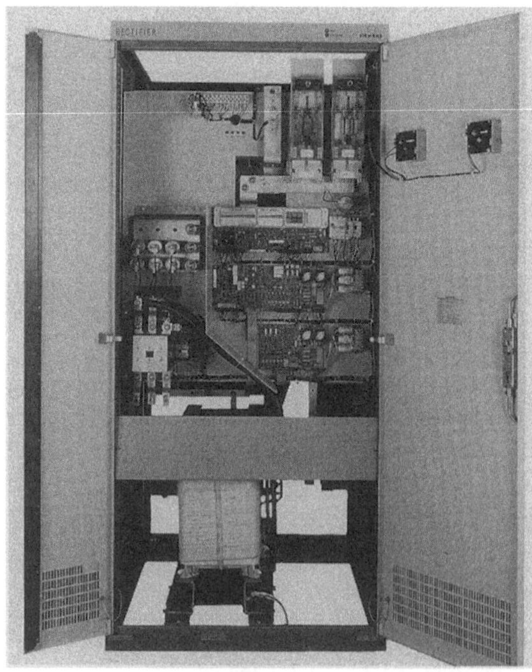

**Fig. 10.4.** Rectifier type 48 V/ 1000 A GR12 with doors open. (Photo by courtesy of Siemens AG)

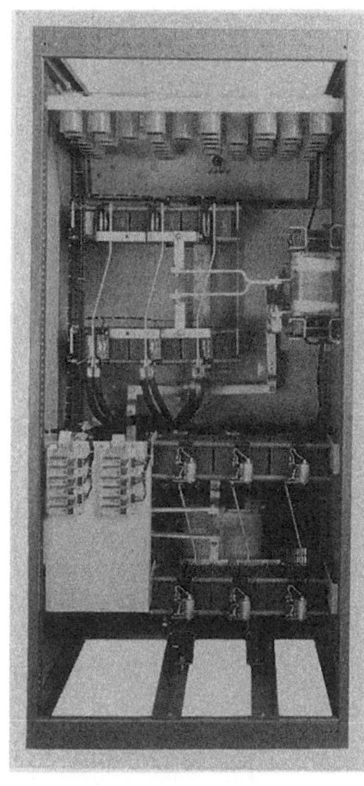

**Fig. 10.5.** Rectifier type 48 V/1000 A GR12, back view. (Photo by courtesy of Siemens AG)

panels BF12 2000 A or 1500 A. Additionally, one or two NF12 mains switch panels are required per system.

### 10.2.2 Modes of Operation

Here, too, the same modes of operation and processes as explained in Sects. 3.1 and 3.3 apply.

### 10.2.3 Survey Diagram of the Power Supply System

Figure 10.6 shows as an example the survey diagram of a power supply system with five rectifier units and three battery switching panels.

### 10.2.4 Survey Diagram and Functioning Principle of the Rectifier Unit 48 V (60 V)/1000 A

Figure 10.7 shows the survey diagram of rectifier unit 48 V(60 V)/1000 A GR12.

*Power section.* The power section of the rectifier unit 1000 A (500 A) GR12 is designed similar to that of the rectifier unit 48 V/(60 V)/200 A GR12 (see Sect. 10.1 and Fig. 10.3). However, unit 1000 A GR12 has an exceptional feature

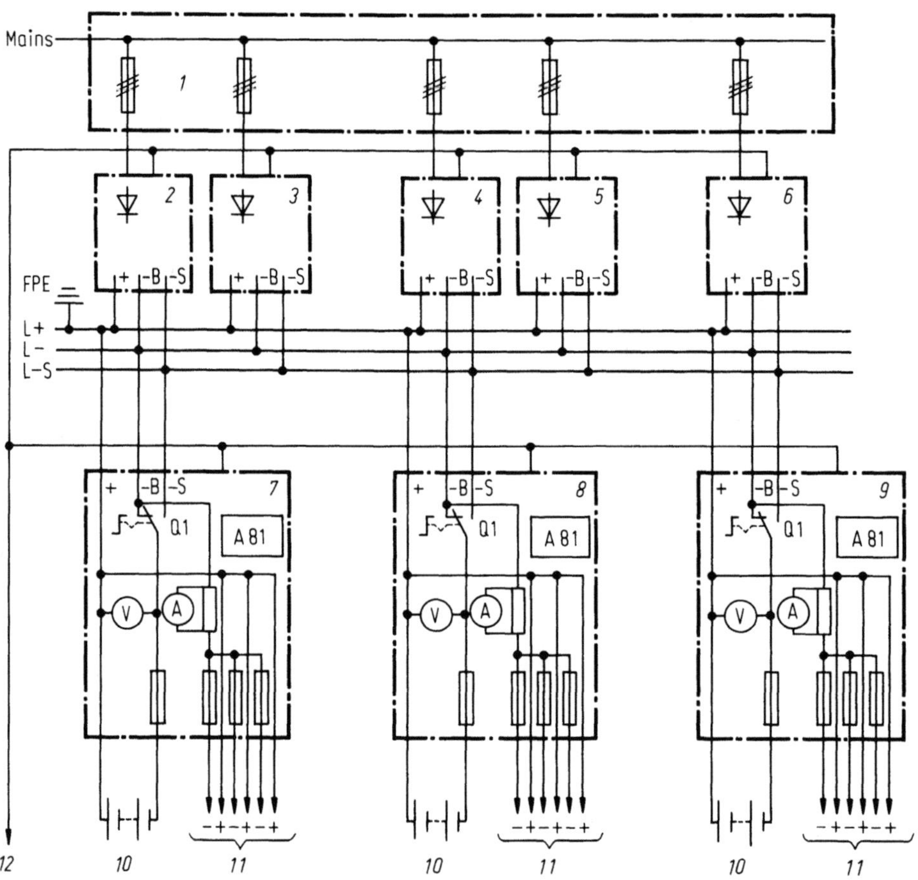


**Fig. 10.6.** Survey diagram of a power supply system for example the 48 V/4000 A system in standby parallel mode. *1* Mains switch panel 1000 A NF12, *2* to *6* rectifier units 48 V/1000 A GR12 with regulation, control, monitoring and measurement, FPE functional earthing and protective earthing conductor, *7* to *9* battery switching panels 48 V/2000 A BF12, Q1 battery switch, A81 fuse monitor, *10* lead-acid battery with e.g. 25 cells, *11* communications system (load), *12* remote signalling

compared to the unit 200 A, which is a double full-wave three-phase bridge configuration. The secondary side of the main transformer T1 is divided into two separate windings: a star winding and a delta winding. This causes a phase shift of each of the two a.c. voltages by 30°. Thus, the system disturbance as well as the interference voltage at the d.c. output are diminished.

Each of the two transformer windings feeds a thyristor set in full-wave three-phase bridge configuration, with load-side filter elements to the common output switched in parallel. The unit 500 A contains the thyristor sets A1 and A2, the variant 1000 A contains the thyristor sets A1.1, A1.2 and A2.1, A2.2. The bridge circuits are synchronized by means of the regulation module, so that a symmetrical current input and a 12 pulse system perturbation may be expected. Compared to a 6-pulse circuit, this way of functioning thus eliminates the fifth and seventh harmonic on the line-side, so that only the eleventh and thirteenth are involved in any distortion of the supply, approximately 9% or 7.7% of the fundamental current.

The d.c. output circuit of the unit 1000 A is built up in a less complicated way because here the circuit distribution is a part of the battery switching panel.

*Regulation, control and monitoring.* In addition to the modules indicated in Fig. 10.3 and Sect. 10.1, (i.e. the 200 A unit) the rectifier unit 1000 A (see Fig. 10.7) contains the modules A23 (12-pulse-regulation) and A21 (trigger-pulse transformer). Module A23 is used together with the regulation module A3 and allows a 12-pulse operation of the two thyristor sets in full-wave three-phase bridge configuration.

Module A23 transmits trigger pulses via module A21 to the thyristor set (A2.1, A2.2) and receives, amongst others, the following functional units and monitors:

– current balancing regulator (for current balancing between the two full-wave three-phase bridge configurations),
– trigger set,
– ripples contents monitor,
– overcurrent cutout monitor and
– short circuit cutout monitor.

The trigger pulse transformer module A21 is designed similar to that of A11. The second auxiliary transformer T32 is designed as T31.

## 10.3 Assemblies

### 10.3.1 Regulation Module A3

The regulation module A3 (Fig. 10.8, see also Fig. 8.12) is used for example in the rectifier unit GR12. The complete regulation for a thyristor set in full-wave three-phase bridge configuration (6-pulse) is housed in the regulation module A3. In addition, this module performs monitoring functions and also contains light-emitting diodes which indicate operation and malfunction which in turn monitor the most important functions and signals (Fig. 10.9).

### 10.3.2 Control Module A9

The control module A9 (see Figs. 10.10 and 10.11) is used for example in the rectifier units GR12. It contains all the control and monitoring functions. The

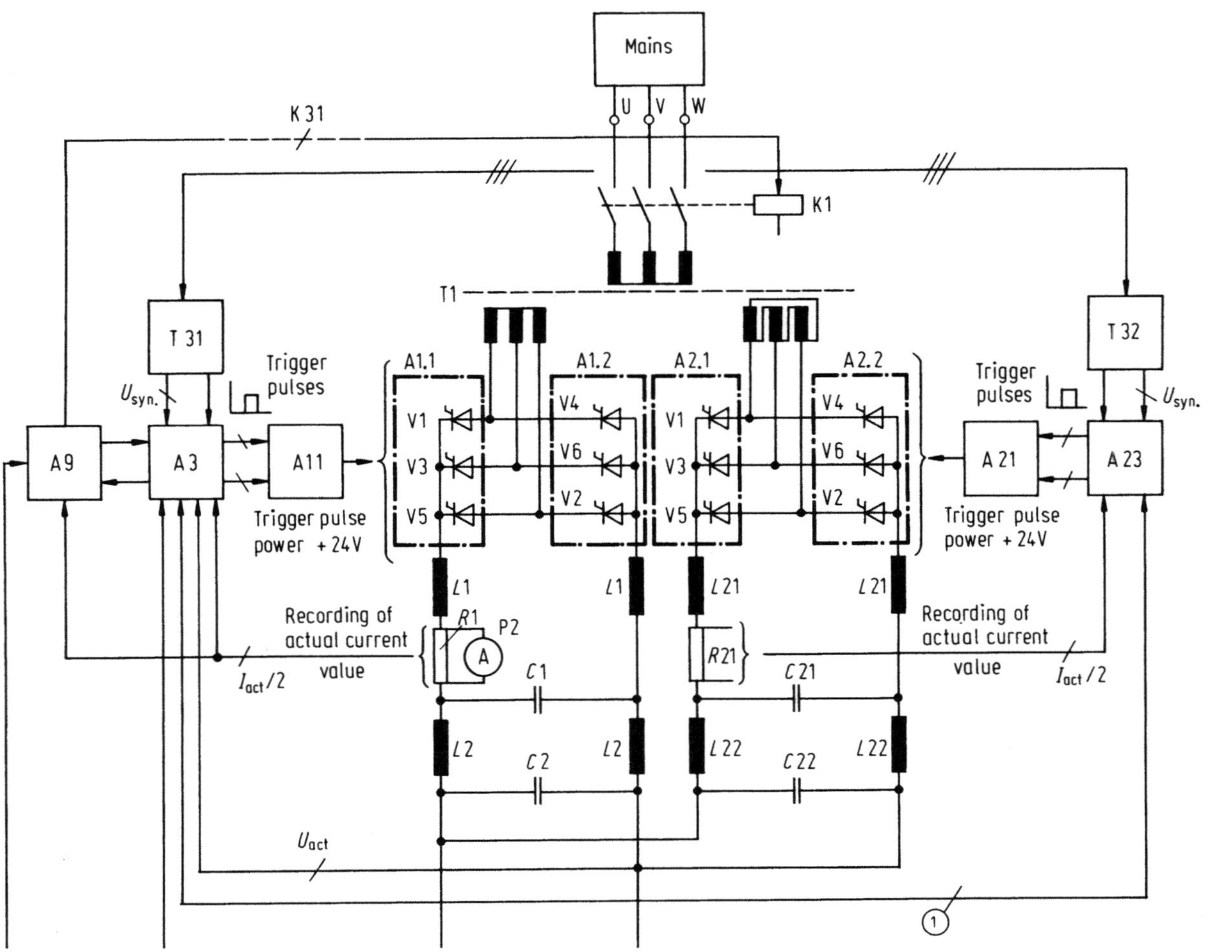

Mains
U V W
K 31
K1
T 31
T1
T 32
U_syn.
Trigger pulses
U_syn.
A9
A3
A11
A21
A23
Trigger pulse power + 24V
Trigger pulses
Trigger pulse power + 24V
A1.1
V1 V3 V5
A1.2
V4 V6 V2
A2.1
V1 V3 V5
A2.2
V4 V6 V2
L1
L1
L21
L21
Recording of actual current value
I_act / 2
R1 P2
A
R21
Recording of actual current value
I_act / 2
C1
C21
L2
C2
L2
L22
C22
L22
U_act

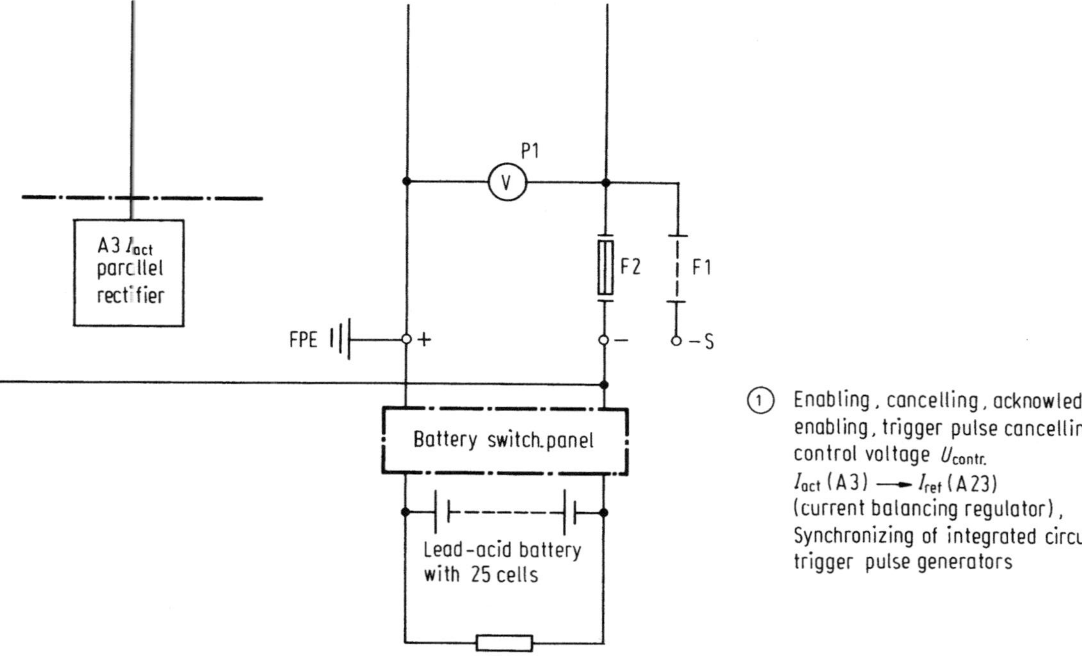


**Fig. 10.7.** Survey diagram of rectifier units 48 V(60 V)/500 A and 1000 A GR12. *1* Enabling, cancelling, acknowledging enabling, trigger pulse cancelling, control voltage $U_{contr.}$, $I_{act.}$(A23) → $I_{ref.}$(A23) (current balancing regulator), synchronizing of integrated circuits of trigger pulse generators. T1 main transformer, T31, T32 auxiliary transformer, FPE function and protective earthing conductor, K1 mains contactor, K31 auxiliary mains contactor, A1.1/A1.2 thyristor set (500 A rectifier A1), A2.1/A2.2 thyristor set (500 A rectifier A2), A3 regulation module, A9 control module, A11, A21 trigger pulse transformer module, A23 12-pulse regulation module; *L1, L2, C1, C2 L21, L22, C21, C22* output filtering

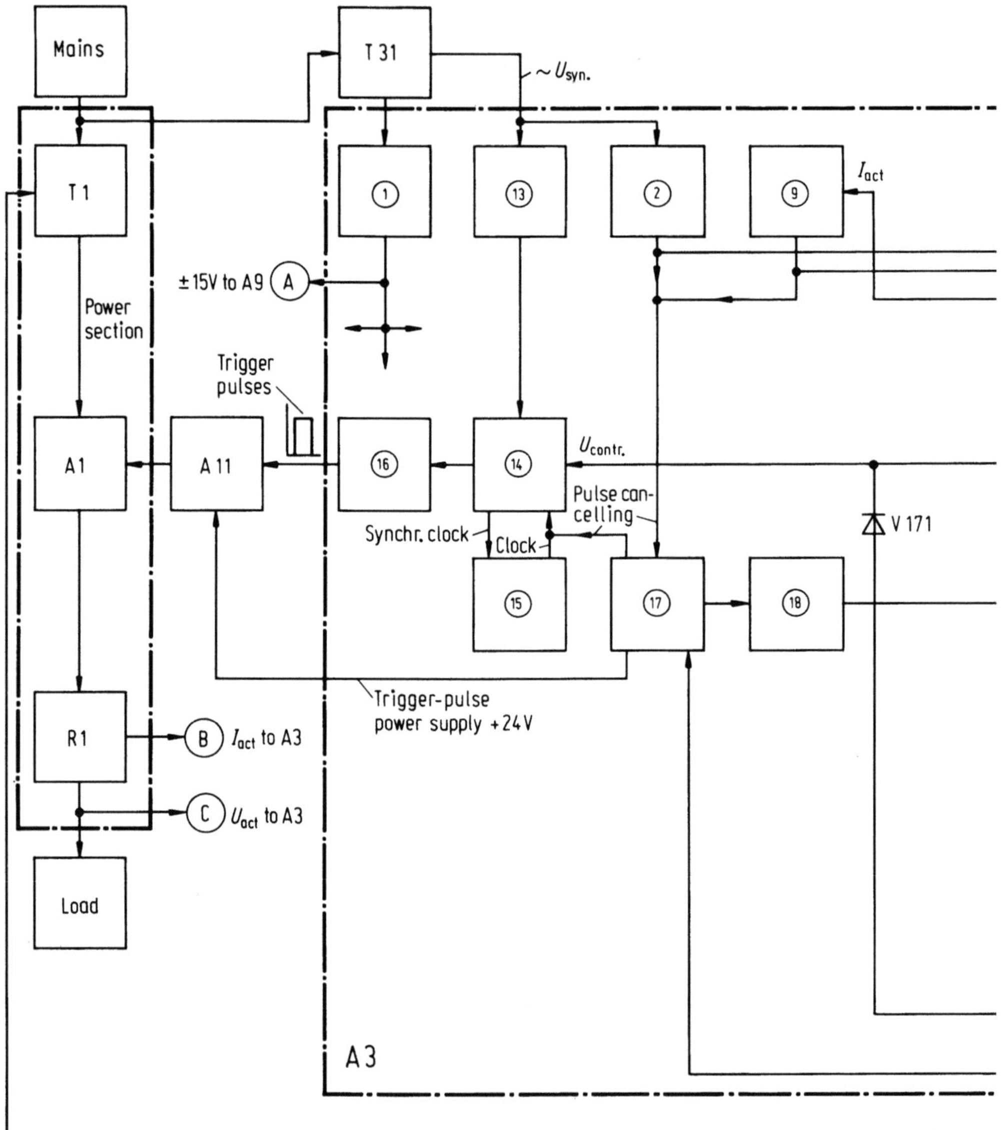



**Fig. 10.8a.** Block diagram of the regulation module A3. T1 main transformer, T31 auxiliary transformer, $R1$ shunt-resistor, A1 thyristor set, A3 regulation module, A9 control module, A11 trigger-pulse transformer module, A23 12-pulse-regulation module, $U_{\mathrm{syn}}$ synchronizing a.c. voltage, $U_{\mathrm{act}}$ actual voltage value, $U_{\mathrm{ref}}$ reference voltage value, $U_{\mathrm{contr.}}$ control voltage, $I_{\mathrm{act}}$ actual current value, $I_{\mathrm{ref}}$ reference current value, K1 mains contactor; *1* Internal power supply: $+/-15$ V stab., $+ 10$ V stab., $+24$ V not stab.; *2* mains monitor (mains voltage supervision and phase cut supervision): $U_{\mathrm{mains}} > 15\% \, U_{\mathrm{mains-nom}}$, $U_{\mathrm{mains}} < 19\% \, U_{\mathrm{mains-nom}}$; *3* current balancing; *4* slope in characteristic (not used in GR12) for diesel generator-operation max. $\pm 3\%$; *5* $I_{\mathrm{act}}$ amplifier and matching; *6* current monitoring (no relevance in rectifiers GR12): $I < 5 \% \, I_{\mathrm{nom}}$ and $I > 95\% \, I_{\mathrm{nom}}$; *7* cur-

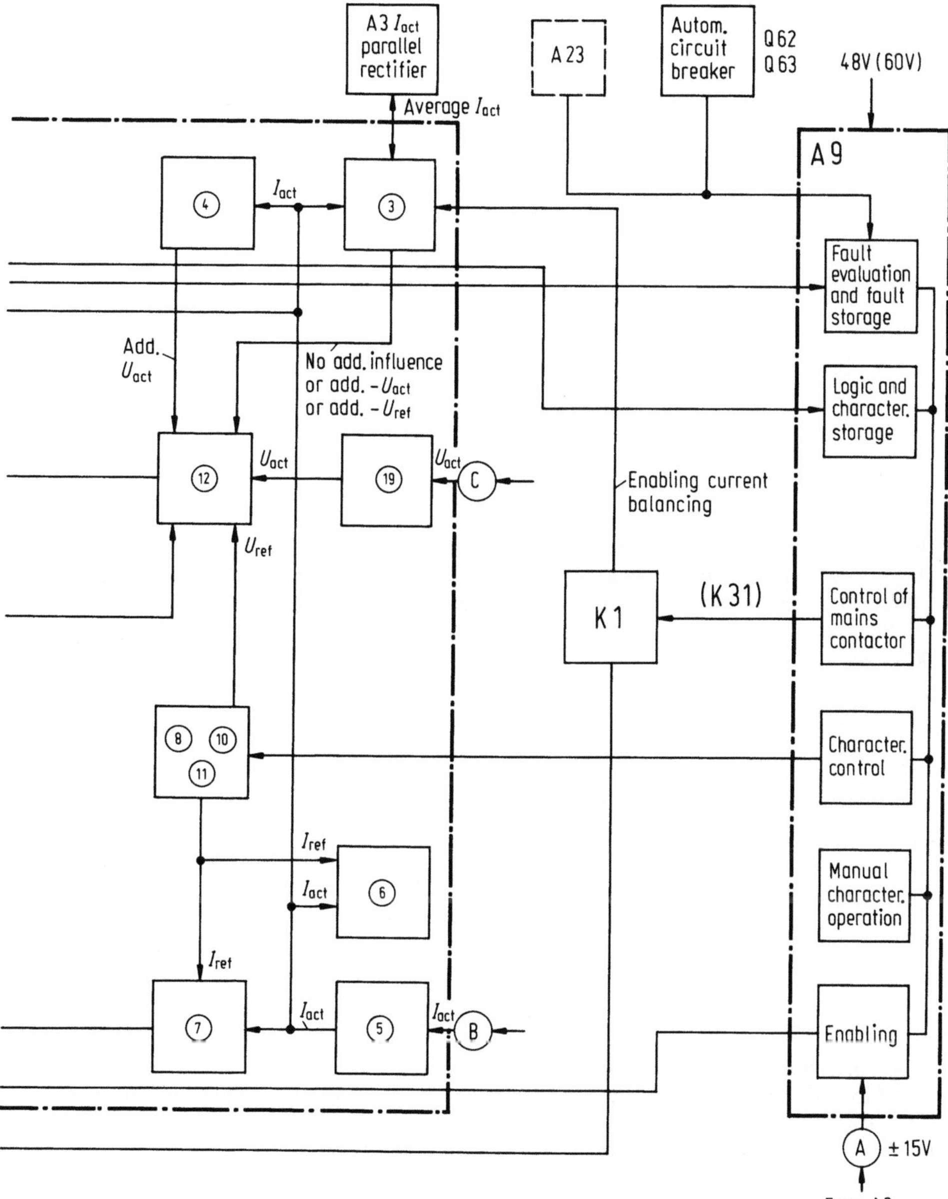


rent limiter regulator; *8* $I_{ref}$ value selector (set point device); *9* ripple contents monitor (The ripple contents monitor detects elevated, superimposed a.c. voltages as occuring for instance in the event of thyristor failure, trigger-pulse failure or trigger control set failure in the first filtering circuit. The monitor responds if the actual superimposed a.c.-voltage exceeds a certain fixed limit value.): $I \sim$ max. 15% $I_{nom}$ or $I \sim$ max. 25% $I_{nom}$ (adjustable); *10* characteristic control; *11* $U_{ref}$ value selector (set point device); *12* voltage regulator; *13* $U_{syn\text{-}filter}$ 60°; *14* trigger control set with integrated circuits of trigger pulse generators; *15* 7-kHz clock generator; *16* trigger-pulse amplifier; *17* pulse cancelling/pulse enabling; *18* rise; *19* $U_{act}$ matching and adjustment

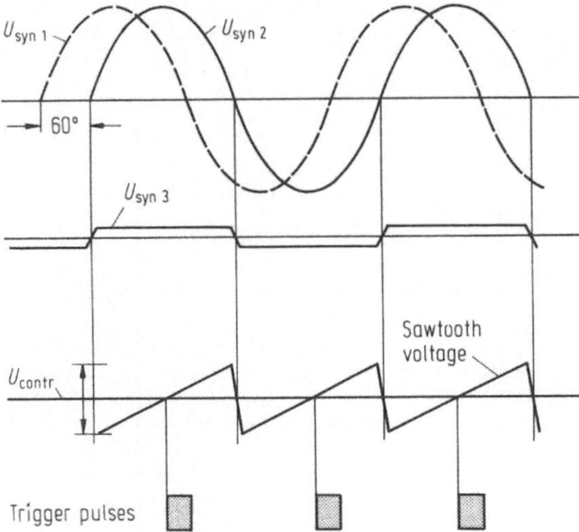



**Fig. 10.8b.** Phase shifting of the synchronization voltage and generating of trigger pulses. $U$syn. a.c. synchronization voltage, $U$syn.1 from transformer T31, $U$syn.2 after $U$syn. -filter 30°, $U$syn.3 input of trigger pulse generator, $U$contr. control voltage at the output of the voltage regulator or current limiter regulator or at the input of the trigger pulse generator

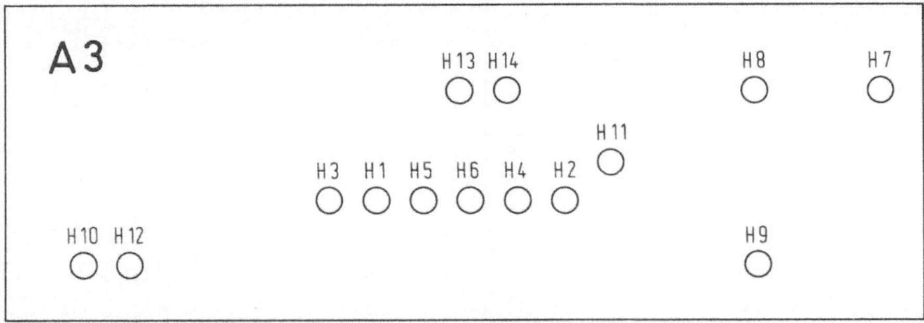



**Fig. 10.9.** Positions and functions of LED regulation module A3. H1 to H6 Trigger pulses, H7 P15 (+15 V) stab., H8 N15 (−15 V) stab., H9 enabling, H10 ripple contents monitor, H11 trigger pulse cancelling (not used in GR12), H12 mains monitor, H13 $I < 5\%\ I_{nom}$. H14 $I > 95\%\ I_{nom}$ no relevance for GR12

following are the main functions integrated into the control module:

– operation of the unit,
– characteristic control,
– timing,
– current and voltage indication,
– monitoring functions and
– signalling.

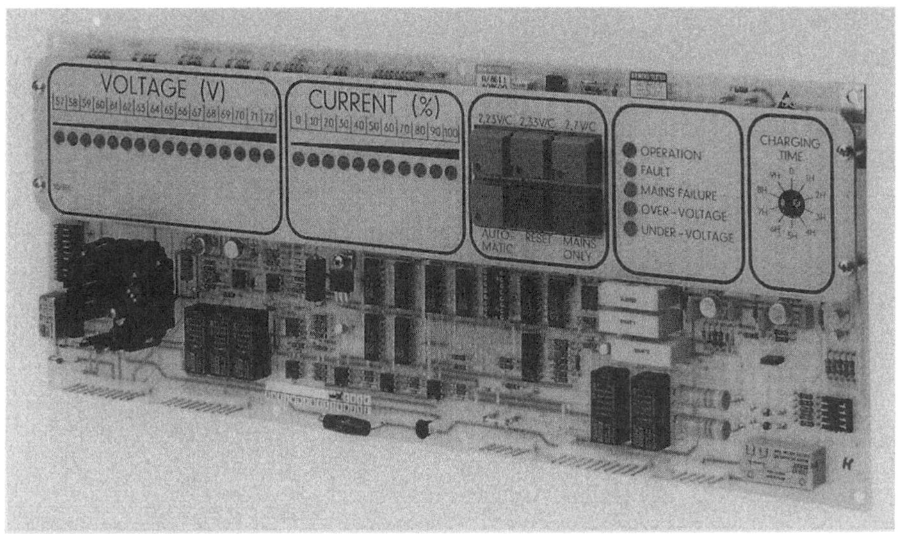

**Fig. 10.10.** Control module A9. (Photo by courtesy of Siemens AG)

Figure 10.12 shows the display and operation panel with the light-emitting diodes of module A9.

For rectifier units operated in the standby parallel mode with batteries it is also necessary to preserve certain control- and monitoring functions in cases of power failure. For this reason the internal power supply ($1$) of module A9 (see Fig. 10.11) operates not only with the line-voltage-dependent supply voltages P15 ($+$ 15 V) and N15 ($-$15 V), which it receives from the regulation module A3, but also with the supply voltages N24 ($-$24 V) and N10 ($-$10 V), formed via in-phase regulators out of the load voltage (battery voltage).

The shunt resistor $R1$ located in the power section provides the actual current value of 30 mV up to 60 mV for the nominal current to the matching $I_{act}$ ($2$) (see Fig. 10.11). The actual current value $I_{act}$ is matched to the shunt resistors of the respective rectifier unit by means of auxiliary switches (DIP-FIX).

In a rectifier unit, where the output voltage increases uncontrollably due to a defective control unit, the current consequently reaches excessively high value. During automatic operation this circumstance allows the elimination of the defective unit without influencing the other units operating in parallel. For this, the criteria overvoltage, $I > 80\% I_{nom}$ as well as automatic operation are evaluated and a (locked) shutdown of the respective unit is caused (*current monitor* $I > 80\% I_{nom}$ ($3$) (see Fig. 10.11)).

In case the current limiter regulator on module regulation A3 is faulty, a two-stage *overcurrent cutout* ($3$) is provided additionally in order to protect the rectifier unit. During the selection of one of the two overcurrent cutout mechanisms, either for $I > 110\% I_{nom}$ (delay approximately 100 ms) or $I > 170\% I_{nom}$

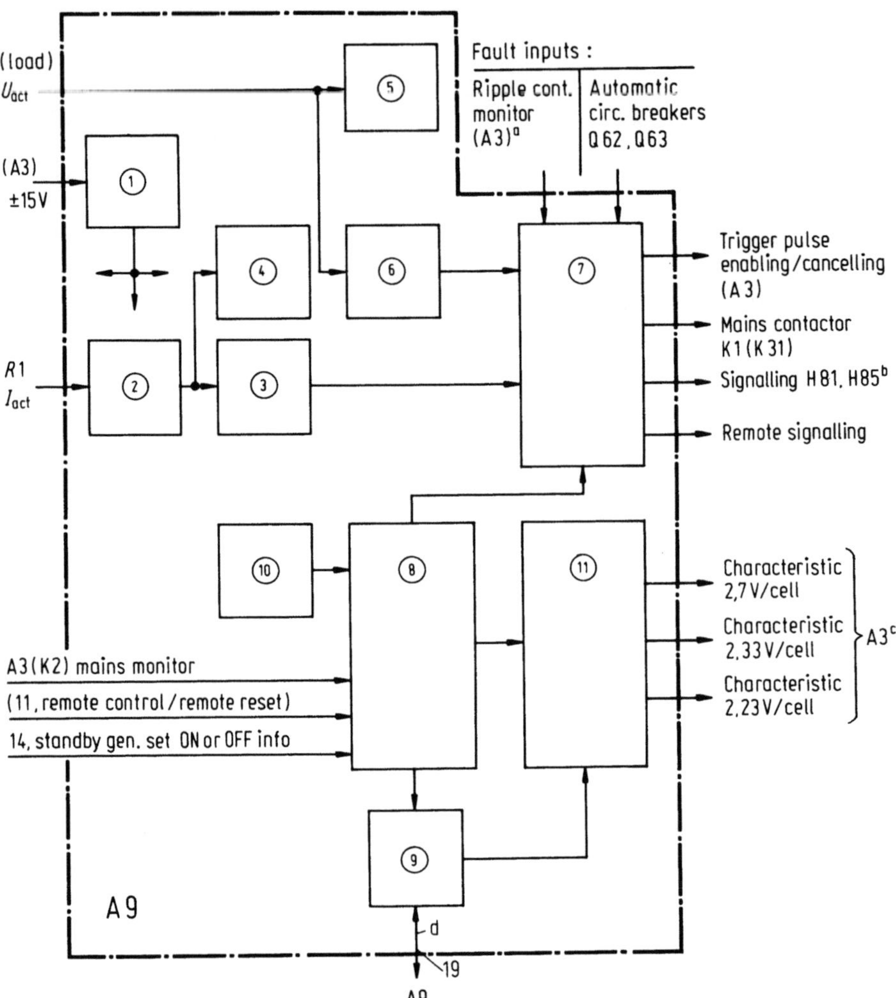


**Fig. 10.11.** Block diagram of the control module A9. *1* Internal power supply $-10$ V stab. and $-24$ V stab., *2* $I_{act}$ amplifier and matching, *3* current monitor $I > 80\%$ $I_{nom}$, overcurrent cutout $I > 110\%$ and short circuit cutout $I > 170\%$ $I_{nom}$, *4* current display (LED) 0 to 100 % $I_{nom}$, *5* voltage display (LED) 45 V to 60 V or 57 V to 72 V, *6* undervoltage monitor $U < 45$ V or 57 V and overvoltage monitor $U > 56$ V, 60 V or 68 V, *7* fault evaluation and fault storage, control of mains contactor incl. switching on delay 1 to 15 sec., signalling, *8* logic control with characteristic control storage, mains failure duration measurement (charging instruction) 3,6 or 9 min and charging time 0 to 9 h, *9* synchronizing of characteristic (charging time synchr.), *10* manual characteristic operation (buttons), *11* characteristic control with display, [a]In application GR12 500 A and 1000 A additional ripple contents monitor (A23), [b]Signalling LED (top panel of the equipment) OPERATION (H81) FAULT (H85), [c]Characteristic control to regulation module A3, [d]From parallel connected rectifier units (A9)

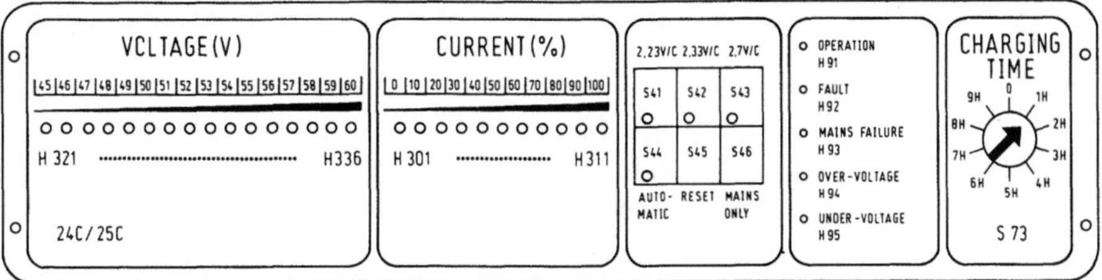

**Fig. 10.12.** Display and operation panel of the control module A9 with LED for certain functioning principles regarding states of operation on control module A9 for 48-V-units (24 or 25 cells of lead acid batteries). *Remark*: The rear side of the metal sheet is designed for 60 V operation (30/31 cells). H321 to H336 LED-voltage $V_{\text{load}}$, H301 to H311 LED-current $I_{\text{output}}$, S41, S42, S43, S44 control buttons with LED, S45, S46 control buttons without LED, S73 switch charging time 0 to 9 h

(delay approximately 10 ms), the rectifier unit is shut down and the signalling system switched on.

An eleven-step LED array H301 to H311 serves as *current display* (*4*). Thus, the current range rating from 0 to 100% is covered in 10% steps.

A sixteen-step LED array H321 to H336 serves as *voltage display* (*5*), so that voltages from either 45 V to 60 V or 57 V to 72 V may be indicated.

If the load voltage gets below $U < 45$ V or 57 V the module has to release a corresponding message for undervoltage. The LED H95 indicates this visually by 'UNDER VOLTAGE' and starts the remote signalling. The delay interval of *undervoltage monitor* (*6*) is approximately 700 ms.

If the load voltage exceeds $U > 56$ V, 60 V or 68 V, the unit is switched off and an 'OVER-VOLTAGE' message is released by means of LED H94 and a remote signalling is started. The delay interval of *overvoltage monitor* (*6*) amounts to approximately 300 ms.

The following criteria are evaluated in the *fault evaluation and fault storage* (*7*):

- $U > 56$ V, 60 V or 68 V for automatic operation simultaneously with $I > 80\%$ $I_{\text{nom}}$,
- $I > 110$ % or $> 170\%$ $I_{\text{nom}}$,
- response of ripple contents monitor on module A3 (A23),
- operation of an automatic circuit-breaker,
- reference trigger-pulse,
- reset,
- remote reset and
- operation of standby generating set.

The selection of the mains contactor, too, is performed by this functional unit after expiration of a programmable switching on delay (adjustable from 1 to 15 s). Thus, the switching on of the rectifier units can be staggered, so that the inrush currents impose the smallest possible load on the mains or on the emergency generator. The trigger pulses are also enabled or disabled and the complete signalling or remote signalling mechanism is started.

The *logic control* (*8*) processes the input signals of the manual characteristic selection and the signals of the mains monitor. It controls the characteristic curve memory and, during automatic operation, the supervision of the power failure interval and charging time.

The *characteristic control storages* (*8*) serve to store the characteristics selected by the control logic until it is altered by hand or automatically (characteristic control changeover). There are three characteristic storages: for the 2.7 V/cell, the 2.33 V/cell and the 2.23 V/cell.

If the power failure lasts less than the programmable period of 3, 6 or 9 min (blocking time) the rectifier unit is reset to float (trickle) charging (normal operation) 2.23 V/cell after mains recovery and a charge of 2.33 V/cell is prevented. If, on the other hand, the outage lasted longer than 3, 6 or 9 min, the unit is switched to 2.33 V/cell after mains recovery. The length of the charging period depends on the setting of the coding switch S73 'CHARGING TIME' which can be set between 0 to 9 hours (1 hour steps). Normally, the charging time is set between 1 to 9 hours. Only for systems with lead-acid batteries having fixed electrolytes i.e. valve-regulated lead-acid batteries (maintenance-free batteries) is charging prevented by setting the timer to 0 hours.

When operating the power supply, the circuit of logic of the module has to be set in a specified state. For this, a reference trigger pulse generator is used (not shown in Fig. 10.11).

In order to ensure that for parallel units operated in the automatic mode all rectifiers work with the same characteristic, the module controls A9 have to be interconnected via a synchronizing link. In this way, all units switch back simultaneously from charging the 2.33 V/cell to the 2.23 V/cell when the 'last electronic timing relay' on the A9 module in the rectifier units has expired (synchronizing of characteristic (*9*)) (see Fig. 10.11).

The *logic control with characteristic control storage* (*8*) is also influenced by the *manual characteristic operation* (*buttons*) (*10*).

In manual operation, the automatism for the charging period and its startup is inactive. All specifications for this mode of operation must be entered manually, including the characteristics, which are set with the same buttons.

– S41   2.23 V/cell parallel mode/float (trickle) charging and
– S42   2.33 V/cell parallel mode/charging.

The characteristic up to 2.7 V/cell (S43) for the initial charging of lead-acid batteries can only be set during manual operation, since the load must be disconnected for this due to the high output voltage; during automatic operation this characteristic is disabled.

If the button 'AUTOMATIC' S44 is switched off, the units operate in the rectifier mode (LED in the button extinct). If the button S44 is on, there is parallel operation mode/automatic operation (LED in the button lights up).

The button S45 ('RESET') serves to unblock a locked shutdown of the unit in case an error occurs. The same effect can be achieved by applying a short-term earth potential ($M \hat{=} 0\,V$) to the remote input. If a constant earth potential ($M \hat{=} 0\,V$) is applied at this input, the unit remains blocked.

If the capacity of the standby generator is not sufficient for all power supply system units, the cut-in priority for the individual units during operation of the standby generator is set with the button 'MAINS ONLY' S46. This requires a potential-free contact (lock for standby operation) from the standby generator control for the corresponding operation of the rectifier unit (earth potential). If the button is ON, the unit can be switched on only for mains operation. If the button is OFF, the unit can be switched on for both, mains and standby operation.

### 10.3.3 Trigger-Pulse Transformer Module A11 (A21)

The trigger pulse transformer module A11 (A21) (not illustrated) is used for example in the rectifier unit GR12. Trigger-pulse transformers T1 to T6 are employed for the selection of each thyristor.

The primary winding of the trigger-pulse transformer carries a constant voltage of $+\,24$ V (P24) through a dropping resistor. When operating the trigger pulse (having M-potential) this winding carries a current which generates a pulse of approximately 2 V on the secondary side and fires the thyristor via the circuit gate with an auxiliary cathode. Thus, the trigger-pulse transformer serves to match the trigger-pulse and for galvanic isolation of the regulation from the power section. Resistors, diodes and capacitors help to prevent the thyristors from being fired by external pulses.

### 10.3.4 Battery- and Thermo Disconnection Module A16

The battery- and thermo disconnection module A16 (Fig. 10.13) is used for example in various power supply cabinets and thyristor-controlled rectifier units (e.g. GR121).

The battery cutoff contactor K10 together with A16 ( the battery undervoltage monitor) serves to protect the lead-acid battery of a possible final discharging voltage.

If the battery voltage is lower than 42 V or 54 V, the operational amplifier N1 (not shown in Fig. 10.13) disconnects the battery from the telecommunications system via the K11 relay and K10 contactor. In this way, the battery is protected against deep discharging. With the potentiometer R16 the response threshold is set.

In case the rectifier unit takes over the system supply again and if the load voltage exceeds 44 V or 56 V, N1 automatically reconnects the battery to the system via the K11 relay and K10 contactor.

The thermo disconnection is intended for air-conditioned systems (e.g. container). In order to protect the exchanges against too high temperatures the power

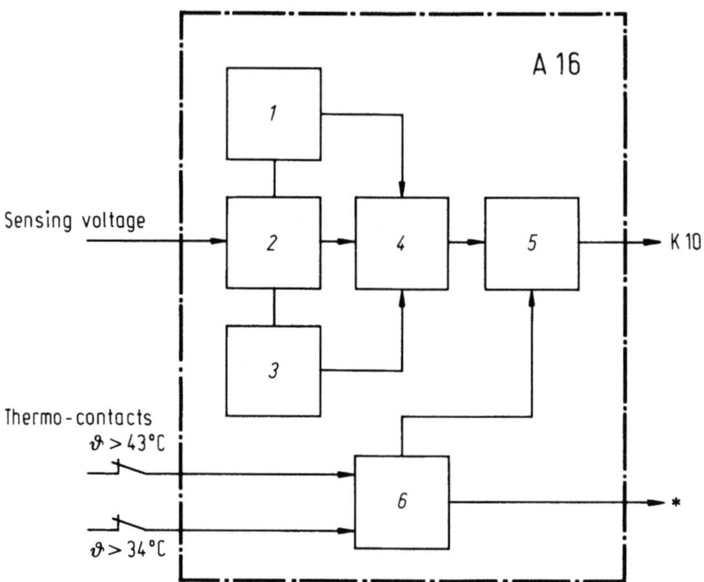



**Fig. 10.13.** Block diagram of the battery and thermo disconnection A16 *1* Reference voltage source, *2* internal power supply −24 V stab., *3* reference value potentiometer *R*16, *4* battery undervoltage monitor, *5* relay circuit K11, *6* thermo disconnection, K10 battery disconnection contactor, * trigger pulse cancelling signal to all rectifier units

supply (battery and rectifier unit) can be switched off via external temperature sensors.

This thermo disconnection is activated by appropriate selection of the auxiliary switch S1 open, S2 closed (on the A16 module) (not shown in Fig. 10.13). For a temperature of > 43°C the power supply is switched off and for a temperature lower than 34°C it is switched on again, in case a line voltage exists.

### 10.3.5 12-Pulse-Regulation Module A23

The 12-pulse-regulation module A23 (see Fig. 10.14) is implemented in the rectifier units with 500 A and 1000 A nominal current (e.g. GR12) together with the regulation module A3. It allows 12-pulse operation of the two parallel switched thyristor sets (fully-controlled-three-phase bridge configuration). The required three-phase voltages, shifted by 30° provide the two secondary windings (a star winding and a delta winding) of the main transformer T1.

The operation mode of the module A23 is similar to that of the regulation module A3 (see Sect. 10.3.1). The only difference concerns the additional 30° phase shift of the synchronizing a.c. voltage. For this, the voltages provided by T32 are interconnected in a triangle configuration after the 60° filtering. Thus, the integrated trigger-pulse generators N200, N300 and N400 (not shown in Fig.

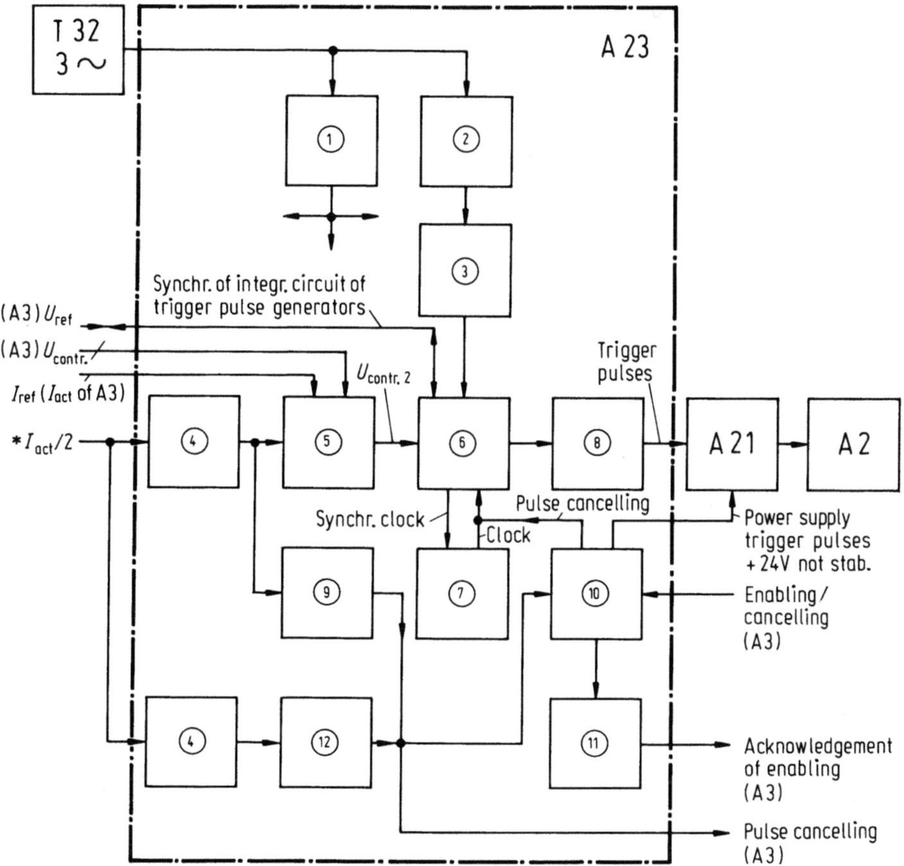



**Fig. 10.14.** Block diagram of the 12-pulse regulation module A23. *1* Internal power supply P (+15 V stab.), N (−15 V stab.) and P 24 (+ 24 V not stab.), *2* $U_{syn}$-filter 60°, *3* phase shifting 30°, *4* $I_{act}$ amplifier and matching, *5* current balancing regulator, *6* trigger control set with integrated circuits of trigger-pulse generators, *7* 7 kHz clock generator, *8* trigger-pulse amplifier, *9* ripple contents monitor, *10* pulse cancelling/pulse enabling, *11* enabling, *12* overcurrent cutout ($I > 110\% I_{nom}$, $I > 170\% I_{nom}$), A21 trigger-pulse transformer, A2 thyristor set, * shunt-resistor in the power section $R21$ (e.g. 1000 A GR12 rectifier units)

10.14) of the trigger control set receive synchronizing voltages shifted by 30° compared to the trigger control sets on the module A3.

The pulse enabling of the module A23 is linked with the pulse enabling of the regulation module A3. This ensures that either both regulations release trigger pulses, or that all trigger pulses are blocked.

The same is valid for the ripple contents monitor. Should one of the two monitors (on the A3 or A23 module) respond, the pulses on both modules are blocked.

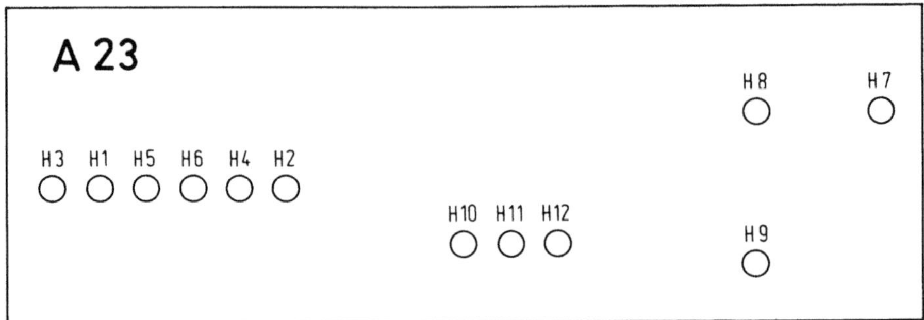


**Fig. 10.15.** Positions and functions of LED-12-pulse-regulation module A23. H1 to H6 Trigger pulses, H7 P15 (+15 V) stab. H8 N15 (−15 V) stab., H9 enabling, H10 ripple contents monitor, H11 on-indicator (normal operation), H12 overcurrent cutout

**Table 10.2.** Technical data for rectifier type 48 V (60 V) / 200 A, GR12 (GR121)

| Main input supply | | | |
|---|---|---|---|
| Voltage (V) | $3 \times 380/400/415$ or $3 \times 220/230$ $+10\% - 15\%$ | | |
| Frequency (Hz) | 50 or 60 $\pm$ 5% | | |
| Fuse protection in mains (A) distribution switch board | 63 (at 220V) 35 (at 380V) | | |
| Degree of radio interference | A (VDE 0878) | | |

| D.C. output | | | |
|---|---|---|---|
| Operating mode or condition | Rated direct voltages (V) (tolerances for 1 to 100% rated current | | |
| | **48 V Systems** | | **60 V Systems** |
| | Equipment voltage for lead-acid batteries with ... cells $\hat{=}$ load voltage (except for initial charging with communications system disconnected) | | |
| | 25 cells | 30 cells | 31 cells |
| Rectifier mode | 51 $\pm$ 0.5% | 62 $\pm$ 0.5% | 62 $\pm$ 0.5% |
| Parallel operation/float (trickle) charging (2.23 V/cell) | 56 $\pm$ 0.5% | 67 $\pm$ 0.5% | 69 $\pm$ 0.5% |
| Parallel operation/charging (2.33 V/cell) | 58.5 $\pm$ 0.5% | 70 $\pm$ 0.5% | 72.5 $\pm$ 0.5% |
| Initial charging | 67.5 (communications system disconnected) | | 84 (communications system disconnected) |
| Rated direct current (A) | 200 | | |
| Interference voltage (mV) | $\leq$ 2 (frequency-weighted with CCITT A-filter) | | |
| Dimensions (H×W×D)  (mm) | cubicle construction 2000 $\times$ 600 $\times$ 600 | | |

The actual current value $I_{act}$ on regulation module A3 serves as reference value $I_{ref}$ for a current balancing between the two fully controlled three phase bridge configurations. For this, that value is compared to the processed actual current value on the A23 module .

The d.c. control voltage supplied from the regulation module A3 is added to the error of the current balancing regulator and then fed into the integrated pulse generators N200 to N400 of the trigger control set (control voltage $U_{contr.\ 2}$).

The module A23 contains an overcurrent cutout mechanism similar to the control module A9 (see Sect. 10.3.2).

The LED on module A23 (see Fig. 10.15) have a similar function as those on module A3 (see Sect. 10.3.1, Fig. 10.9). The LED H13 and H14 are not to be found on module A23. The importance of LED (on module A23) is explained in detail below.

## 10.4 Technical Data

The principal technical data for the rectifier types 48 V (60 V)/200 A GR12 (GR121) and 48 V/1000 A GR12 are listed in Tables 10.2 and 10.3.

**Table 10.3.** Technical data for rectifier type 48 V/1000 A, GR12

| Main input supply | |
|---|---|
| Voltage (V) | 3 × 380/400/415 or 3 × 220/230 <br> +10% − 15% |
| Frequency (Hz) | 50 or 60 ± 5% |
| Fuse protection in mains switching panel (A) | 250 (at 220 V) <br> 160 (at 380 V) |
| Degree of radio interference | A (VDE 0878) |
| **D.C. output** | |
| Operating mode or condition | Rated direct voltages (V) (tolerances for 1 to 100% rated current) |
| | Equipment voltage $\hat{=}$ load voltage (except for initial charging with communications system disconnected |
| Rectifier mode | 51 ± 0.5% |
| Parallel operation/float (trickle) charging (2.23 V/cell) | 56 ± 0.5% |
| Parallel operation/charging (2.33 V/cell) | 58.5 ± 0.5% |
| Initial charging | 67.5 (communications system disconnected) |
| Rated direct current (A) | 1000 |
| Interference voltage (mV) | $\leq$ 2 (frequency-weighted with CCITT A-filter) |
| Dimensions (H×W×D)(mm) | cubicle construction <br> 2000 × 900 × 1200 |

# 11 Magnetically Controlled Rectifiers

## 11.1 Types 24V to 220V/10A to 400A

### 11.1.1 General and Application

Magnetically controlled rectifiers have been used in power supply systems for more than 40 years.

A brief introduction to the new magnetically controlled rectifiers (see Fig. 11.1)[1] now follows. The main advantages of these rectifiers are

- insensitive to mains power line input distortion,
- good current waveform factor,
- low disturbance feedback to mains power line (to draw a sinusoidal current from the mains with low conducted EMI),
- very high input power factor,
- low inrush current,
- good dynamic response,

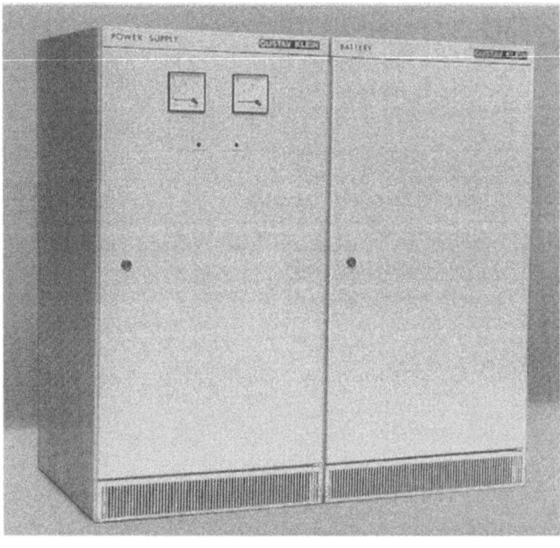


Fig. 11.1. Magnetically controlled rectifier e.g. type LGDM-IU 48 V/100 A (left) with battery cabinet (right). (Photo by courtesy of Gustav Klein GmbH & Co. KG)

---

[1] *Source*: Gustav Klein GmbH & Co. KG.

- short circuit proof,
- low radio interference and
- high MTBF.

The magnetically controlled rectifier unit is very often used in unmanned mains-independent power supply stations with transmission systems.

### 11.1.2 Modes of Operation

The rectifier unit can be used in the rectifier mode (see Sect. 3.1), standby parallel mode (see Sect. 3.3) or changeover mode (see Sect. 3.4).

### 11.1.3 Survey Diagram of the Power Supply System

Figure 11.2 shows a basic survey diagram of a power supply system with magnetically controlled rectifiers.

During normal operation with mains supply available both rectifiers supply the communication system and the battery with float (trickle) charging in parallel mode.

On power failure the battery is discharged and takes over the supply of the communications system without interruption. The voltage relay is deenergized and so the diode (5) bridged (see Fig. 11.2).

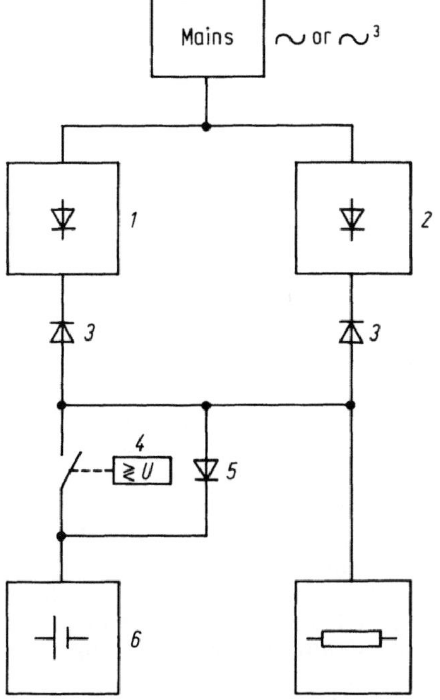


Fig. 11.2. Survey diagram of the power supply system. (1) Rectifier 1, (2) rectifier 2, (3) decoupling diodes, (4) voltage dependent control of the battery contactor, (5) bridging diode, (6) lead-acid battery with e.g. 24 cells, (7) load 40 to 75 V

If one of the rectifiers is faulty (e.g. after a overvoltage on the output) the rectifier switches off automatically and the non-faulty rectifier increases the current and takes over the whole supply of the load.

### 11.1.4 Survey Diagram and Functioning Principle of the Rectifier Unit 24V to 220V/10A to 400A

The rectifier unit 24V to 220V / 10A to 400A LGM-*IU* and LGDM-*IU* (see Fig. 11.3) both use the magnetic regulation principle which keeps the unit to a large extent insensitive to mains power line input distortion ensuring low disturbance feedback to the mains power input.

In addition to such inherent filtering capability a specially designed mains interference filter A4 is installed at the a.c. input side of the unit and an A6 filter at the output side.

Magnetically regulated rectifier units consist of a transformer T1, a capacitive reactance ($C1$) and an inductive reactance ($L1$) (reactive impedance), two rectifier diodes V1, a control element A3, filter, output sensors, a control circuit A1 and

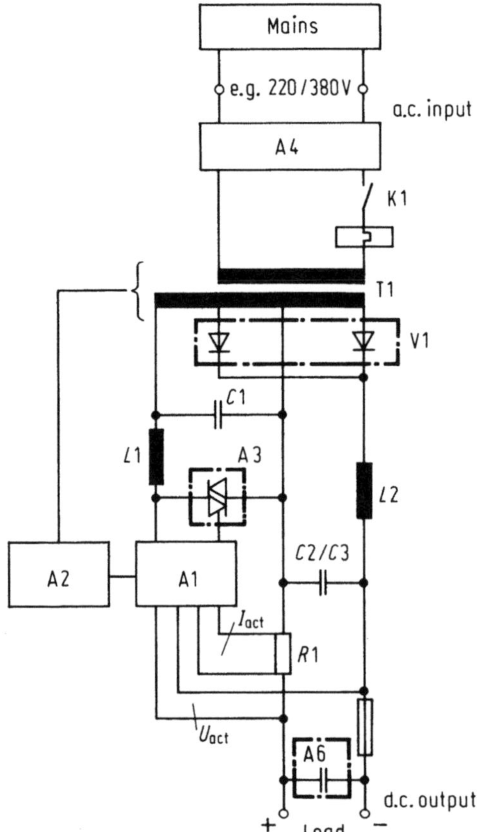


**Fig. 11.3.** Survey diagram of magnetically controlled rectifier type 24 V to 220 V/10 A to 400 A, LGM-*IU* and LGDM-*IU*. A1 Control amplifier, A2 supervision, A3 triac (control element), T1 main transformer, C2, C3, L2 output filter, A4,A6 HF interference filter, L1, C1 reactance, V1 rectifier, K1 mains contactor, R1 measuring shunt resistor

a supervision A2. The a.c. input voltage is converted by the transformer to a voltage determined by the nominal d.c. output voltage, rectified by the diodes and filtered by a capacitor. The secondary side of the transformer includes a control winding which is loaded under the control of the control element. A capacitive and an inductive reactance are connected across the control winding, the inductive reactance being under the control of the control element. $C1$ and $L1$ have been dimensioned so that, depending upon the value of the drive to the control element, the transformer can be loaded either inductively or capacitively.

The construction and dimensioning of the input transformer are such that the input power factor of the rectifier unit is greater than 0.95 over a very wide load range despite the variable loading of the control winding (capacitive and inductive).

Together with the available inductance (choke integrated in the transformer), a voltage divider is formed from reactive elements, allowing the secondary voltage, and therefore the rectifier d.c. output voltage, to be regulated almost without loss.

The control element is driven by the control circuit in such a way that a d.c. voltage, constant to within $\pm 1\%$, is available at the output terminals. The a.c. input choke provides a good decoupling from the input supply line. To a large extent this prevents feedback of disturbances from the rectifier diodes into the a.c. supply so that very clean a.c. input current waveforms are obtained. The high attenuation in the other direction also protects the rectifier diodes from overvoltages on the a.c. input supply. The rectifier diodes are also unaffected by a.c. input voltages with undefined zero-crossing.

This optimal primary current waveform enables the rectifier to be fed from a standby power supply without any restriction or additional equipment.

**Table 11.1.** Technical data for magnetically controlled rectifier units

| Mains input supply | |
|---|---|
| Voltage (V) | LGM-*IU*: 220 $\pm$ 10 %<br>LGDM-*IU*: 220/380 $\pm$ 10 % |
| Frequency (Hz) | 50 $\pm$ 5% |
| Input current (A) | 3.1 to 130.0 |
| Degree of radio interference | A(VDE 0878) |
| **D.C. output** | |
| Nominal d.c. voltage (V) | 24 to 220 |
| Constant voltage (V/cell) | 2.23 to 2.40 $\pm$ 1 %<br>(lead acid) |
| | 1.40 to 1.65 $\pm$ 1%<br>(nickel-cadmium) |
| Rated direct current (A) | 10 – 400 |
| Interference voltage (mV) | $\leq$ 2 (frequency-weighted with CCITT<br>A - filter) |

The integral input choke produces a large angle of current flow in the rectifier diodes and they can thus be operated optimally. The resultant low input current increase contributes to the good efficiency achieved with this equipment technique.

Using only electrolytic capacitors for d.c. output filtering results in a good dynamic response.

In the case of a fault in the control circuit, the rectifier is still in operation, it then works like an uncontrolled rectifier.

### 11.1.5  Technical Data

The principal technical data for the magnetically controlled rectifier units are listed in Table 11.1.

# 12 D.C./D.C. Converter

D.C./D.C. converters are designed either as single-height or double-height power supply units or as complete built in equipment; they all deliver constant controlled (component) supply voltages.

At this point a selection of the equipment will be described below. In general they all employ the principle of pulse-width control and use the single-ended forward converter circuit.

## 12.1 Type 50 to 200 W

The double-height d.c./d.c. converters 50 to 200 W[1] (see Fig. 12.1) in SIVAPAC® construction operate, in general, on the single-ended forward converter principle (60 kHz, with pulse-width control). From an input supply of 48 or 60 V (40 to 75 V), they produce an output voltage of 5, 5.4, 12, 24, 60 or 93 V (of either polarity).

Some of those assemblies employ the single-ended fly back-converter principle. The grip boards carry sockets for voltage measurement, a light-emitting diode (indicating operation) and a switch.

**Fig. 12.1.** Double-height d.c./d.c. converter DCCCL (EZL) 5 V/25 A. (Photo by courtesy of Siemens AG)

---

[1] *Source*: Siemens AG.

### 12.1.1  Application

The d.c./d.c. converters can be used as decentralized power supplies to loads in 48 and 60 V-communications systems as for example:

- digital electronic switching system EWSD,
- communications system ISDN,
- data switching system EWSP,
- optical broad-band systems,
- communication network KN system (PABX),
- electronic modular (telephone) system EMS (PABX) and
- office telecommunications system HICOM (PABX).


### 12.1.2  Modes of Operation

The d.c./d.c. converters are used for the rectifier mode or standby parallel mode (see Fig. 3.7).


### 12.1.3  Survey Diagram and Functioning Principle

The basic circuit of the double-height d.c./d.c. converters is shown in Fig. 12.2.

The input is applied periodically to the transformer Tr through the input filter and the switching transistors (Power MOS-transistors T4/T6), transformed to the required output level, rectified by the diode G20 and smoothed by the output filter $L2$ and $C27$ to $C30$. The switching frequency is 60 kHz. The conduction period of the transistors T4/T6 is determined by the closed-loop and open-loop control.

The output is controlled, short-circuit-proofed and monitored for undervoltage and overvoltage.

Operation is indicated by a green light-emitting diode, which lights when the output voltage is at its rated value.

The *power section* of the d.c./d.c. converter consists of:

- combined input filter (h.f. and l.f.),
- switching transistors T4/T6 (Power MOS-transistors),
- power transformer Tr,
- rectifier diode G20,
- free-wheeling diode G21,
- output filter $L2$ and $C27$ to $C30$ and
- current-measuring resistor $R33$ for current limiting.

The transistors T4/T6 applies the voltage periodically to the primary winding of the transformer Tr. During the conduction period of the transistors T4/T6 current flows through the rectifier diode G20. The transformer Tr provides electrical separation and transforms the input voltage to the level required in the output circuit. In the non-conducting period the energy stored in the inductor $L2$ flows

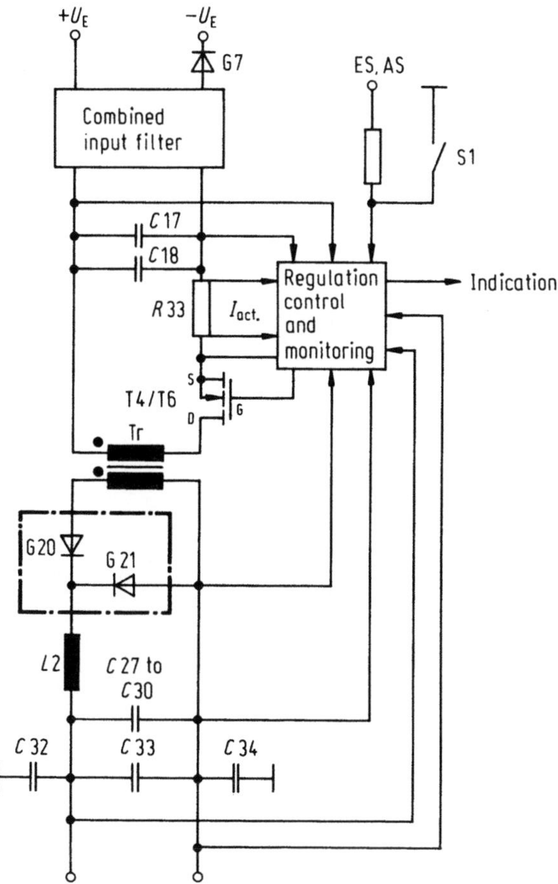



**Fig. 12.2.** Survery diagram of a d.c./d.c. converter (single-ended forward converter). ES Remote control, AS common synchronized off command of many different d.c./d.c. converters.

through the diode G21. The output voltage is filtered by the inductor $L2$ and the capacitors $C27$ to $C30$. Capacitors $C32$ to $C34$ are for EMI suppression.

The resistor $R33$ is used to obtain the actual-current feedback signal in the input circuit and the transistors T4/T6 are thereby protected from overcurrent spikes.

The auxiliary supply of 12 V is obtained from the input and stabilized by a control.

The switching frequency of 60 kHz is produced by an integrated control circuit. The maximum switching ratio is 0.5.

The integrated control circuit controls the output voltage. It performs the actual-reference comparison and effects the conversion into a corresponding pulse-width controlled trigger pulse. If the voltage across the resistor $R33$ exceeds a predetermined value due to overload, the integrated control circuit reduces the conduction period.

**Table 12.1.** Technical data for d.c./d.c. converter 50 to 200 W

| D.C./D.C. converter | Type | | | | | | |
|---|---|---|---|---|---|---|---|
| | 5V/14A | 12V/6A | 12V/2.5A 5V/4A | +12V/0.8A − 5V/0.3A | +12V/1.1A −12V/0.5A +5V/4A −5V/1.2A | 24V/3A | 5 V/25A |
| Abbreviated designation | DCCCA (EZA) | DCCCB (EZB) | DCCCC (EZC) | EZG | DCCCH (EZH) | DCCCF (EZF) | DCCCL (EZL) |
| Input | | | | | | | |
| Rated direct voltage (V) | −48 or −60 (−40 to −75) | | | | | | |
| Rated direct current | | | | | | | |
| at 48 V (A) | 1.90 | 1.94 | 1.32 | 0.45 | 1.61 | 1.93 | 3.39 |
| at 60 V (A) | 1.54 | 1.56 | 1.08 | 0.37 | 1.29 | 1.58 | 2.68 |
| Degree of radio interference | Limit class B (VDE 0871) | | | | | | |
| Output | | | | | | | |
| Rated direct voltage (V) | 5±3% | 12±3% | 12±3%/ 5±3% | +12±3%/ − 5±3% | +12±3%/ −12±3% + 5±3%/ − 5±3% | 24±3% | 5±3% |
| Rated direct current (A) | 14 | 6 | 2.5/4 | 0.80/0.3 | 1.1/0.5 4/1.2 | 3 | 25 |
| Polarity | Either | Either | Either | Prescribed (rel. to 0 V) | Prescribed (rel. to 0 V) | Either | Either |
| Dimensions (L×W) (mm) | SIVAPAC® construction, double-height: 277×230 | | | | | | |

| D.C./D.C. converter Type | 60V/2.5A | +12V/2.1A<br>+5V/10A<br>−12V/0.8A<br>−5V/0.8A | 5V/40A | 5.4V/16A | +12V/3A<br>−12V/3A | 93V/0.6A | +12V/2.1A<br>+5V/10A<br>−12V/0.8A<br>−5V/0.8A | 5V/46A |
|---|---|---|---|---|---|---|---|---|
| Abbreviated designation | DCCCK | DCCCR | DCCCS | DCCCT | DCCCU | DCCCV | DCCMR | DCCMS |
| **Input** | | | | | | | | |
| Rated direct voltage (V) | −48 or −60(−40 to −75) | | | | | | | |
| Rated direct current | | | | | | | | |
| at 48 V (A) | 4.1 | 2.7 | 5.49 | 2.48 | 2.01 | 1.41 | 2.7 | 6.35 |
| at 60 V (A) | 3.3 | 2.1 | 4.38 | 1.96 | 1.62 | 1.13 | 2.1 | 5.1 |
| Degree of radio interference | Limit class B (VDE 0871) | | | | | | | |
| **Output** | | | | | | | | |
| Rated direct voltage (V) | 60±3% | +12±5%<br>+ 5±5%<br>−12±5%<br>− 5±5% | 5±3% | 5.4±3% | +12±8%<br>−12±8% | 93±3% | +12±5%<br>+ 5±5%<br>−12±5%<br>− 5±5% | 5±3 % |
| Rated direct current (A) | 2.5 | 2.1/10<br>0.8/0.8 | 40 | 16 | 3 | 0.6 | 2.1/10<br>0.8/0.8 | 46 |
| Polarity | Either | Prescribed (rel. to 0 V) | Either | Either | Prescribed (rel. to 0 V) | Either | Prescribed (rel. to 0 V) | Either |
| Dimensions (L×W) (mm) | SIVAPAC® construction, double-height: 277 × 230 | | | | | | SAS(SIPAC®) construction, double-height 287 × 230 | |

The output voltage is monitored by the undervoltage and overvoltage monitor. In operation within the rating limits the green light-emitting diode on the board of the assembly is lit, the 'potential-free' contact is closed and a high signal is produced. If an undervoltage is detected at the output, it is indicated through the contact (relay contact opened) and by a low signal. If the rated voltage is exceeded by up to 40%, the output is switched off by suppression of the trigger pulses. The switched-off condition is maintained, since the auxiliary supply continues to be produced. The power section can be restored to operation by a brief application of the earth potential ($>$ 50 ms) to the remote control input ES, or by switching the input supply off and on.

A particular converter can be switched on and off by means of the switch S1, while the control point AS enables a number of converters to be switched off simultaneously.

### 12.1.4 Technical Data

The principal technical data for the 50 to 200 W d.c./d.c. converters are listed in Table 12.1.

### 12.2 Type 750 W

The built-in or wall-mounting equipment d.c./d.c. converter $750^2$ or 1000 W (see Fig. 12.3) operates on the single-ended forward converter principle (approximately 60 - 75 kHz constant, with pulse width control). From an input supply of 48 V (+ 20 % / -10 %), it produces an output voltage of 24 or 60 V (e.g. ± 1%).

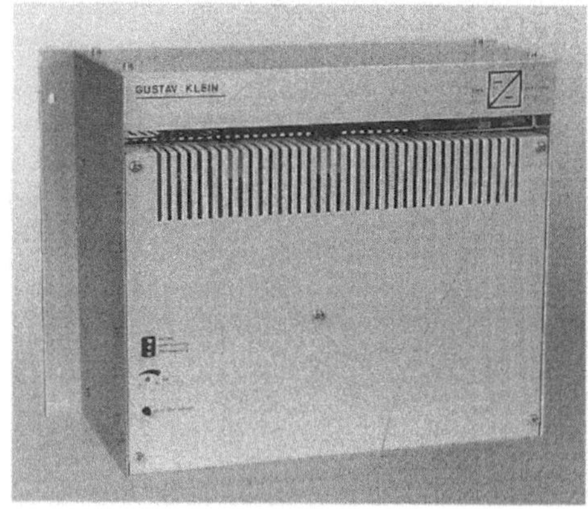


**Fig. 12.3.** D.C./D.C. converter 1000 W. (Photo by courtesy of Gustav Klein GmbH & Co. KG)

---

[2] *Source*: Gustav Klein GmbH & Co. KG.
The equipment 750 W is an example; the power range of this kind of d.c./d.c. converters is approximately 100 – 5000 W.

### 12.2.1 Application

The d.c./d.c. converter can also be used with rectifiers 48 V and batteries as centralized power supplies to loads in 24 or 60 V communications systems.

### 12.2.2 Modes of Operation

The d.c./d.c. converter is used for the standby parallel mode (see Fig. 12.4, and also Fig. 3.7).

In order to attain greater safety of the supply, several units can be connected in parallel.

### 12.2.3 Survey Diagram and Functioning Principle

The basic circuit of the d.c./d.c. converter is shown in Fig. 12.5.

The d.c./d.c. converter is used to transform 48 V d.c. into 24 or 60 V d.c. The output d.c. voltage is galvanically separated from the input d.c. voltage.

The output voltage is electronically regulated which means it is independent from load changes and variations of the input voltage.

At normal operation an equal, stable distribution of the current occurs for all units (see Fig. 12.4). A decoupling diode V6/7 in the output of the converter prevents a feedback.

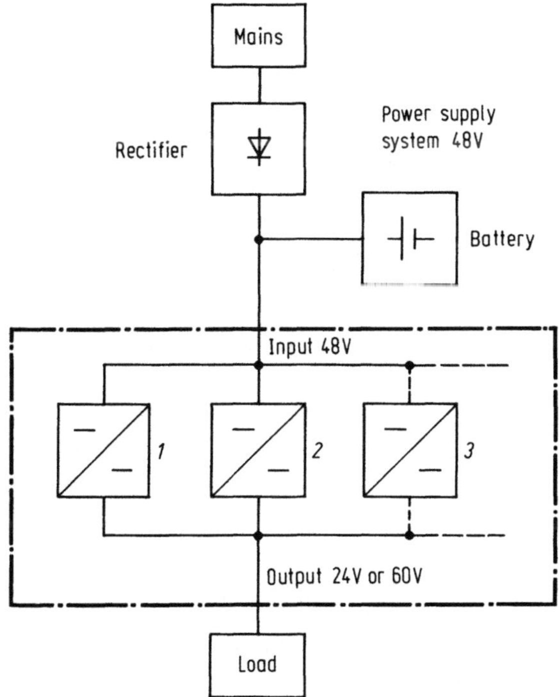


**Fig. 12.4.** Standby parallel mode – parallel connection of d.c./d.c. converters 750 W (e.g. n + 1)

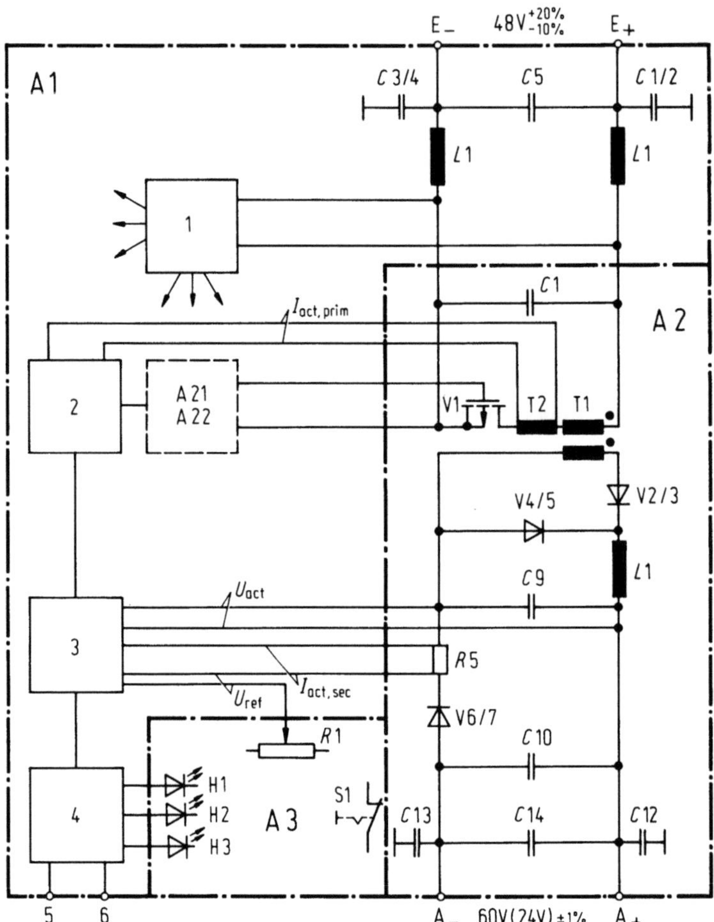

**Fig. 12.5.** D.C/D.C. converter 750 W. *1* Auxiliary power converter (internal power supply) 2 × ± 12 V stab., *2* pulse-width control, *3* regulation, *4* voltage monitoring (output over- and undervoltage), *5* remote control, *6* remote signalling, E–/E+ d.c. input, A–/A+ d.c. controlled output, A1, A2 pc-board of power section, regulation, control and monitoring circuit, A3 signalling board, A21/A22 drive stage and protection circuit board, LED H1 operation, LED H2 undervoltage, LED H3 overvoltage, S1 button reset

Each d.c./d.c. converter incorporates an over- and undervoltage supervision in the output. At normal operation, overvoltage and undervoltage are signalled by LED's. For the remote signalization 'failure' a potential-free changeover contact is available. Additionally an external LED for 'operation' can be connected.

The d.c. input voltage connects to the single-ended forward converter via the interference suppression filter $C1$ to $C5$, $L1$, A1) and the input filter (e.g. $C1$, A2). The input voltage is applied periodically to the main transformer T1 through the switching transistor V1, transformed to the required output level.

The output voltage is rectified V2/3, A2) and smoothed via a *LC* filter (*L*1, *C*9, *C*14, A2). Small components can be used for the power transmitter and output filter because of the high switching frequency. The regulation dynamic is essentially better than with units using the conventional technique. Additional radio interference suppression devices are also available in the output (e.g. *C*10, *C*12, *C*13, A2). An integrated circuit of pulse-width control (e.g. Siemens TDA 4718) (2) (see Fig. 12.5) is used for the control of d.c./d.c. converter.

The auxiliary voltage for the regulation and supervision is generated via the auxiliary converter (1) (incorporated as single-ended flyback (blocking) converter).

The output voltage is kept constant to $\pm$ 1 % within the load range of 0 – 100 %. From rated current (e.g. approximately 31 A) on the output voltage is controlled down so that the output current remains constant.

Figure 12.6 shows an example of an oscilloscope picture measured at the output of pulse-width control (2) (see also Fig. 12.5) and Fig. 12.7 shows flow between the source and draining of power switching transistor V1 at rated power.

### 12.2.4 Technical Data

The principal technical data for the 750 W (1000 W) d.c./d.c. converter are listed in Table 12.2.

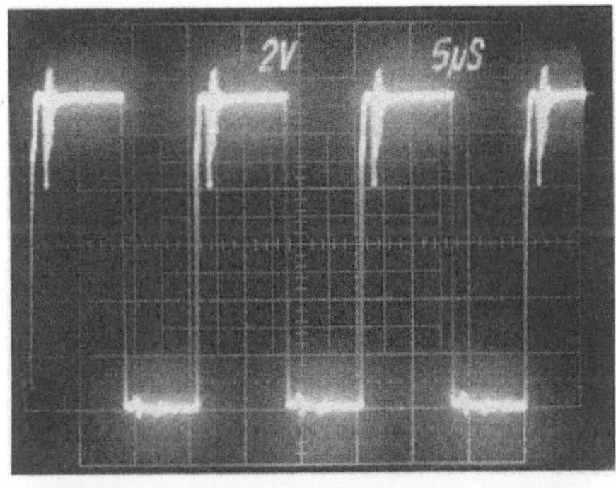


**Fig. 12.6.** Pulse width controlled trigger pulse (output of TDA 4718)

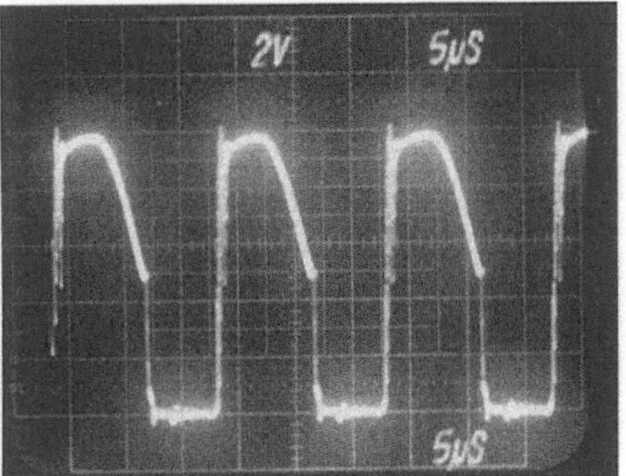


**Fig. 12.7.** Pulses between drain and source of transistor V1 (approximately 66.6 kHz $\hat{=}$ 1 period T $\hat{=}$ 15 $\mu$s)

**Table 12.2.** Technical data for d.c./d.c. converter 750 and 1000 W

| Input | |
| --- | --- |
| Rated direct voltage      (V) | 48 + 20 %<br>    − 10 % |
| Degree of radio interferences | Limit class B (VDE 0871) |
| Interference noise voltage  (mV) | $\leq$ 1 (with A-filter) |

| Output | | |
| --- | --- | --- |
| Rated direct voltage      (V)<br>(before decoupling diode) | 24 $\pm$ 1 % | 60 $\pm$ 1 % |
| Rated capacity      (W) | 750 | 1000 |
| Dimensions (H $\times$ W $\times$ D) (mm) | 316.5 $\times$ 446 $\times$ 252 | |

# 13 D.C./A.C. Inverters

At this point a selection of equipment will be described.

## 13.1 Type 2.5 kVA

In the pulsed modulated 2.5 kVA inverter module (WR20[1], see Figs. 13.1 and 13.2) with transistor power section the principle of the forward converter (20-kHz-pulse width control) is employed. From the 48- or 60 V input voltage (40 to 75 V), a single-phase a.c. voltage of 220 V/50 Hz (60 Hz) is produced.

### 13.1.1 Application

The d.c./a.c. inverter modules serve to supply 220-V loads, such as

- memories,
- keyboard printer terminals and
- printer

in 48- and 60-V communications systems (e.g. system EWSD, text- and data switching systems as well as solar-power supplied loads). Generally, the units are housed in inverter connecting panels (e.g. AS 124, see Figs. 13.3 and 13.4).

### 13.1.2 Type Designation

The d.c./a.c. inverter modules are characterized as shown in Fig. 13.2.

Fig. 13.1. D.C./A.C. inverter module 2.5 kVA WR20. (Photo by courtesy of Siemens AG)

---

[1] *Source*: Siemens AG.

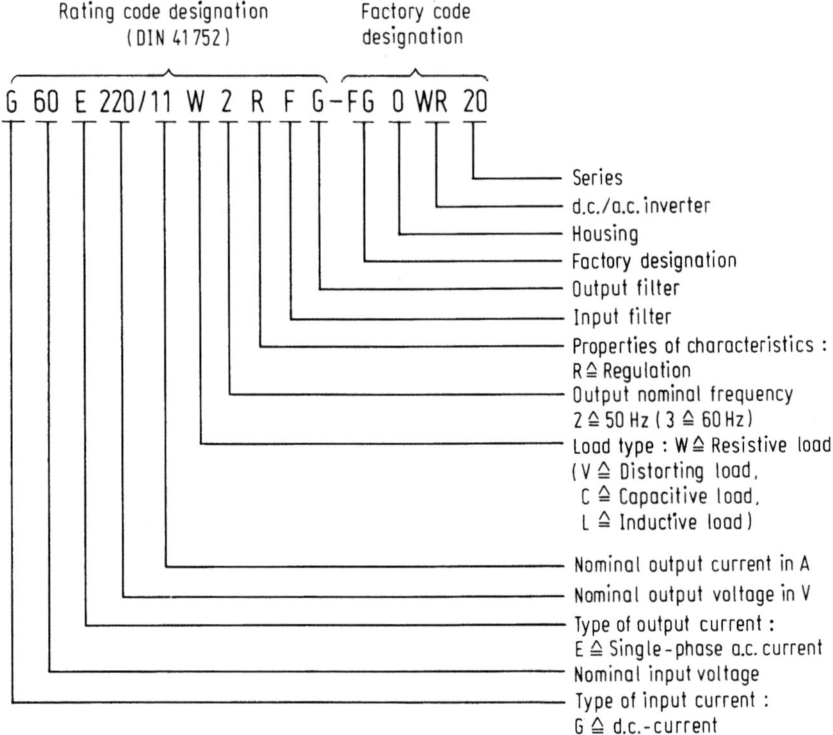



**Fig. 13.2.** Type designation d.c./a.c. inverter module WR20

### 13.1.3 Modes of Operation

In general the d.c./a.c. inverter modules are operated in the standby parallel mode (continuous operation).

### 13.1.4 Survey Diagram of the Power Supply System

The connecting cabinet AS124 (see Figs. 13.3 and 13.4) is equipped with a maximum of four (+ one spare) inverter modules 220 V/50 Hz (60 Hz) 2.5 kVA WR20.

The d.c. input is connected with the fail-safe d.c. power supply system (rectifiers and battery, as in Fig. 4.11).

### 13.1.5 Survey Diagram, Block Diagram and Functioning Principle of the d.c./a.c. Inverter Module 2.5 kVA

According to Fig. 8.28 in chapter 8 the input direct voltage $U_E$ (−40 to −75 V) passes through a filter (NF) which serves to reduce the interference voltage to

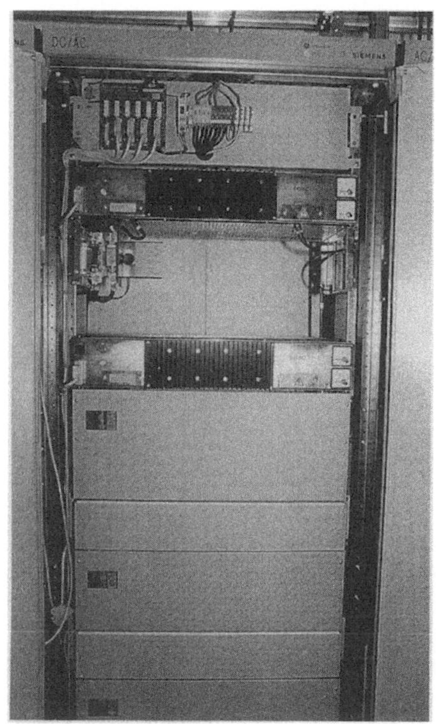

**Fig. 13.3.** Connecting cabinet AS124 equipped with two inverter modules 2.5 kVA WR20. (Photo by Hans Gumhalter and by courtesy of Siemens AG)

the converter which is triggered by the regulator 1 at a constant frequency (20 kHz). The sinusoidal reference voltage is rectified by a rectifier circuit and input to the regulator 1 as reference value $U_{RM}$. The modulated converter voltage $U_M$ corresponds largely to a rectified mains voltage. Subsequently, using the polarity-inverting bridge circuit, the output a.c. voltage $U_A$ (see Fig. 8.29) is formed from it; here the bridge control switches alternately the bridge branches S1/S4 or S2/S3 at every zero crossing of the sine-wave generator.

For ambient temperatures of up to 45 °C the d.c./a.c. inverter can carry a load of up to 2.5 kW (effective power) or 2.5 kVA[2] (apparent power) for a power factor of cos phi 0.7 inductive to 0.85 kapacitive. The current limitation is set to approximately 13.6 A. A loading which exceeds this value results in voltage drops, for a loading current of more than approximately 24 A, there is a disconnection of the device (i.e. by pulse disabling).

When designing the systems it is important to keep the maximum current value of the load below the value of the current limitation. Otherwise, there is a

---

[2] The continuous rating of 2.5 kVA for 45° C is only valid with a purely linear load; with a distorting load (e.g. switched mode power supply units which are not able to draw a sinusoidal current from the input with low conducted EMI) a limitation of approximately 1.5 kVA is required.

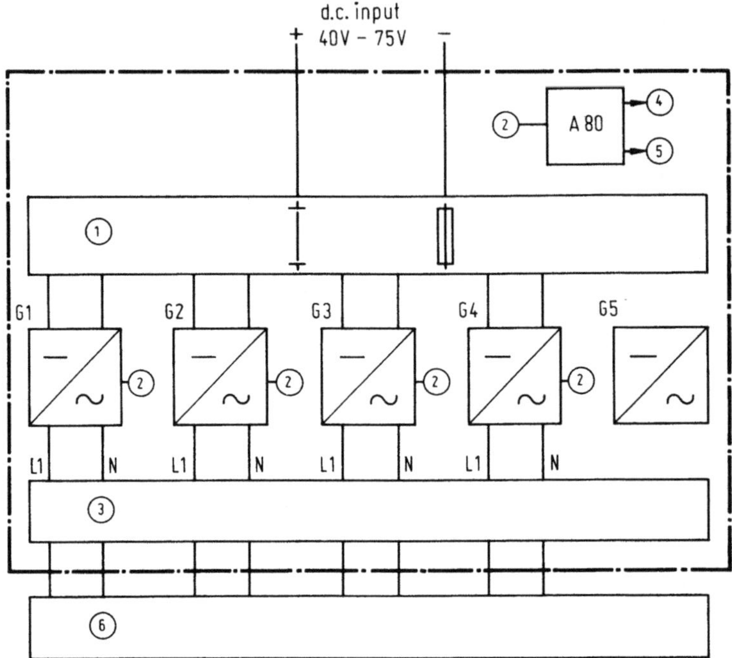



**Fig. 13.4.** Connecting cabinet AS124 for one to four inverter modules 220 V/50 Hz (60 Hz) 2.5 kVA WR20. *1* d.c. distribution, *2* signal connector, *3* a.c. distribution, *4* signalization: LED H81 operation, LED H85 fault, *5* remote signalization, *6* a.c. load, G1 to G4 d.c./a.c. inverter modules 2.5 kVA WR20, G5 inverter 2.5 kVA WR20 (not connected i.e. spare equipment), A80 signalling board

corresponding drop in the output voltage and when the value of approximately 24 A is exceeded, the disconnection of the d.c./a.c. inverter unit is performed with automatic reclosure after the current decreases below this value. This is necessary in order to protect the semi-conductor; the device is protected against overload and short circuits.

During normal operation and with automatic reclosure the device is running with 'softstart', which means that after operation the output voltage is increased slowly (during an interval of 0.8 s) from 0 to the rated value. The softstart with current limiting effect ensures smooth operation even for loads with higher current values.

The modules of the d.c./a.c. inverter WR20 2.5 kVA and their interaction can be seen from the block diagram (see Fig. 13.5).

*Logic module A1.* The circuits for signalling and monitoring as well as for logic and auxiliary power generation are housed on the logic module A1. The input of the module is located just above the filter of the trigger module A3 at the input voltage of 40 to 75 V. An auxiliary converter generates four output voltages, two

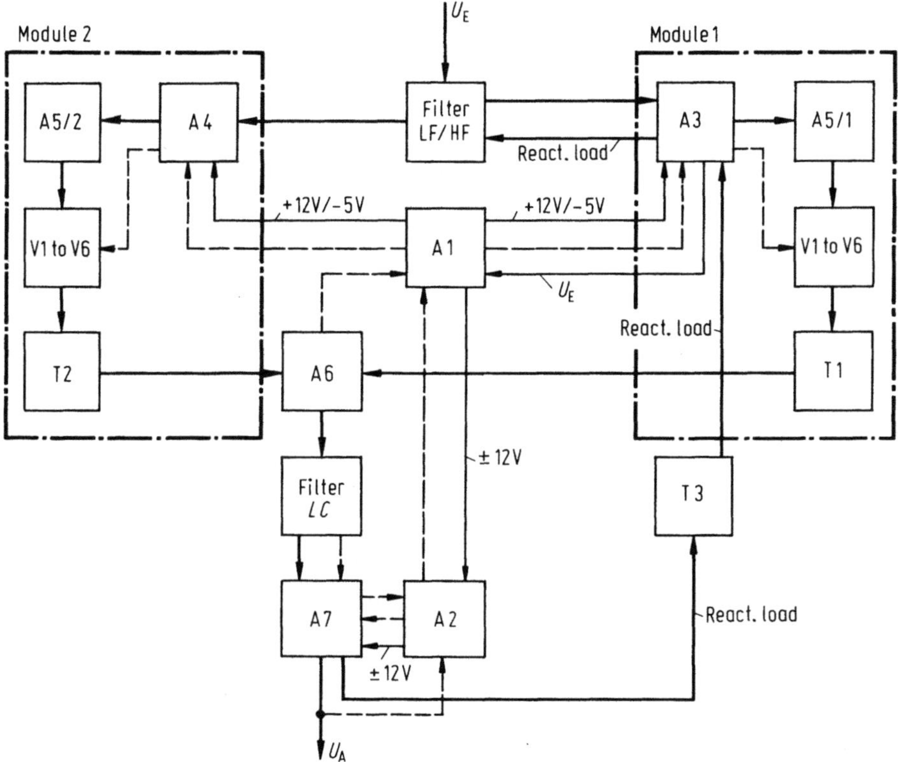



**Fig. 13.5.** Block diagram of the pulsed, d.c./a.c. modulated inverter module 2.5 kVA WR20. $U_E$ Input d.c. voltage 40–75 V, $U_A$ output a.c. voltage 220 V/50 Hz (60 Hz), T1, T2, T3 transformers, V1 to V6 power transistors, ———▶ energy flow, - - - - -▶ signal flow, ⟶ internal power supply, A1 logic module, A2 regulation module, A3 trigger module 1, A4 trigger module 2, A5/1 emitter module, A5/2 emitter module, A6 pulse circuit module, A7 polarity inverting module

of which supply the logic and trigger equipment on the input-side (+12 V, −5 V), whereas the other two (+ 12 V, −12 V) supply the output side of the inverter.

*Regulation module A2.* The regulation module A2 comprises the components for the following functions: quartz oscillator, reference value generator, zero indicator, voltage regulator and monitors. The circuits are at the level of the output voltage and are supplied with +12 V and −12 V from the logic module A1.

The quartz oscillator works with a fundamental frequency of 6.55 MHz (50 Hz d.c./a.c. inverter) which is then divided into two operating frequencies of about 50 kHz and 50 Hz by means of binary counters. Out of the rectangular 50 Hz voltage a sine-wave is generated with an active filter. The energy recovery equipment (reactive load converter) is synchronized with the 50 kHz frequency. The rectified 50 Hz sinusoidal voltage is the reference value for the voltage regulator. It is also used for the zero control of the polarity-inverting bridge circuit A7.

The regulation module performs the regulation and three monitoring functions: for overvoltage, undervoltage, and a thermic monitoring for polarity-inverting transistors. All results are evaluated here and then forwarded to the optocoupler of the logic control module A1.

*Trigger module A3 and A4.* The two trigger modules A3 and A4 are PCBs of equal electrical design. They contain all components for triggering the power transistors. The elements affected during the triggering process are housed on a common module beneath the rapidly switching power transistors in order to hold the switching paths short and thus ensure low induction. The PCB A3 also comprises the elements performing the rectification of the feedback reactive current occurring with complex loading of the d.c./a.c. inverter.

*Emitter module A5/1 and A5/2.* The emitter modules A5/1 and A5/2 are located between the power transistor and the two trigger modules. Each of them contains six power transistors switched in parallel. Every PCB also houses the capacitor and the diode of the RCD circuit. The trigger module A3 and the emitter module A5/1 as well as the trigger module A4 and the emitter module A5/2, including the corresponding power transistors and heat sinks, make up two modules. The electric isolation required for power transformation is achieved through the two main transformers T1 and T2. The output power to be transformed has a value of 2.5 kVA and the switching frequency is of about 20 kHz. For the transport of energy there are two modules: the pulse circuit module A6 and the polarity-inversing module A7.

*Pulse circuit module A6.* The pulse circuit module A6 contains all power diodes necessary for the rectification of the pulsating intermediate circuit voltage. For this, every branch of the intermediate circuit has two diodes connected in series and two free-wheeling diodes connected in parallel. Each of the six power diodes is connected to one varistor between anode and cathode, which ensures an equal distribution of the blocking voltage to all power diodes affected and thus leads to a symmetrical voltage stress.

In addition, module A6 contains the elements needed for reactive current balancing (energy-recovery), for the power transistor control and the current limiting equipment. The transformer T3 is located in the energy recovery circuit.

*Polarity inverting module A7.* A *LC* filter transforms the sine-modulated triangular voltage into a sine-wave loop voltage which is forwarded to the polarity-inverting transistor module on A7. Apart from the Darlington transistors all other elements required for triggering are housed on A7. The field effect transistors are introduced here as low-capacity elements. The sine-wave loop voltage is transformed by means of the polarity-inverting transistors into a 50 (or 60) Hz a.c. voltage and passed on via a filter to the output of the inverter.

In general, a complex load is to be expected. The energy stored in the reactive load distorts the form of the output voltage and impairs the distortion

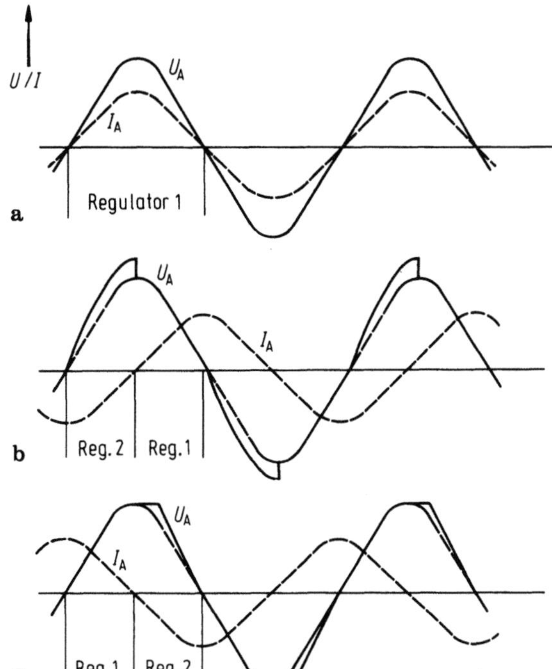



**Fig. 13.6a–c.** Area of operation, regulator 1 and 2 of the inverter modules for different types of load. $U_A$ Output voltage, $I_A$ output current, **a** ohmic load ($\varphi = 0°$), **b** inductive load ($\varphi = -90°$), **c** capacitive load ($\varphi = +90°$)

**Table 13.1.** Technical data for d.c./a.c. inverter module, 2.5 kVA WR20

| D.C. input | |
|---|---|
| Nominal voltage (V) | $-48$ or $-60$ ($-40$ to $-75$) |
| Maximum input current approximately (at 40 V d.c. input voltage and rated load) (A) | 73 |
| Conducted interference/ voltage (mV) | $\leq 1.8$ (in the range up to approximately 1.5 kVA, frequency-weighted with CCITT A -filter) |
| RFI suppression (input and output) | 10 kHz to 30 MHz as per VDE 0878/limit class B |
| **A.C. output** | |
| Nominal voltage (V) | $220 \pm 5\%$ |
| Frequency (Hz) | $50 \pm 0.5\%$ (or $60 \pm 0.5\%$) |
| Rated capacity (continuous) (kVA) | 2.5 (at 45° C[a]) 1.5 (at 60° C) |
| Displacement factor | 0.7 lagging to unity to 0.85 leading |
| Total harmonic distortion factor (%) | $\leq 2.5$ (at non-reactive load) |
| Dimensions (H×W×D) (mm) | 118×660×370 |

[a]For purely linear loads; for distorting loads reduced to approximately 1.5 kVA

factor. That is why a reactive load compensation is provided which is realized as reactive load converter feeding the excessive energy back into the input source. Figure 13.6 shows the basic voltage behaviour related to the type of load.

### 13.1.6  Technical Data

The principal technical data for the d.c./a.c. inverter module 2.5 kVA WR20 are listed in Table 13.1.

## 13.2  Type 15 kVA (3 × 5 kVA)

### 13.2.1  General and Application

The three phase 15 kVA (3 × 5 kVA) d.c./a.c. inverter in the centre-tap circuit U-SFUIEK[3] 60 V d.c. (see Fig. 13.7, illustrating the 10 kVA type, and also Fig. 7.17) is used for example in communications systems, e.g. text- and data switching systems.

**Fig. 13.7.** Inverter U-SFUIEK 60 V d.c./10 kVA. (Photo by courtesy of Gustav Klein GmbH & Co. KG)

---

[3] *Source*: Gustav Klein GmbH & Co. KG.

From the 60 V input voltage a three phase a.c. voltage of 360 V/60 Hz is produced.

### 13.2.2  Type Designation

The d.c./a.c. inverter is characterized as follows (see Fig. 13.8).

### 13.2.3  Modes of Operation

In general the d.c./a.c. inverter operates in the standby parallel mode (continuous operation) (see Fig. 4.10).

### 13.2.4  Survey Diagram of the Power Supply System

In addition to a fail-safe d.c. power supply system (±60 V) with rectifier and batteries, d.c./a.c. inverters 15 kVA (3 × 5 kVA) are also used (see Fig. 13.9 and also Fig. 4.10).

### 13.2.5  Survey Diagram and Functioning Principle of the d.c./a.c. Inverter 15 kVA (3 × 5 kVA)

Figure 13.10 shows a survey diagram of the d.c./a.c. inverter U-SFUIEK 60 V/15 kVA (3 × 5 kVA) with static transfer switch. The principal function has already been explained as shown in Fig. 4.10.

The electronic revert to mains unit (static transfer switch) supervises the inverter and the mains. The inverter is steadily synchronized as long as the mains frequency and voltage is within the permissible tolerance. At a failure of the inverter or at overload (consumer short circuit) an uninterruptible change-over to the mains is effected (Fig. 13.11). The static transfer switch operates automatically (Fig. 13.12).

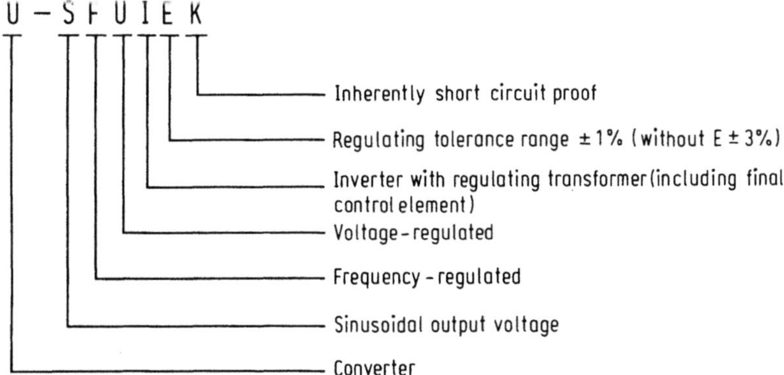


**Fig. 13.8.** d.c./a.c. inverter U-SFUIEK 60 V d.c./3 × 5 kVA - type designations

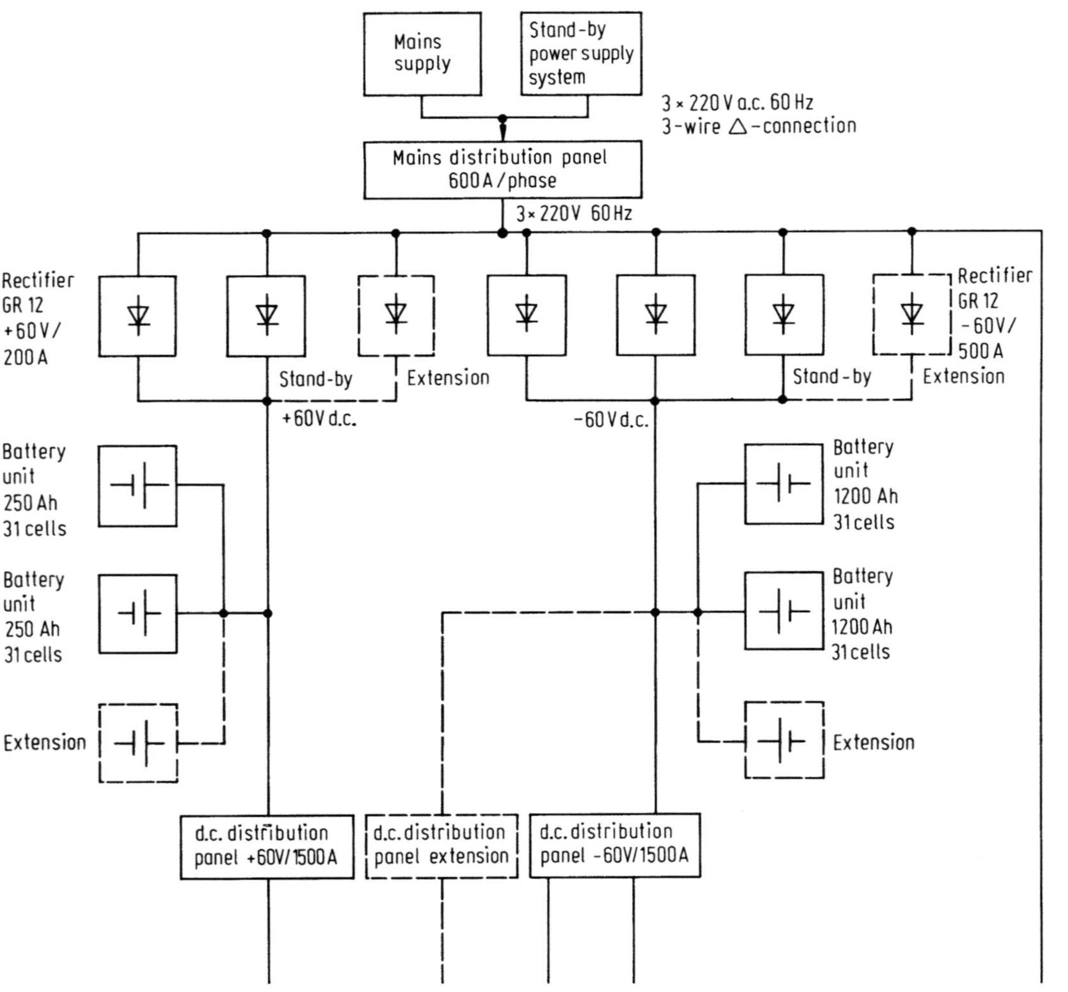

Mains supply
Stand-by power supply system
3 × 220 V a.c. 60 Hz
3-wire △-connection
Mains distribution panel 600A/phase
3 × 220V 60Hz
Rectifier GR 12 +60V/ 200A
Stand-by
Extension
+60V d.c.
Stand-by
Rectifier GR 12 -60V/ 500A
Extension
-60V d.c.
Battery unit 250 Ah 31cells
Battery unit 250 Ah 31cells
Extension
Battery unit 1200 Ah 31cells
Battery unit 1200 Ah 31cells
Extension
d.c. distribution panel +60V/1500A
d.c. distribution panel extension
d.c. distribution panel -60V/1500A

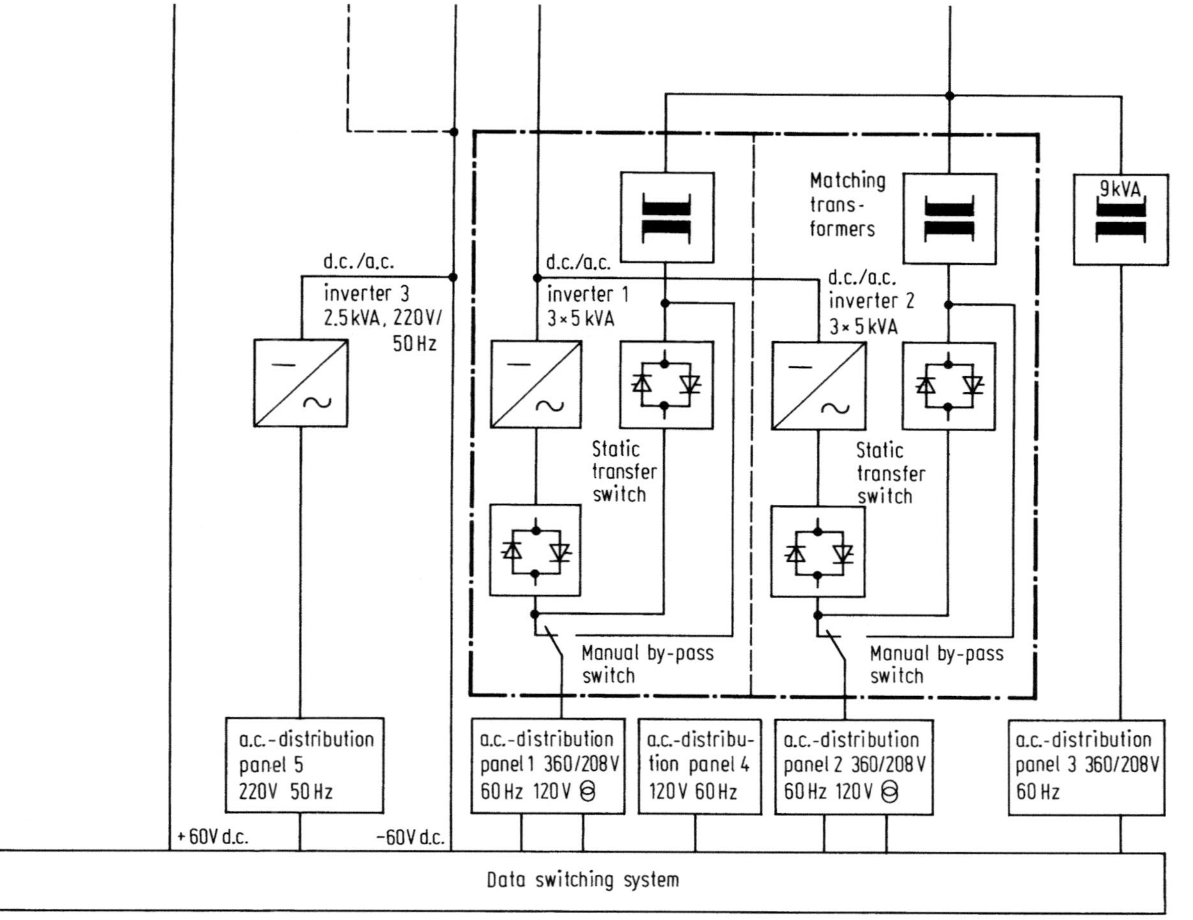


**Fig. 13.9.** Survey diagram of a power supply for a data switching system

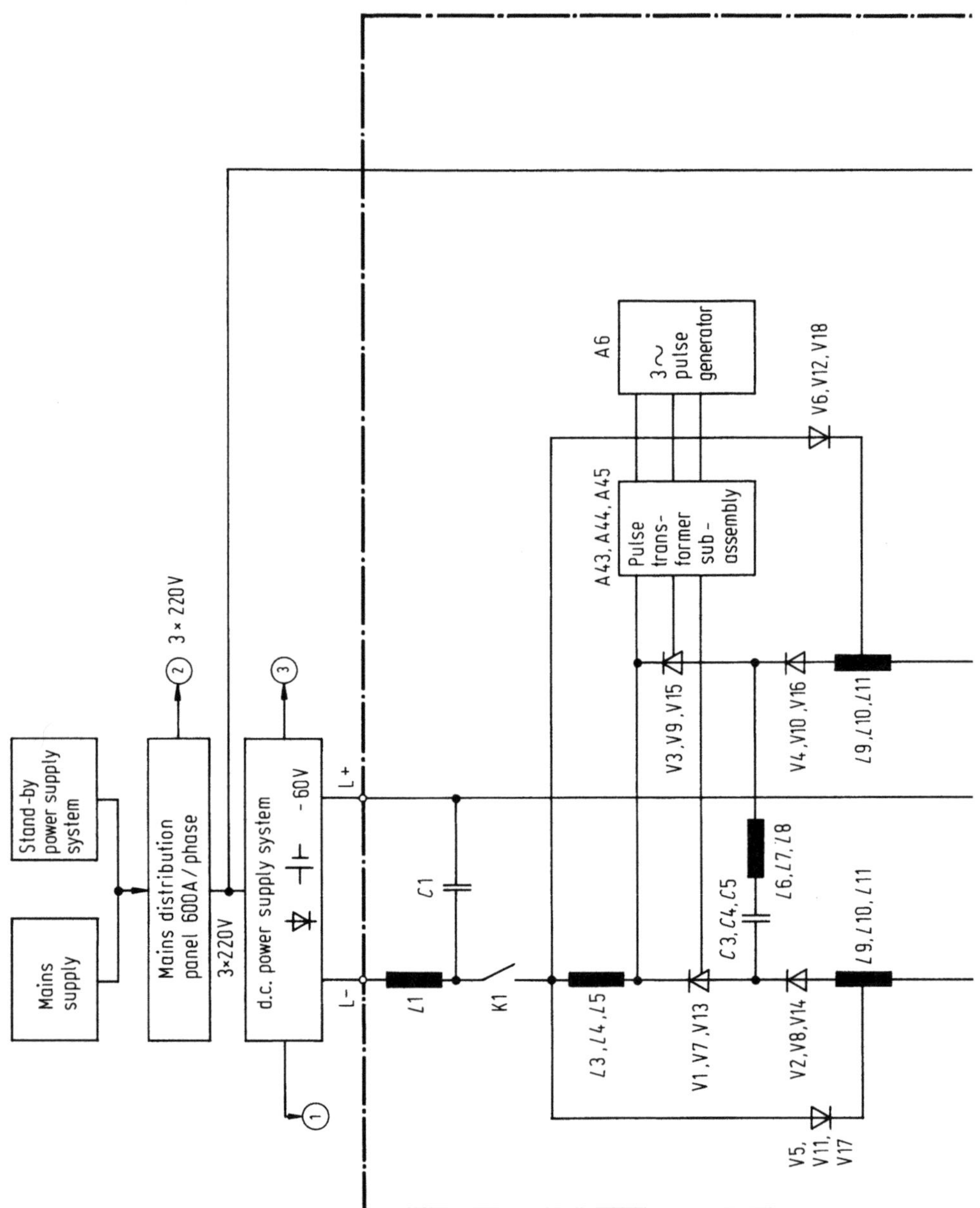

Mains supply
Stand-by power supply system
Mains distribution panel 600A / phase
3×220V
d.c. power supply system
−60V
L+
L−
C1
L1
K1
A6
3~ pulse generator
A43,A44,A45
Pulse trans-former sub-assembly
V3,V9,V15
V4,V10,V16
L9,L10,L11
L6,L7,L8
C3,C4,C5
V1,V7,V13
V2,V8,V14
L3,L4,L5
L9,L10,L11
V5, V11, V17
V6,V12,V18
3 × 220V

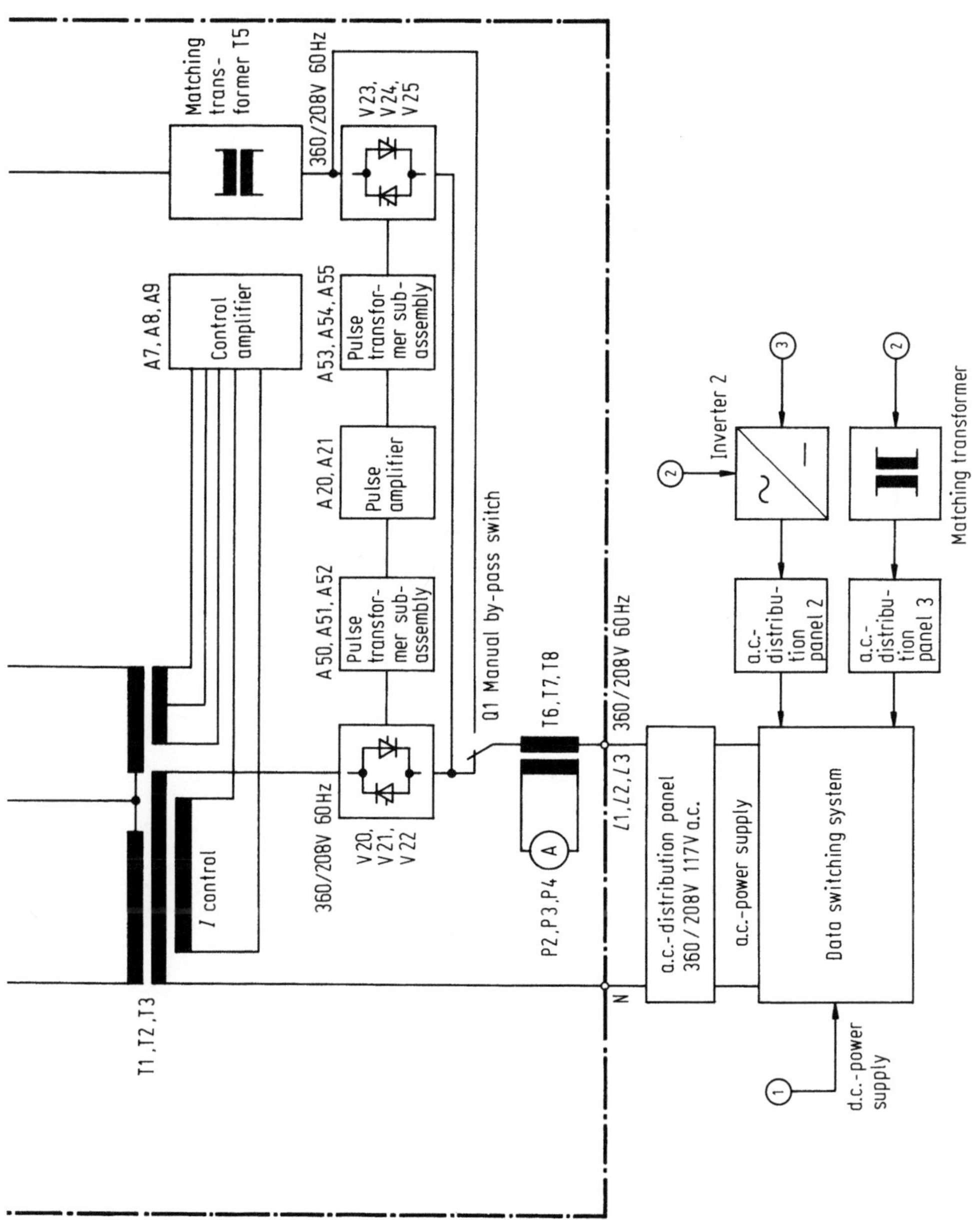


Fig. 13.10. For caption see next page

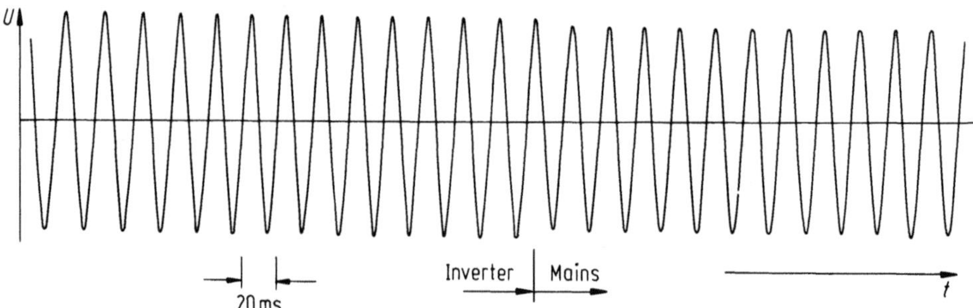

**Fig. 13.11.** Changeover from inverter to mains supply (without interruption of the load), inverter off, type U-SFUIEK 15 kVA (3 × 5 kVA) 60 Hz

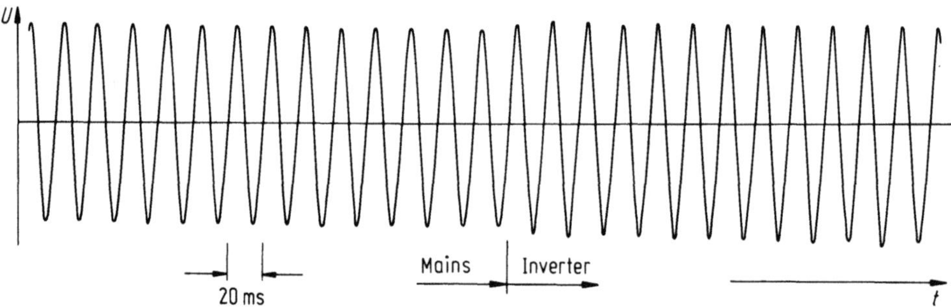

**Fig. 13.12.** Switch-back from mains to inverter supply (without interruption of the load), inverter on, type U-SFUIEK 15 kVA (3 × 5 kVA) 60 Hz

The inverter incorporates three thyristor centre-tap circuits, the rectangular a.c. voltages of which are proportional to the d.c. input voltage. The a.c. voltage is smoothed to a sinusoidal voltage by a stabilizer.

In the centre-tap circuit the thyristors are fired alternately at a clock rate of 60 Hz. The forced extinguishing from the thyristors is effected by commutation capacitors. Free wheeling diodes render circular currents possible as well as energy feedback in case a thyristor is blocked and thereby prevents high voltage peaks in the corresponding current circuit.

**Fig. 13.10.** Survey diagram of the d.c./a.c. inverter U-SFUIEK 60 V/15 kVA (3 × 5 kVA) (one phase circuit shown) with static transfer switch. L+, L− d.c. input (60 V), $L1, C1$ battery (input)-filter, K1 inverter input contactor (inverter on/off), V1, V7, V13, V3, V9, V15 thyristors of the centre-tap d.c./a.c. inverter circuit, V2, V8, V14, V4, V10, V16 series-decoupling diodes against a reverse discharging of capacitors $C3, C4, C5$; $C3, C4, C5$ commutations capacitors, $L6, L7, L8$ commutations coils, V5, V11, V17, V6, V12, V18 reactive current diodes, T1, T2, T3 control transformers (magnetically controlled stabilizers), V20, V21, V22 thyristor contactor 'inverter' of the static transfer switch, V23, V24, V25 thyristor contactor 'mains' of the static transfer switch, T6, T7, T8 current transformers, N, L1, L2, L3 a.c. output (360 V/208 V/60 Hz)

**Table 13.2.** Technical data for d.c./a.c. inverters U-SFUIEK

| D.C. input | |
|---|---|
| Nominal voltage (V) | 24, 48, **60,** 110, 220 + 20% /−10% |
| Tolerable ripple feedback to d.c. bus (% rms) | $\leqq$ 5 (with input filter) |
| RFI suppression | Limit class G as per VDE 0875 |
| **A.C. output** | |
| Nominal voltage (V) | 360/208 ± 5% (in general 220 ±5%) |
| Frequency (Hz) | 60 ± 0.5% (50 ± 0.5%) |
| Rated capacity (continuous) (kVA) | 15 (3×5) (in general 1.5 to 15) |
| Displacement factor | 0.7 lagging to unity to 0.9 leading |
| Total harmonic distortion factor (%) | $\leqq$ 5 |
| Dimensions (H×W×D) (mm) | 1800×1200×700 |

The battery filters $L1, C1$ limits the feedback of the inverter to the batteries and rectifier.

### 13.2.6  Technical Data

The principal technical data for d.c./a.c. inverters types U-SFUIEK, e.g. 15 kVA (3 × 5 kVA) are listed in Table 13.2.

# 14 Static UPS Systems

At this point the introduction and basic circuits of static UPS systems (as in Sects. 1.7, 4.3 and 4.3.5) are recalled. A selection of the equipment will be described in Fig. 14.1. Different examples of types, series and power ranges of UPS-systems are listed in Table 14.1.

## 14.1 Type 10 to 500 kVA

The power range of the UPS system 40, 41 and 42[1] is 10 to 500 kVA (see Table 14.1). Figure 14.2 shows for an example a three-phase UPS system type 42 (4233) with a rated output power of 330 kVA.

### 14.1.1 Application

UPS systems are employed wherever a fully-uninterruptible, independent and secure power supply is required.
A UPS system:

- protects users from all types of mains disruptions,
- bridges mains failures and
- provides constant voltage with close tolerances for amplitude and frequency.

Typical applications:

- telecommunication systems,
- DP equipment and systems,
- process controls,
- control centres,
- drives in continuous production processes,
- life-support systems in hospitals
- air traffic control systems and
- safety systems in power stations and many others.

### 14.1.2 Modes of Operation

The continuous mode of operation is used in the UPS system.

---

[1] *Source*: Siemens AG.

**Fig. 14.1.** UPS-System series – types 40, 41 and 42, exterior view (Photo by courtesy of Siemens AG)

### 14.1.3 Overview

System 42 UPS units are used for uninterruptible supply of three-phase loads. A UPS system consists of at least a UPS unit and a battery. To provide redundancy and/or to increase output two in the case of types 4212 to 4233 or up to four in the case of type 4250 or even up to eight UPS units can be connected in parallel.

*Features:*

- high efficiency (up to 94%), low thermal loading of environment,
- UPS units up to and including type 4212 can be installed in the telecommunications system room close to the loads, thanks to DP-oriented design and low space requirement,
- low noise level,
- high reliability,
- transistor inverter,
- parallel operation possible to increase redundancy and/or system power rating,
- excellent dynamics,

**Table 14.1.** Types, series and power ranges of static UPS systems (Source: Siemens AG)

| Type | Number of phases/a.c. output voltages[a](V) | Output fre-quence (Hz) | Rated power | | | | | | | | | | | | | | | | | | | | | |
| --- | --- | --- | --- | --- | --- | --- | --- | --- | --- | --- | --- | --- | --- | --- | --- | --- | --- | --- | --- | --- | --- | --- | --- | --- | --- |
| | | | (VA) | | | | (kVA) | | | | | | | | | | | | | | | | |
| | | | 500 | 1000 | 2000 | 3000 | 1 | 1.2 | 2.5 | 3.5 | 5 | 7 | 7.5 | 10 | 15 | 20 | 25 | 30 | 40 | 60 | 80 | 120 | 160 | 220 | 330 | 500 |
| 40 CP05 | 1/N,~230 | 50 | × | | | | | | | | | | | | | | | | | | | | | | | |
| 40 CP10 | 1/N,~230 | 50 | | × | | | | | | | | | | | | | | | | | | | | | | |
| 40 CP20 | 1/N,~230 | 50 | | | × | | | | | | | | | | | | | | | | | | | | | |
| 40 CP30 | 1/N,~230 | 50 | | | | × | | | | | | | | | | | | | | | | | | | | |
| 4010 | 1/N,~220 | 50/60 | | | | | × | | | | | | | | | | | | | | | | | | | |
| 4012 | 1/N,~220 | 50/60 | | | | | | × | | | | | | | | | | | | | | | | | | |
| 4025 | 1/N,~220 | 50/60 | | | | | | | × | | | | | | | | | | | | | | | | | |
| 4035 | 1/N,~220 | 50/60 | | | | | | | | × | | | | | | | | | | | | | | | | |
| 4050 | 1/N,~220 | 50/60 | | | | | | | | | × | | | | | | | | | | | | | | | |
| 4070 | 1/N,~220 | 50/60 | | | | | | | | | | × | | | | | | | | | | | | | | |
| 4105 | 1/N,~230 | 50/60 | | | | | | | | | × | | | | | | | | | | | | | | | |
| 4107 | 1/N,~230 | 50/60 | | | | | | | | | | | × | | | | | | | | | | | | | |
| 4110 | 1/N,~230 | 50/60 | | | | | | | | | | | | × | | | | | | | | | | | | |
| 4115 | 1/N,~230 | 50/60 | | | | | | | | | | | | | × | | | | | | | | | | | |
| 4120 | 1/N,~230 | 50/60 | | | | | | | | | | | | | | × | | | | | | | | | | |
| 4130 | 1/N,~230 | 50/60 | | | | | | | | | | | | | | | | × | | | | | | | | |
| 4201 | 3/N,~400 | 50/60 | | | | | | | | | | | | × | | | | | | | | | | | | |
| 4202 | 3/N,~400 | 50/60 | | | | | | | | | | | | | | × | | | | | | | | | | |
| 4204 | 3/N,~400 | 50/60 | | | | | | | | | | | | | | | | | × | | | | | | | |
| 4206 | 3/N,~400 | 50/60 | | | | | | | | | | | | | | | | | | × | | | | | | |
| 4208 | 3/N,~400 | 50/60 | | | | | | | | | | | | | | | | | | | × | | | | | |
| 4212 | 3/N,~400 | 50/60 | | | | | | | | | | | | | | | | | | | | × | | | | |
| 4216 | 3/N,~400 | 50/60 | | | | | | | | | | | | | | | | | | | | | × | | | |
| 4222 | 3/N,~400 | 50/60 | | | | | | | | | | | | | | | | | | | | | | × | | |
| 4233 | 3/N,~400 | 50/60 | | | | | | | | | | | | | | | | | | | | | | | × | |
| 4250 12pulse | 3/N,~400 | 50/60 | | | | | | | | | | | | | | | | | | | | | | | | × |

[a]In single-phase UPS (e.g. Type 41) systems in addition to the 230 V a.c. output voltages of 220 or 240 V are also possible
In three-phase UPS (e.g. Type 42) systems in addition to the 400 V a.c. output voltages of 380 or 415 V are possible

Remark: UPS units (systems 41 or 42) can also operate as 50/60 Hz or 60/50 Hz frequency converters. If it is not necessary to bridge line failures, the frequency converters can be supplied without battery.

**Fig. 14.2.** UPS system 42 (4233) 330 kVA with 12 pulse rectifiers, single block unit with revert to mains unit, view of the interior of the equipment, covers removed. (Photo by courtesy of Siemens AG)

- switched mode power supply load compatible by virtue of instantaneous value regulation, fast control response,
- compact design,
- easy to instal,
- maintenance free,
- attractive and informative control panel with LC display and membrane keyboard,
- easy to operate, automatic operation,
- menu-driven monitoring and diagnostic system,
- interfaces for connection of personal computer, printer and modem (option),
- interfaces for connection to networks (option),
- the UPS units meet the specifications of all well-known computer manufacturers and
- world-wide service organization.

*Environmental considerations.* The excellent efficiency of the equipment contributes to energy saving. The unavoidable losses of modern converters are low, and so there is only a slight extra load on the air conditioning system.

The latest state of the art and environmental protection legislation are taken into account at factories throughout the production process.

*Quality assurance (QA).* A comprehensive QA system ensures that all processes, from design and development through the construction, procurement and manufacturing to sales, installation and service, are fully covered.

The QA system conforms to international standard ISO 9001.


*Frequency converter.* The UPS units can also be used as frequency converters (conversion from 50 to 60 Hz or vice versa) with or without a battery. The information given in this description also applies when they are used in this way, but not to the static bypass switch (SBS) and service bypass functions.


### 14.1.4 Survey Diagram and Functioning Principle

*Power and Electronics Section.* The *power section* consists of five main components:

– rectifier for rectifying the mains voltage,
– d.c. link circuit to which the battery is connected,
– inverter which converts the d.c. voltage into a sinusoidal a.c. voltage,
– static bypass (transfer) switch (SBS) (revert-to-mains unit) which can provide an immediate direct connection between the output and the mains feed and
– service bypass (manual override) which can be used to isolate the UPS equipment manually without interrupting the supplied load (from type 4222 to type 4250 not part of unit).

The SBS and the service bypass can be installed in a separate cubicle for certain plant configurations.


*Rectifier.* A fully controlled three-phase thyristor bridge circuit (6-pulse type) with mains commutation choke $L10$ is connected to the mains via switch Q10 (Fig. 14.3). It converts the mains voltage into d.c. voltage.

In order to diminish significantly the effects on the supply network, the 220 kVA and 330 kVA units can be provided optionally with 12-pulse rectifiers. The 500 kVA units are always of 12-pulse design. The 12-pulse rectifier consists of two fully controlled 3-phase bridges which operate by mutually displacing 30° el. with the aid of a special transformer (similar to GR12, see Sect. 10.2.4).

Open and closed-loop control of the rectifier includes the following main functions:

– software-controlled battery-saving automatic charger with *IU* characteristics and charge current limiting,
– current limiting, prevents rectifier overload (important for multi-unit equipment with common d.c. link circuit),
– reduced current limiting and sloping voltage characteristic for operation with a standby power supply,
– controlled runup: slow rise in voltage and current, no power-up surges,

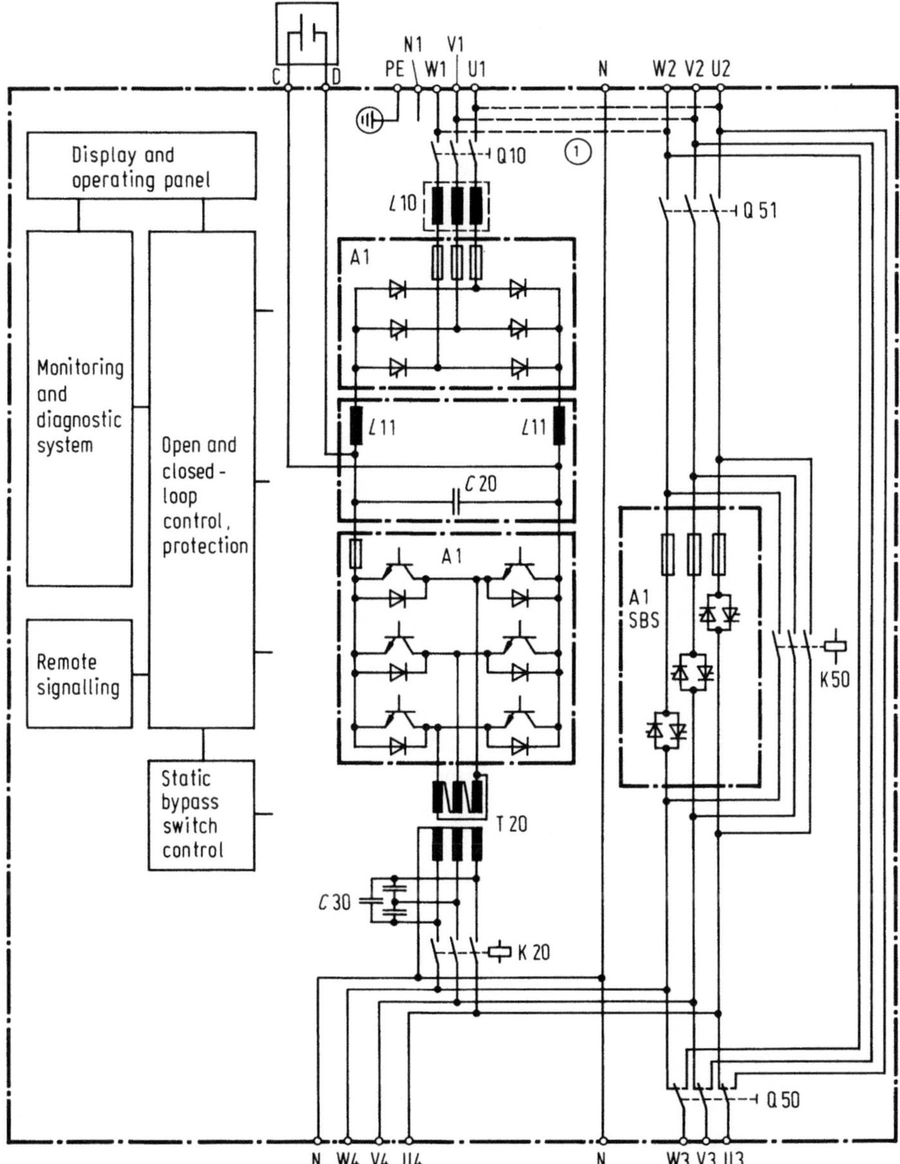


**Fig. 14.3.** Survey diagram of UPS system series 42 (80/120 kVA), other units have slight deviations (e.g. various switchgear combinations according to rating). A1 Power semiconductor module, consisting of thyristor controlled rectifier, transistor controlled inverter and thyristor switch (static bypass switch SBS), $C20$ d.c. link capacitors, $C30$ filter capacitors, K20 inverter contactor, K50 SBS contactor, $L10$ mains commutation choke, $L11$ smoothing choke, Q10 rectifier input switch, Q50 service bypass switch (manual), Q51 SBS input switch, T20 inverter output transformer, C,D battery connection, U1, V1, W1, N1 rectifier power connection, U2, V2, W2, N SBS mains connection, U3, V3, W3, N connection for load, U4, V4, W4, N connection for parallel unit 1) links, remote for separate SBS feed (units 10 kVA to 330 kVA)

– controlled current division for parallel operation of two rectifiers (dual-unit system with common d.c. link circuit) prevents uneven loads and
– sequenced runup of the rectifiers for multi-unit system.

*D.C. link circuit.* A filter circuit, consisting of choke $L11$ and capacitor $C20$ reduces the natural ripple of the d.c. voltage and eliminates any voltage peaks. The d.c. link circuit voltage is smoothed and the ripple of the battery current becomes very small. This protects the battery.

The UPS principle is based on the fact that the d.c. link circuit voltage is maintained either by the mains or, if this fails, by the battery, within a range which allows the inverter to keep the output voltage constant.

*Inverter.* The main components of the power section are the transistor bridge circuit, transformer $T20$ capacitors $C30$.

In simple terms this is a changeover switch, consisting of power transistors, which uses a switching cycle determined by the electronics to alternatively apply the individual phases of the transformer to the plus and minus busbars of the d.c. link circuit (compare Fig. 8.27, Chapter 8).

The leakage inductances of the transformer, in conjunction with the capacitors, form a filter circuit which produces a sinusoidal output voltage with a low distortion factor from the pulse train (Fig. 14.4a).

Figure 14.4b shows a load changing of 100% and the a.c. output voltage.

The output voltage is controlled by gating the inverter phases using pulse width (control) modulation (PWM) with supersine. Because of the high pulse frequency control can take effect very rapidly (instantaneous value control). Since this also actively affects the wave form of the output voltage, feedback effects from loads of non-sinusoidal currents (non-linear load) are compensated.

The salient features of open and closed-loop control of inverters are as follows:

– highly -dynamic voltage control ($<$ 5 ms).

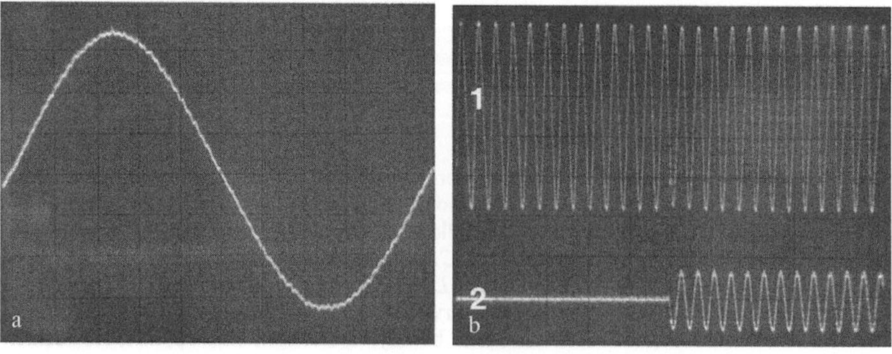


**Fig. 14.4a, b.** Oscillogram of output voltage (**a**) and (1), e.g. 100% load changing (**b**); (**b**) (1) output voltage of the inverter, (**b**) (2) load current. (Photo by courtesy of Siemens AG)

- current limiting:
  prevents equipment becoming overloaded, gives short-circuit protection.
- frequency control:
  highly-accurate quartz-controlled natural frequency (tolerance: $\pm$ 0.1 Hz);
  soft synchronization without frequency jumps (0.1 Hz/s);
  wide synchronization range for load transfer in the case of unstable mains.
- angle control:
  precise synchronization with the mains and with parallel UPS units;
  no compensating currents;
  no current surges during switchover;
  full mains redundancy;
  highly dynamic ($<$ 5 ms).
- instantaneous value control:
  low output voltage distortion,
  even for non-linear load;
  high peak currents;
  highly dynamic ($<$ 5 ms).
- pulse generation:
  constant monitoring of the switching status of power;
  transistors;
  no error pulses possible;
  optimum utilization of power section in borderline situations.
- peak current recording:
  reliable protection of power section in the case of short circuits;
  rapid response of load fuses thanks to high r.m.s value of short-circuit current
  (square-wave current blocks).

*Static bypass (transfer) switch (SBS) (revert-to-mains unit).* The thyristor
switch and the contactor K50 are the main components of the SBS (see Fig. 14.3).
The thyristor switch can accept the load current with no interruption. The contactor then accepts the continuous current after about 50 ms. The 10 kVA and
25 kVA units are designed without contactor K50.

The SBS provides uninterrupted switching under the following circumstances:

- inverter overloading,
- short-circuit,
- inverter fault and
- when the inverter is switched off by hand.

**Open and closed-loop control, protection, monitoring and diagnostic, display
and operation system**
The *electronic components* for open and closed-loop control, protection, monitoring, display and operation are plugged into a subassembly rack and accommodated
in function or component-related peripheral modules.

The main features are:

- largely digital with microprocessors and microcontrollers,
- separate actual values for SBS and for parallel units: fully redundant, highest possible reliability and operational integrity,
- separate power supply for the SBS electronics,
- SMD, multilayer and hybrid technologies and
- industrial-grade electronics with high reliability and immunity to electromagnetic interference.

In addition to the modules necessary for unit functions an independent *monitoring and diagnostics system* is provided as standard. The LC display on the control panel gives comprehensive information on the equipment status and assists fault diagnosis.

The system:

- holds edited data for current measured values which can then be retrieved,
- calculates the currently available battery bridging time using the characteristic curve method (takes account of charge state and present load),
- provides statistics,
- continuously stores analog and digital values,
- records events occurring before and after a fault and
- records faults and serves as an aid to diagnosis.

The monitoring and diagnostic system contains a battery-backed process signal memory which continuously records numerous signals from the UPS unit. In the event of a fault the storage process continues for a few more cycles so that the events preceding and following the fault are recorded.

The operating data stored can be read later.
Other functions are available as an optional extra. The *control panel display* then offers:

- display of unit and plant statuses,
- selection of an automatic battery test and
- display of other up-to-date measurements and derived variables.

This option also allows a data report of an unit fault and a listing of measurements to be output via a printer interface. A further option enables the user to produce a thorough analysis on an IBM-compatible PC.

*Performance.* This section describes the behaviour of the UPS unit in the situations which occur most frequently in practice. The equipment is designed so that the loads connected to it as well as the equipment itself are protected in all situations. Even extreme situations are safely dealt with.

### Normal operation

*Mains power available, inverter ON.* The rectifier converts the three-phase mains voltage into a d.c. voltage which the inverter then converts back into a three-phase system in order to supply the load with constant current and frequency. The battery draws only a very small trickle charging current.

The inverter operates synchronously with the mains, if mains voltage and mains frequency at the static bypass switch (SBS) input are within the defined limits (mains ready):

– mains voltage      $U_N \pm 10\%$ (adjustable) and
– mains frequency   $f_N \pm 1\%$ (settable to 2%).

These values are determined by the loads placed on the mains. In contrast the UPS unit controls the output voltage with an accuracy of $\pm$ 1%.

*Inverter OFF.* The inverter can be switched off by hand. In units without SBS the load is switched off, but the rectifier is still on and continues to charge the battery.

In units with SBS the SBS takes over the load current provided that the 'mains ready' message is displayed (mains within tolerance range $U_N \pm 10\%$; $f_N \pm 1\%/2\%$, SBS automatic mode).

*Fluctuations in mains voltage.* The UPS equipment operates with no restriction to its rated performance data at deviations up to $\pm 10\%$ from rated mains voltage $U_N$ (DIN VDE 0558). At $U_N - 10\%$ the battery charging voltage is still reached.

The rectifier operates down to $U_N - 15\%$ without the battery being discharged. If the mains voltage drops below this value the rectifier switches off. The UPS equipment then behaves as it would be for mains failure.

The SBS remains operable in the range $U_N \pm 10\%$.

*Mains overvoltages.* The rectifier and SBS input of the UPS unit are equipped with noise suppression capacitors to ground and with varistors (overvoltage limiter with nanosecond response times) between phases and N. In conjunction with the *LC* filter in the d.c. link circuit this protects the unit from mains overvoltages.

*Mains frequency fluctuations.* Mains frequency fluctuations of as much as $f_N \pm 5\%$ have no effect on the rectifier.

To take account of the demands made by sensitive loads the inverter only operates synchronously with the mains within a range of $f_N \pm 1\%$ (synchronization range). For larger fluctuations it changes over to self-clocked operation. The SBS is not ready in this case. Depending on the load tolerances the synchronization range can be extended.

## Battery operation

*Mains failure, battery discharging.* If there is a mains failure the battery immediately provides current for the inverter. The changeover to battery operation has no effect on the output voltage. The battery arrangement governs the duration of this state (bridging time). The bridging time is increased if the inverter is operating with part load. The bridging time can also be extended by turning off noncritical loads. The available bridging time relative to the actual load can be interrogated via the LC display.

If the final discharge voltage is reached before mains power is restored, the inverter switches itself off and a tripping signal for an external battery circuit-breaker is output in order to avoid an exhaustive discharge and thereby damage to the batteries. The same applies to the electronics power supply.

*Restoration of mains supply, battery charging.* After mains power is restored the rectifier starts automatically, supplies the inverter and simultaneously charges the battery. Rectifier start-up current surges are avoided by a controlled start with slow rate of current rise. Start-up takes place in about 1 s, according to load and the remaining battery capacity. The rectifier current is limited to 125% $I_N$ (factory setting). Provided that the static bypass switch SBS input voltage are within the permissible tolerances ($f_N \pm 1\%$, $U_N \pm 10\%$), the inverter is synchronized with the mains supply (not applicable to units without SBS, or to frequency converters).

Battery recharging follows the *IU* characteristic. The automatic charging system can be matched to the requirements of the given battery type by appropriate setting of the charging voltage, trickle charge voltage, boost charge voltage, switchover of charging stages and charging current limit.

### Service bypass

When the service bypass switch is actuated, an uninterrupted switchover of the loads to the mains supply is effected. The construction of the bypass is such that it is possible to work on the interior of the equipment while supply to the loads is maintained. The service bypass switch is supplied as standard on units with ratings up to and including 160 kVA.

### Special operating conditions

*Operation without battery.* Despite the absence of buffering, the dynamic performance data are adhered to during operation without a battery.

*Load unbalance.* A 100% load unbalance is permitted. This means that in extreme cases one phase may be loaded with rated current while the other two are not loaded at all. Similarly, two phases can be driven with rated current while there is no load on the third. When this occurs, there are only slight voltage and angle deviations and so large single-phase loads can be connected.

*Non-linear load.* Modern equipment (such as computers with switched-mode power supplies with an exception of new switched mode power supplies such as those mentioned in Sect. 9.1 and converter-fed drives) is seldom 'mains-friendly'. It does not load the mains with a continuous sinusoidal current, but with a periodic square-wave or pulse-shaped current. This can result in voltage distortions and subsequent equipment malfunctions.

Due to the particular control principle employed by the UPS 42 System (instantaneous value control), such voltage distortions occur only to an insignificant extent. These UPS units can therefore be operated continuously up to their rated load with non-linear loads of this type.

*Dynamic load changes.* Sudden changes in load can occur when individual loads or groups of loads are switched on and off. Because of the excellent dynamic performance of the inverter, load changes cause only insignificant short-duration changes to the output voltage. Even when there is a sudden 100% load change, the output voltage does not deviate by more than $\pm$ 4% from its rated value, and the recovery time is less than 5 ms.

*Overload, short circuit.* The inverter can deliver 1.5 times the rated current (three-phase) for up to 30 seconds while keeping to its rated output voltage. If this limit is exceeded, the current limiter regulator acts and reduces the voltage.

If there is a short circuit in one of the load circuits, the inverter delivers a short-circuit current.

The SBS combines with the inverter to optimize equipment performance under transient-load-peak and short-circuit conditions. The dynamic control capabilities of the inverter and the precise current and voltage monitoring circuits allow the mains to be connected briefly via the SBS in parallel with the inverter to utilize the high power of the mains. Control is so precise that no compensation current flows between the mains and inverter this occurs.

The SBS provides uninterrupted switching in the following cases:

– immediately on the inverter output voltage dropping below the $U_N - 10\%$ limit,
– immediately on the load current rising above 1.5 $I_N$
  (inverter and SBS in parallel) and
– after approximately 200 ms if overcurrent > 1.05 $I_N$ (adjustable).

When the overload has died down or the relevant overcurrent protection device of the load circuit has been tripped, the unit automatically returns to inverter operation.

*Overtemperature.* Critical UPS components are equipped with thermoswitches to provide protection against thermal overloading. A prewarning is given when the temperature first starts to rise. If the critical threshold is exceeded, the unit automatically switches itself off and the SBS takes over to provide an uninterrupted supply.

Since the ventilation system is of redundant design, failure of one fan does not result in failure of the UPS unit.

*Inverter fault.* On tripping of the inverter by an internal monitoring circuit, the SBS instantly switches over the supply without interruption. If desired, the control can be set so that the SBS switches over to a 'non-ready' mains supply.

*D.C. voltage fault in d.c. link.* The d.c. link (and the battery) are monitored for over- and undervoltage:

- if the d.c. link voltage rises to 470 V d.c. the inverter cuts out instantly (signal and display: INV fault),
- the battery undervoltage, prewarning signal is output as soon as the d.c. link voltage falls to 350 V d.c. (factory setting) and
- if the d.c. link voltage falls to 300 V d.c. the inverter cuts out. The inverter starts up again when the voltage rises above 350 V (factory setting).

*Start-up under unstable mains conditions.* When starting up the UPS unit, after a long break in operation, for example, the user should normally first start the SBS and then the inverter. The UPS can accept the load even if the mains frequency happens to be outside the inverter's synchronization range. In this event, frequency monitoring is temporarily switched off by hand until the inverter has accepted the load and has gradually adjusted the frequency to the rated value (this is important where motors are connected).

The UPS unit can also be started directly from the battery with no mains power.

*Operation with standby power supplies.* A standby power supply, such as a diesel generator set, generally cannot meet the requirements of sensitive loads. It is therefore good policy to signal to the UPS unit via an external contact that a standby supply is in operation. The static bypass switch is then locked out to avoid repeated synchronizing operations being set in train by generator frequency fluctuations. The inverter operates self-clocked. The rectifier current limit is cut from 125% $I_N$ to 105% $I_N$ (factory setting) and reduced still further if the rectifier input voltage falls (sloping voltage characteristic), due to loading of the standby supply, for instance. Stable operation is assured in this way.

Standby supply start-up is facilitated by prolonging the rectifier ramp-up time (to about 10 s, depending on load and battery rating) and by sequenced starting of multi-unit UPS systems (time interval between one UPS and the next: 1.5 s).

### UPS Unit Variants
Various types of UPS units are available which facilitate the supply of single and multi-unit systems, as well as the frequency converter operation (see Fig. 14.5)

### Multi-Unit Systems
*Principle.* Multi-unit systems can be used in order to:

- improve availability (redundancy) and/or
- increase the power output of the system.

To achieve *redundancy*, the system must contain at least one more unit than would be necessary for supplying the loads. The principle involved is that upon

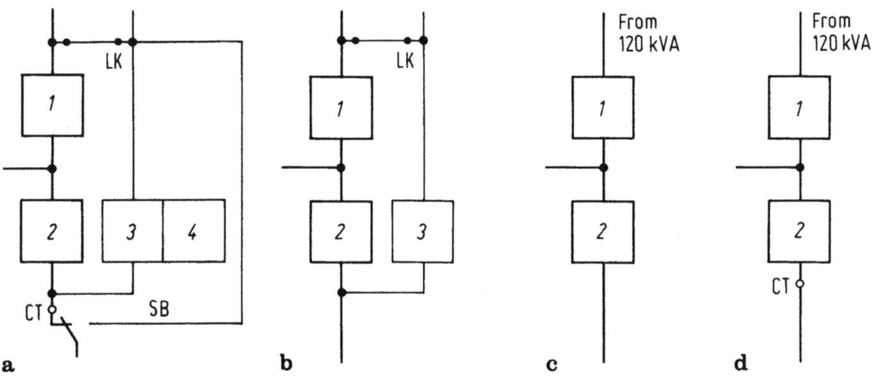

**Fig. 14.5a–d.** UPS unit types

**Table 14.2.** Different configurations of multi-unit systems

| USV type | 4201 | 4202 | 4204 | 4206 | 4208 | 4212 | 4216 | 4222 | 4233 | 4250 |
|---|---|---|---|---|---|---|---|---|---|---|
| Rating (kVA) | 10 | 25 | 40 | 60 | 80 | 120 | 160 | 220 | 330 | 500 |
| Maximum no.of units | 2 | 2 | 2 | 2 | 2 | 4 | 4 | 4 | 4 | 8 |

non-availability of a unit (e.g. because of maintenance, or a fault) the other unit or units continue to supply the loads without interruption.

Multi-unit systems always consist of UPS units of the same rating which share the overall load. The most frequent variant is the two-unit system (Table 14.2).

*Parallel operation.* The concept of parallel operation employed in the UPS System 42 guarantees maximum independence for the individual units so that they do not adversely affect each other. It ensures that when faults occur the faulty unit is detected and switched off. This is achieved by:

– precise parallel control of the inverters,
– separate actual value measurement for the individual units,
– potential-isolated signal traffic via optocouplers and
– precise parallel control of the rectifiers.

*Batteries.* The UPS units can be operated with either a common battery or with individual batteries, depending on requirements. For redundant systems individual batteries are recommended to achieve explicit segregation of the individual units. This variant also permits subsequent modular expansion.

*Integrated static bypass switch (SBS) service bypass.* In two-unit systems, the SBS power modules of the two UPS units can be connected in parallel, with the

SBS control and load current measurement equipment being installed only in the main unit ( $\hat{=}$ version (a)/Fig. 14.5). The integrated service bypass which is also in the main unit is rated for double the unit rating and is used in installations with unit ratings up to and including 160 kVA.

In UPS units with ratings above 160 kVA, a service bypass has to be installed as an external device (see Fig. 14.8). The external service bypass (option) is fitted in a cubicle of its own to match the UPS cubicles.

The use of a central SBS represents another way of installing a service bypass.

*Central static bypass switch (central SBS)*. Installations of more than two units are equipped with a central SBS. This also applies to installations intended for later expansion. Installations of this type have unit ratings of 120 kVA and upwards.

*Expansion capability*. A UPS unit with integrated SBS (type a, see Fig. 14.6) can subsequently be expanded if necessary by the addition of a parallel unit (type b, see Fig. 14.6). In this way, either the UPS system rating can be doubled or the new unit can serve in a redundant capacity. This also applies, as appropriate, to systems without SBS (see Fig. 14.7).

If future expansion is expected, a system with central SBS should be installed at the outset. Systems with central SBS can be expanded step by step to the extent allowed by the central SBS rating (see Fig. 14.8).

### Operator Control, Monitoring, Diagnosis
*Control panel*. The clearly laid out control panel in the door (see Fig. 14.9) allows simple monitoring and operation of the UPS unit. An LED display on the

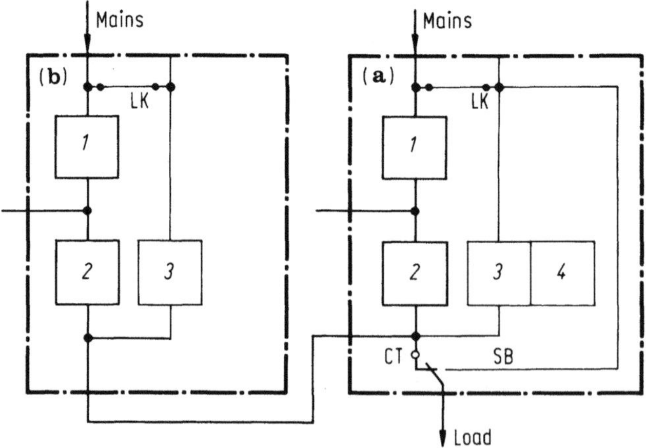


**Fig. 14.6.** Two-unit system with integrated SBS and integrated service bypass (for explanation of abbreviations see Fig. 14.5)

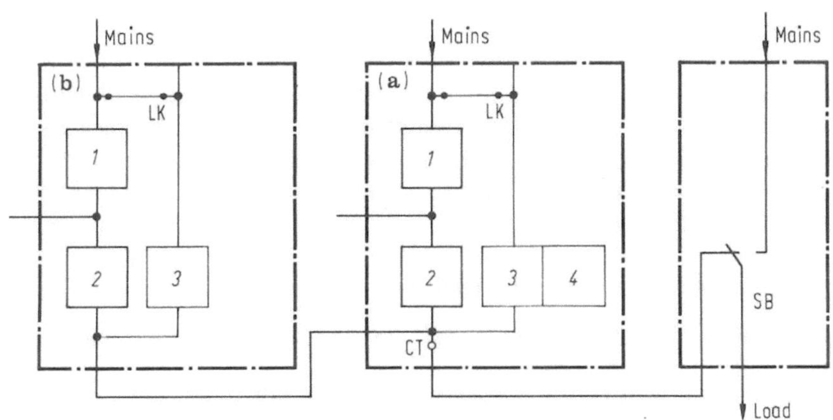


Mains
Mains
Mains
(b)
(a)
LK
LK
1
1
2
3
2
3
4
CT
SB
Load


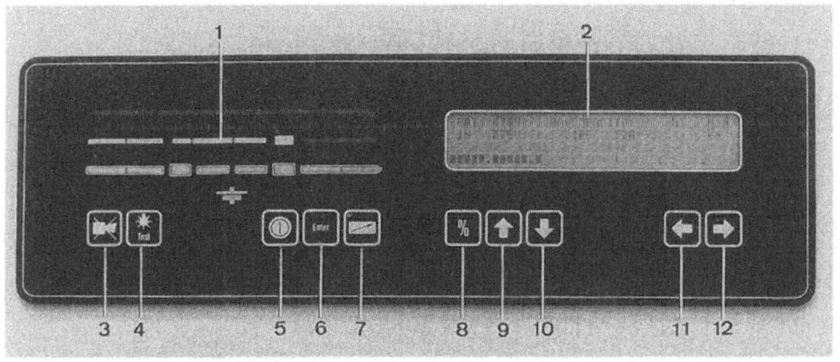


1
2
Enter
Test
%
3 4
5 6 7
8 9 10
11 12

left of the panel shows the UPS operating state at a glance (see Fig. 14.10). An LC display on the right can show information from the monitoring and diagnostic system. Warnings and fault indications are accompanied by an audible alarm.

The controls are in the form of a membrane keyboard. To prevent operator errors the 'enter key' must be pressed after the inverter on/off command is input.

### Additional Equipment

*Remote control panel.* The remote control panel allows remote control and monitoring of the UPS system. The remote control panel is available with functions and design identical to the UPS control panel. Two optional interfaces are added for the UPS-unit where necessary.

*Battery cubicles with valve-regulated (maintenance-free) lead-acid batteries.* Battery cubicles with valve-regulated batteries provide an economical, space-saving solution since they fit in with the arrangement and design of the UPS units and are supplied ready for connection and can be sited in the immediate vicinity of the UPS unit.

*Other addition equipment.* For the UPS unit:

- exchangeable EPROMs allowing the UPS unit for display messages at the control panel in other languages,
- PC software for monitoring and diagnosis (using MS Windows),
- security service (remote monitoring)
- increased degrees of protection (IP 21, IP 31) and
- cubicles of different colours.

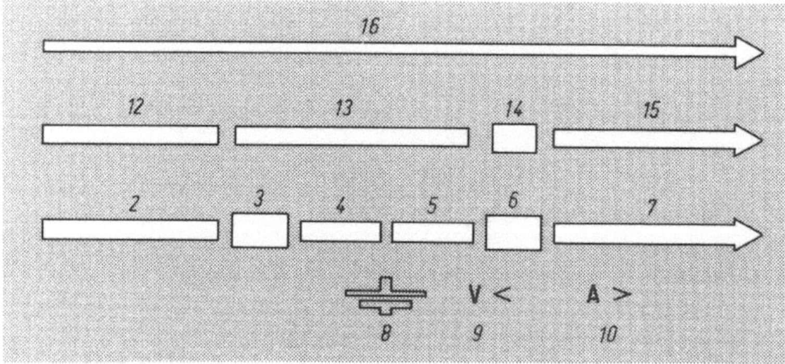


**Fig. 14.10.** Meanings of the LED symbols. *2* Rectifier supply present, *3* rectifier on (green)/fault (red), *4* rectifier operating, *5* rectifier/battery operating, *6* inverter on (green)/fault (red), *7* inverter operating, *8* battery switch on, *9* d.c. link undervoltage warning, *10* inverter/SBS overcurrent, *12* SBS supply present, *13* SBS ready, *14* SBS blocked (red), automatic/mains operation (green), *15* SBS operating, *16* service bypass on

For the energy storage equipment:

- NiCd batteries available as alternative,
- batteries on racks, with accessories and
- battery fuse boxes.

Further equipment:

- matching transformers,
- filter cubicles and
- mains and load distribution boards.

### 14.1.5 Technical Data

The principal technical data for UPS systems series 42 are listed in Table 14.3 (see page 298).

**Table 14.3.** Technical data of UPS system series 42 (three-phases)

| Type | 4201 | 4202 | 4204 | 4206 | 4208 | 4212 | 4216 | 4222 | 4233 | 4250 |
|---|---|---|---|---|---|---|---|---|---|---|
| No. of pulses | 6 | 6 | 6 | 6 | 6 | 6 | 6 | 6 or 12 | 6 or 12 | 12 |

**A.C. input**

| | |
|---|---|
| Rated voltage (V) | $3/N \sim 380$, **400**, $415 \pm 10\%$ |
| Rated frequency (Hz) | 50 or 60 $\pm$ 5% |

**A.C. output**

| | |
|---|---|
| Rated voltage (V) | $3/N \sim 380$, **400**, $415 \pm 1\%$ (stedy-state) with balanced load, $\pm 3\%$ with 100% unbalanced load (single-phase), transient $\pm 4\%$ for 100% load step change |
| Rated frequency (Hz) | 50 or 60 $\pm$ 0.1% self-clocked, $\pm 1\%$ or $\pm 2\%$ (selectable) line-clocked |
| Harmonic distortion | $\leq 2\%$ (with linear load, single harmonic $< 2\%$) |
| Overload capability (three-phase ) | 1.5 $I_N$ for 30 s<br>1.25 $I_N$ for 15 min |

**D.C. link circuit**

| | |
|---|---|
| Permitted range of cell numbers (lead-acid battery) | 174 to 192 |
| Recommended cell number (lead-acid battery) | 192 |
| Float (trickle) charge (V) voltage (adjustable) | 388 to 440 (voltage tolerance $\pm 1\%$ ) |
| Charging voltage (V) (adjustable) | 400 to 440 (voltage tolerance $\pm 1\%$) |
| Final discharge (V) voltage | 307 |

**UPS-unit**

| Mode of operation | continuous mode | | | | | | | | | |
|---|---|---|---|---|---|---|---|---|---|---|
| Multi-unit operation (units) | 2 | 2 | 2 | 2 | 2 | 4 | 4 | 4 | 4 | 8 |
| Cooling | boosted by built-in fans (F) | | | | | | | | | |
| Output rated power kVA | 10 | 25 | 40 | 60 | 80 | 120 | 160 | 220 | 330 | 500 |
| Overall efficiency (%) (at 100 % load) | 90.0 | 90.0 | 91.0 | 92.5 | 92.5 | 93.0 | 93.2 | 93.5/ 93.0 | 94.5/ 93.5 | 93.5 |
| Dimensions / (mm) (H) | 1700 | | | | | | | 1900 | | |
| (W) | 650 | | 950 | | 1250 | | | 1850/2150 | | 2750 |
| (D) | 650 | | | | | | 850 | | | 1050 |

# 15 Diesel Generating Sets

An introduction to standby power supply systems (diesel generating sets) was presented in Sect. 4.2.1.

In the event of failures of the a.c. line supply, batteries provide the necessary power to the telecommunication equipment. Air conditioning systems supplied from the a.c. line supply (to extract heat from the operating rooms) must remain operational during long a.c. line failures.

This can only be achieved with the aid of automatically starting emergency generating sets which can then also supply the rectifiers and other important equipment.

The design of the generating sets takes into account the fact that the generator must be suitable for supplying rectifiers. Emergency generating sets are manufactured to comply with quality specifications as per DIN, VDE and IEC, so that the power necessary to ensure the required high availability of the telecommunications equipment is guaranteed.

## 15.1 Standby Power Supply System with Fully Automatic Mains Failure Control

Fully automatic mains failure control cubicles[1] (see Fig. 15.1) are employed to safeguard the supply to consumers when the mains system fails. They contain the electric equipment required. In line with the amperage and/or environmental conditions available, mains failure systems are used as cubicles for wall-mounting or cabinets for floor-mounting.

### 15.1.1 Motor and Generator

This is composed of

- air- or water-cooled diesel motor (see Fig. 15.2) with exhaust-sound-proofing installation (partly insulated) and air inlet and outlet louvers (motor driven),
- daily fuel tank for up to 8 hours on wall bracket with filling pumps (manual and motor-driven), and fuel pipes,
- three-phase synchronous generator, brushless with static voltage stability $\pm$ 1%,

nominal voltage e.g. 3×400 V/230 V; 50 Hz (60 Hz), (see Fig. 15.3).

---

[1] *Source*: Strüver KG (GmbH & Co.).

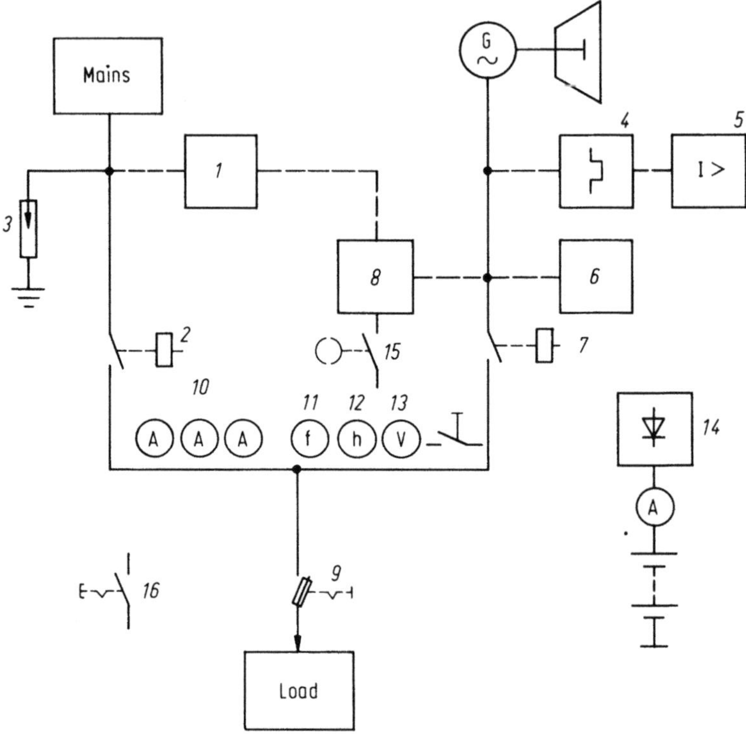

**Fig. 15.1.** Survey diagram of fully automatic mains failure control. *1* Mains monitor $\leqq U$, *2* mains contactor, *3* overvoltage protector, *4* thermal overcurrent release, *5* electronic short-circuit release, *6* generator over- and undervoltage monitor $\leqq U$, *7* generator contactor, *8* automatic electronic control system, *9* load fuse connector/disconnector, *10* amperemeter, *11* frequencymeter, *12* sevice running hour meter, *13* voltmeter, *14* starter battery charger, *15* manual start, *16* emergency stop

### 15.1.2 Switchgear Control Cubicle

This is composed of

- switchgear for separate installation (floor/wall) with diesel control unit,
- three contact fuse load-break switches for outputs to load,
- fully electronic control unit with automatic stop/start,
- with selectable operating mode:
  Automatic,
  Manual,
  Test and
  Off.
- protection against transient overvoltages by means of protective equipment at the a.c. line connection,

**Fig. 15.2.** Water cooled diesel motor with generator. (Photo by courtesy of AD. Strüver KG, GmbH & Co.)

**Fig. 15.3.** Three-phase synchronous generator, brushless with static voltage stability ±1%. (Photo by courtesy of Stamford A.C. Generators NewAge Internationals Company)

– with overvoltage and undervoltage monitoring for the a.c. line,
– phase sequence monitoring (clockwise rotating field) with indication,
– settable transfer time following a.c. line failure and a.c. line return,
– with overvoltage and undervoltage monitoring of the generator,
– remote signals for a.c. line operation, generator operation, disturbance (collective signal) and lack of fuel (daily tank and/or storage tank) via floating potential changeover contacts and

– floating-contact control contacts in generator protective equipment to control the rectifier.

*Operation.* Fully automatic mains failure systems have the following basic functions:

– supply of consumer load entitled to mains failure supply by the mains system,
– in case of mains fault or failure – automatic starting of the mains failure generating set,
– reversing of consumer load from mains supply to mains failure generating set,
– monitoring during operation and in case of fault or failure automatic shut-down,
– when mains returns – delayed reversing of consumer load from mains failure set to mains supply
(optionally direct, manual reversing),
– follow-up operation of the generating set idling for cooling down,
– shut-down,
– possibility for test operation with manual connecting and disconnecting of load, with changeover to fully automatic mains failure operation in case of mains fault or failure and
– possibility for manual operation with manual starting and manual connecting and disconnecting of load.

The measuring equipment of the standard version is suitable for monitoring the generator under normal operating conditions.

The monitoring equipment of the standard version is designed to include the following monitoring criteria:

– starting fault,
– lubricating oil pressure,
– engine temperature and
– overload/overcurrent

Triggering of a diesel fault alarm will effect an immediate tripping of the generator contactor/breaker as well as the shut-down of the generating set by the engine stopping solenoid.

Simultaneously, the visual and audible alarm will be triggered. Thermal overcurrent releases and short-circuit fuses, or electronic short-circuit releases for currents of 63 A and upwards will secure the protection of the plant and the generator by rapid tripping of the generator contactor/breaker.

*Standard Equipment.*
Power section:

– 1 mains contactor, 3–pole/mains circuit breaker with electric drive, 3–pole (for 1000 A upwards),
– 1 generator contactor, 3–pole/generator circuit breaker with electric drive, 3–pole (for 1000 A upwards),

- 3 generator short-circuit fuses (up to 63 A),
- 1 thermal overcurrent release and
- 1 electronic short-circuit release (above 63 A)/Cu. bus bar/cable connections.

Measuring section:

- 1 voltmeter,
- 1 voltmeter selector switch, 7 positions,
- 3 ammeters,
- 3 current transformers (above 63 A),
- 1 frequency meter, reed-type,
- 1 operating hours counter,
- 1 battery charging voltmeter and
- 1 battery charging ammeter.

Control and monitoring section:

- 1 fully automatic electronic control system,
- 1 control contactor for generator–mains reversing,
- 1 horn relay,
- 1 horn,
- 1 operation relay,
- 1 manual start blocking relay,
- 1 electronic-controlled battery charging device 4 A for maintaining the battery
  charging condition (additional charging by set-mounted dynamo),
- 1 set automatic cut-outs and
- 1 emergency–off push-button.

Starter and control section:

- 1 starter (glow) key switch,
- 1 glow control[2],
- 1 preheating device[2],
- 1 starting relay and
- 1 STOP–relay.

Additional Equipment:

- 1 fault alarm fuel shortage,
- 1 kW–meter.

Additional equipment to VDE 0108:

- 1 kW–meter,
- 1 stronger battery charging device with $IU$ characteristics,
- 1 monitoring battery voltage.

---

[2]If required.

**Table 15.1.** Technical data of fully automatic mains failure control system

| | |
|---|---|
| Operation voltage (V) | 220/127–440/254 |
| Permissible voltage stability (%) | + 10 / − 15 |
| Frequency (Hz) | 50 or 60 |
| Rated currents (A) | 25–6000 |
| Ambient temperature / environmental conditions (°C/%) | + 35 / 90 rel. humidity |
| System | Three-wire, three-phase a.c. with neutral point, 3–pole main switching elements |

Additional equipment for automatic reverse synchronizing when mains supply has returned:

- 1 key switch synchronizing I/O,
- 1 lamp synchronizing,
- 1 quick-acting synchronizing device[3] either
- 1 synchronizing device[4],
- 1 two point regulator[4],
- 4–pole mains and generator contactors (breakers),
- 4–pole mains contactor (breakers) enlarged electronic control system and
- electronic control system in plug-in units,
- electronic control system in memory programmable technique,
- enlarged battery charging system, 8 or 16 A and
- mains overvoltage controller.

The principal technical data for the fully automatic mains failure control system are listed in Table 15.1.


## 15.2 Mains Independent Island Power Supply System with Fully Automatic Alternating Operation

### 15.2.1 Switchgear Control Cubicle

*General.* Control cubicles[5] for fully automatic alternating operation (see Fig. 15.4) are employed for the base load power supply to consumer supply systems which demand increased safety of supply.

The power generating system consists of 2 or 3 generating sets of equal output which automatically take over the power generation at fixed intervals (for example every 24 hours). Should one of the sets fail, the second set will automatically start and accept load.

---

[3] $P \leq 250\,\text{kVA}$.
[4] $P > 250\,\text{kVA}$.
[5] *Source:* Strüver KG, GmbH & Co.

Control cubicles for fully automatic alternating operation are available for wall-mounting or cabinets for floor-mounting depending on the amperage and/or environmental conditions.

Figure 15.5 shows a mobile standby power supply system with a diesel generating set and control cubicle.

*Operation.* Systems for fully automatic alternating operation will have the following basic functions:

- supply of consumer load by the generating set selected as priority set,
- automatic changeover of priority to the second set by adjustable time switch clock, taking over load by this set with momentary interruption of < 0.2 sec,
- follow-up operation of the first generating set idling for cooling-down and shut-down,
- in case of fault or failure affecting the set in operation, changeover of priority to the second set, automatic start and accepting load by this set,
- manual change of generating set operation by priority push-button with momentary interruption of < 0.2 sec,
- possibility of test operation of the second generating set for the purpose of maintenance while the first set is fed through to the consumers in automatic operation.
- possibility of manual operation of both generating sets with manual starting and manual connecting and disconnecting of load.

**Fig. 15.4a.** Control cubicle for fully automatic alternating operation. (Photo by courtesy of AD. Strüver KG, GmbH & Co.)

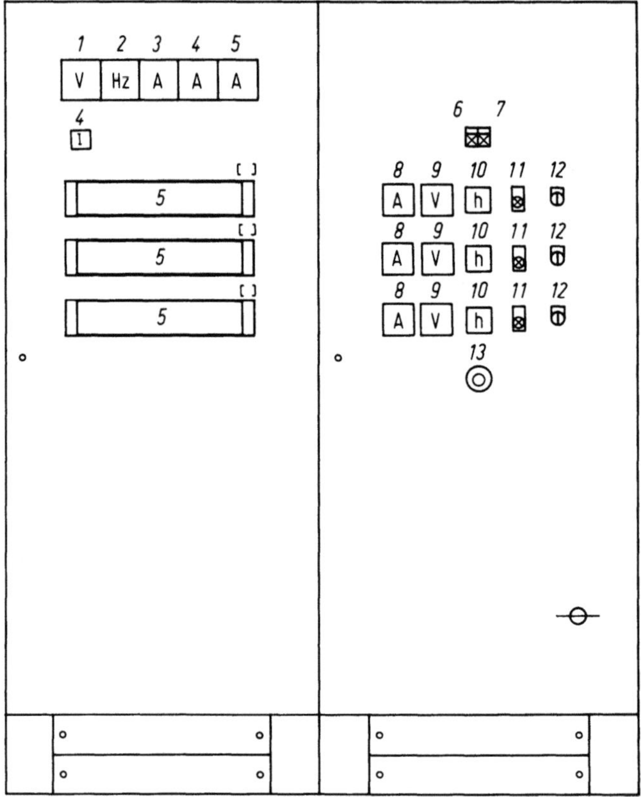

**Fig. 15.4b.** Control cubicle for fully automatic alternating operation. *1* Voltmeter, *2* frequency meter, *3* ammeter, *4* voltmeter selector switch, *5* full automatic electronic control system, *6* signalling lamp-mains available, *7* signalling lamp-mains operation, *8* charging ammeter, *9* battery voltmeter, *10* operating hours meter, *11* priority push button with lamp, *12* start switch, *13* emergency off

The measuring equipment of the standard version is suitable for monitoring the generator under normal operating conditions.

The monitoring equipment of the standard version is designed to include the following monitoring criteria:

- engine fault,
- lubricating oil pressure fault,
- excess engine temperature,
- overload/overcurrent, and
- fuel shortage.

**Fig. 15.5.** Mobile standby power supply system (diesel generating set with control cubicle). (Photo by courtesy of AD. Strüver KG, GmbH & Co.)

Triggering of a diesel fault alarm will effect an immediate tripping of the generator contactor/breaker as well as the shut-down of the generating set by the engine stopping solenoid.

Simultaneously, the visual and audible alarm will also be triggered. Thermal overcurrent releases and short-circuit fuses for 63 A and upwards for electronic short-circuit release will secure the protection of the plant and the generator by rapid tripping of the generator contactor/breaker.

*Standard Equipment*

Power section of generator panels:

- 1 generator contactor, 3–pole,
    generator circuit breaker with electric drive, 3–pole,[6]
- 3 generator short-circuit fuses[7],
- 1 thermal overcurrent release and
- 1 electronic short-circuit release[8]
    Cu bus bar/cable connections.

---

[6] 900 A and upwards.
[7] up to 63 A.
[8] above 63 A.

Measuring section of generator panels:

- 1 operating hours counter,
- 1 battery charging ammeter and
- 3 current transformers[8].


Control and monitoring section of generator panels:

- 1 fully automatic electronic control system,
- 1 control contactor for generator connecting,
- 1 horn relay,
- 1 horn,
- 1 operation relay,
- 1 collective fault relay,
- 1 auxiliary relay AUT / TEST,
- 1 priority selection push-button,
- 1 priority relay,
- 1 priority sequence relay,
- 1 electronic-controlled battery interval charging device 4 A and
- 1 set automatic cut-outs.


Starter and control section of generator panels:

- 1 starter (glow) key switch,
- 1 glow control,[9]
- 1 preheating device,[9]
- 1 starting relay and
- 1 STOP–relay.


Measuring section of the central panel:

- 1 voltmeter,
- 1 voltmeter selector switch, 7 positions,
- 3 ammeters and
- 1 frequency meter, reed-type.


Control section of the central panel:

- 1 time switch clock and
- 1 emergency–off push-button.

---

[8] above 63 A.
[9] If required.

**Table 15.2.** Technical data

| | |
|---|---|
| Operation voltage (V) | 220/127–440/254 |
| Frequency (Hz) | 50 or 60 |
| Rated currents (A) | 25–6000 |
| Ambient temperature / environmental conditions (°C/%) | + 35 / 90 relative humidity |
| System | Three-wire, three-phase a.c. with neutral point, 3–pole main switching elements |

Additional equipment:

- 1 kW–meter,[10]
  additional equipment for automatic transfer synchronizing during generating set changeover:
- 1 quick-acting synchronizing device[10] either
- 1 precision synchronizing device,[11]
- 1 two point regulator,[11]
- 4-pole generator contactors/breakers,[12]
- enlarged electronic control system (maximum 8 fault alarms),
- electronic control device in plug-in units (version with flashing lights),[12]
- electronic control device in memory programmable technique.

Additional measuring, monitoring and operating devices, control of ancillary equipment, special class of protection and standard specifications etc. are available.

Auxiliary panels for consumer load supply are available. The principal technical data for the fully automatic alternating operation control system are listed in Table 15.2.

## 15.3 Fully Automatic Electronic Control System

The control device[13] (see Fig. 15.6) performs all the fully automatic controlling actions required by an alternating operation of a mains independent power supply system (see Sect. 15.2).

The control device for systems with public mains connections is very similar (see Sect. 15.1).
- monitoring mains voltage,[14]

---

[10] $P \leqq 400\,\text{kVA}$.
[11] $P > 400\,\text{kVA}$.
[12] for each generator panel.
[13] *Source*: Strüver KG (GmbH & Co.).
[14] Only for systems with public mains connections.

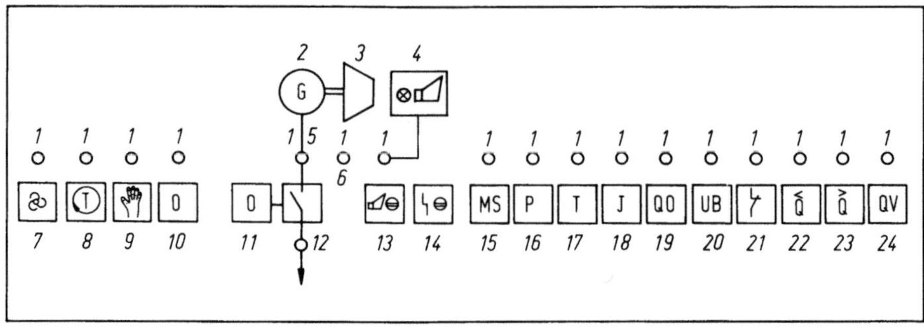

**Fig. 15.6.** Fully automatic electronic control system (compact automatic) in the application for alternating operation of mains independent power supply systems. Operation selection position: *1* LED–indication, *2* generator, *3* combustion engine, *4* fault alarm system device (activated), *5* generator voltage ON, *6* ON instruction engine, *7* AUTOMATIC*, *8* TEST*, *9* MANUAL*, *10* OFF*, *11* generator contactor OFF–ON (manual)*, *12* generator contactor ON, *13* Horn acknowledgement/signal testing*, *14* fault acknowledgement*, *15* engine start fault, *16* lube oil pressure fault, *17* excess temperature, *18* overcurrent, *19* lube oil level low, *20* starter battery voltage low, *21* automatic circuit breaker tripped, *22* fuel level low, *23* fuel level high, *24* fuel storage tank level low

* Button switch

– automatic starting in response to mains fault or failure and monitoring of the generating set [14],
– changeover of consumer load to the generating set,
– delayed reversal of consumer load to the mains system when mains supply returns[14],
– diesel follow-up run and automatic shut-down,
– possibility of test operation with manual connecting and disconnecting of consumer load; transition to automatic mains failure operation in the case of mains fault or failure[14] and
– possibility for manual operation by external starter switch with manual connecting and disconnecting of consumer load.

All fault alarm circuits are operating with flashing light. The device is constructed for front-face mounting into a switchgear panel.

Tactile bubble switches are available as operating elements and luminescent diodes for indication, mounted in a front plate, protected against dust and water.

A main assembly contains all the functions and time circuits in digital C–MOS control technique including command devices and connecting plugs.
Separately arranged in the device are two further module plates containing the mains voltage[14] and generator voltage monitors.

All fault alarm circuits are available in plug-in module technique form. The device is also available with the following modifications:

– extension to 10 fault alarm functions,

---

[14]Only for systems with public mains connections.

– remote starting automatic without mains voltage monitor (an example is shown
in Fig. 15.3),
– additional devices for fault alarm extensions for any number of fault alarms
and
– fault alarm functions with coding to open-circuit or closed-circuit system.

The principal technical data and available settings for the fully automatic control
system (as in mains failure control) are listed in Tables 15.3 and 15.4. The flow
chart (programme sequence plan) of the same device is shown in Fig. 15.7.

**Table 15.3.** Technical data

| | |
|---|---|
| Battery voltage (V) | 12 or 24 d.c. |
| Ambient temperature ($^0$C) | $-$ 20 to $+$ 65 (operation)<br>$-$ 30 to $+$ 85 (storing) |
| Permissible variation of<br>  supply voltage (V) | $10-32$ (static)<br>$6-100$ ($<$1s) (dynamic) |
| Meas. voltage and frequency mains[a] (V/f) | $380/220-416/240$ ;<br>50/60<br>$208/120-240/138$ ;<br>50/60 |
| Meas. voltage and frequency generator (V/f) | 208–240 ; 50/60<br>280–416 ; 50/60 |

[a]For voltages $<$ 208 V or $>$ 416 V use separate matching transformers

**Table 15.4.** Available settings

| Function | Setting range | Setting by factory |
|---|---|---|
| Delayed release (s) | 6–16 | 12 |
| Starting time (s) | 1–15 | 5 |
| Start interval time (s) | 1–15 | 5 |
| Preglowing time (s) | 0.2–7 | 3 |
| Closing delay generator<br>  circuit breaker (s) | 0.5–6 | 2 |
| Tripping delay generator<br>  circuit breaker (s) | 1.5–6 | 2 |
| Reversing gap generator/mains<br>  (mains/generator) as required (s) | – | 1 |
| Mains failure time lag (s) | 0.2–3.5 | 2 |
| Reverse switching time (s) | 20–100 | 60 |
| Follow-up cooling run (s) | 130–800 | 180 |
| Stop time diminishing (s) | – | 6 |
| Stop time (s) | 20–35 | 25 |

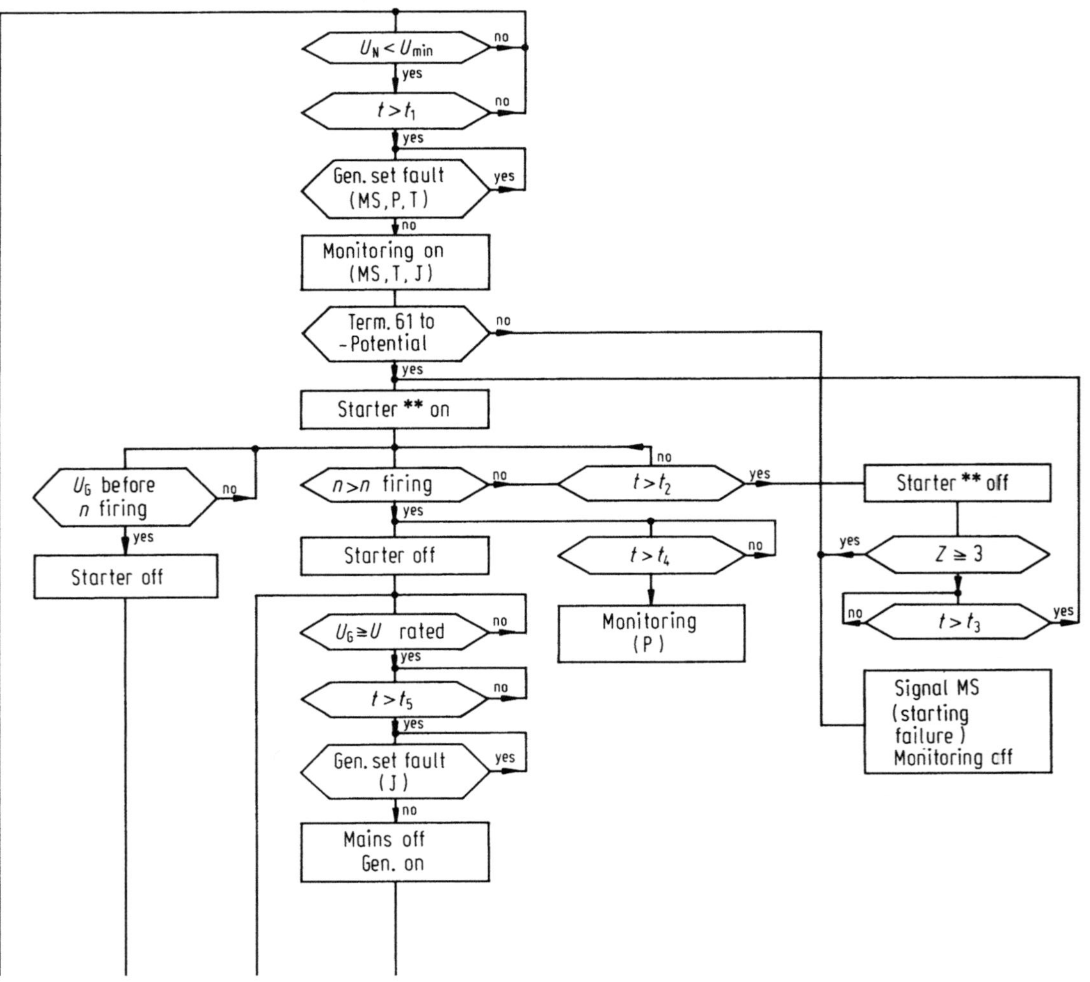

$U_N < U_{min}$
no
yes
$t > t_1$
no
yes
Gen. set fault (MS, P, T)
yes
no
Monitoring on (MS, T, J)
Term. 61 to –Potential
no
yes
Starter ** on
$U_G$ before $n$ firing
no
$n > n$ firing
no
$t > t_2$
yes
Starter ** off
yes
Starter off
yes
Starter off
$t > t_4$
no
$Z \geqq 3$
no
$U_G \cong U$ rated
no
yes
Monitoring (P)
$t > t_3$
yes
$t > t_5$
no
yes
Signal MS (starting failure) Monitoring off
Gen. set fault (J)
yes
no
Mains off Gen. on

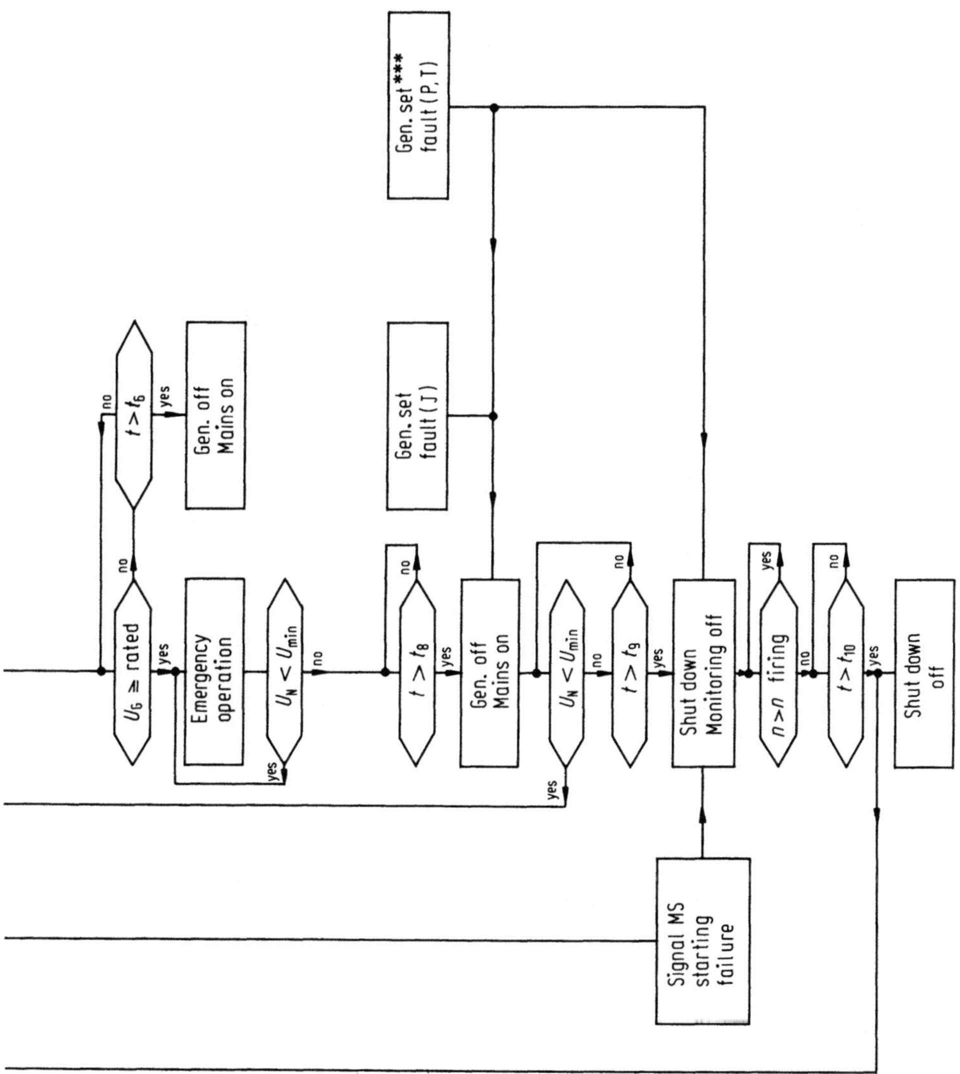


**Fig. 15.7.** For caption see next page

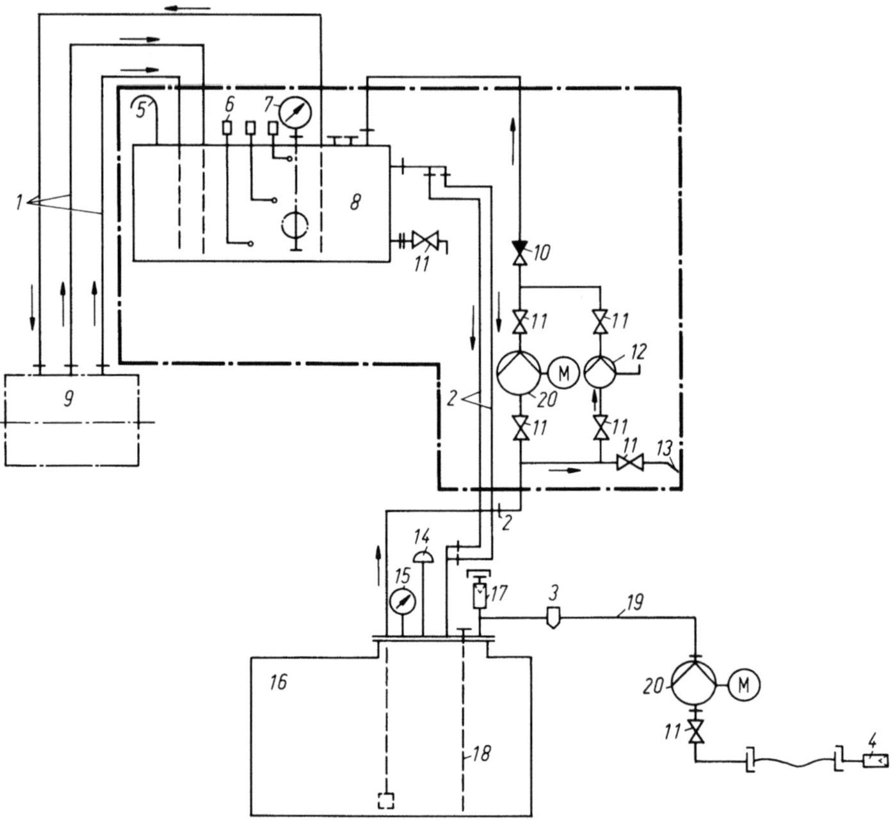

**Fig. 15.8.** Fuel flow diagram. *1* Pipe 10×1, *2* pipe 28×2.75, *3* filter, *4* strainer, *5* vent pipe, *6* float switch, *7* level indicator, *8* service tank, *9* diesel engine, *10* return valve, *11* valve, *12* hand pump, *13* from barrel, *14* vent pipe, *15* level indicator, *16* storage tank, *17* tank cap, *18* pipe 22×2, *19* pipe 28×2, *20* electro-pump

**Fig. 15.7.** Flow chart for the operation selection 'AUTOMATIC' of the fully automatic electronic control system (example of mains failure control). $t_1$ Mains failure time lag, $t_2$ starting time, $t_3$ start interval time, $t_4$ release time for pressure monitoring, $t_5$ connecting time lag gen. circuit breaker, $t_6$ switching off time lag gen. circuit breaker, $t_8$ reverse switching time, $t_9$ follow-up running time, $t_{10}$ stop time, $U_N$ mains voltage, $U_{MIN}$ min. permissible mains voltage, $n_{zünd}$ firing speed, $U_G$ generator voltage, $U_{Soll}$ permissible generator voltage, $Z$ number of starting intervals, MS engine starting fault (starting failure), P lube oil pressure fault, T lack of temperature, J overcurrent

** Possibly necessary glow programme not considered
*** Fault alarm during emergency operation warning signal only!

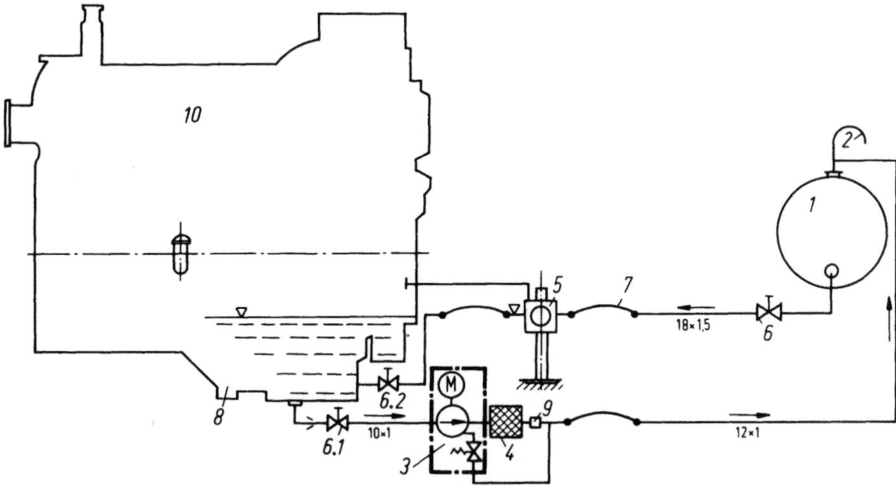


**Fig. 15.9.** Lubrication oil diagram. *1* lube oil tank, *2* tank vent, *3* electro-pump set, *4* filter, *5* oil level regulator, *6* shutoff valve, *7* flexible hose, *8* oil pan, *9* flow switch, *10* diesel engine

## 15.4 Fuel Distribution System

Figure 15.8[15] shows a fuel diagram of a system with one diesel engine, service (day) tank and storage tank.

## 15.5 Lubrication Oil Distribution System

Long service intervals of up to 1500 running hours can be reached by following this system (Fig. 15.9[15]). The fresh oil from the lubricating oil tank automatically changes that already used in the diesel engine.

---

[15] *Source:* Strüver KG (GmbH & Co.).

# 16 Special Features
## for Transmission System Power Supplies

In considering power supplies for long-range transmission systems it is necessary to distinguish between the following:

- power supply systems in telecommunication towers,
- power supply systems in ground communications stations and
- power supply systems independent of mains supplies.

Radio link stations and repeater stations often have to be built in places where public mains supplies are not available. For these situations there are power supply systems which are independent of mains supplies, constructed in shelters. These can be hybrid systems, consisting, for example, if a wind generator and a solar generator together with a diesel generating set and a battery.

## 16.1 Power Supply for Radio Link Apparatus

Since telecommunications towers (Fig. 16.1) are particularly vulnerable to lightning, special measures have to be adopted in connection with the power supply systems for earthing and lightning protection (see Chapter 18). As regards the energy source, there are installations operating in the standby parallel mode.

The power supply system (standby parallel mode) is usually installed in a building adjacent to the telecommunications tower.

Power supply for transmission systems for small until medium power range consists of:

- a.c. distribution panel,
- transistor controlled rectifier modules,
- d.c. distribution panel and
- batteries.

Power supply for transmission systems for large power (e.g. 100 kW) consists of:

- a.c. mains distribution switchboard or mains switch panel,
- thyristor controlled rectifier units,
- battery switching panels,
- distribution panels and
- batteries.

**Fig. 16.1.** Telecommunications tower. (Photo by courtesy of Siemens AG)

The basic circuit of a d.c. fail-safe power supply system is shown in Fig. 16.2.

The principal components are

- standby generating set (NEA),
- rectifiers.
- lead-acid batteries (B1, B2).
- battery switching panel with control unit BS and
- d.c. distribution board with lightning protection assemblies.

For the sake of security, at least two rectifiers and two batteries are provided.

In normal operation power is obtained from the supply mains. In the event of a supply failure the rectifier switches off and the battery takes over the load without interruption. At the same time the standby generator is started automatically; it provides power after a few seconds and the rectifiers switch on again. When

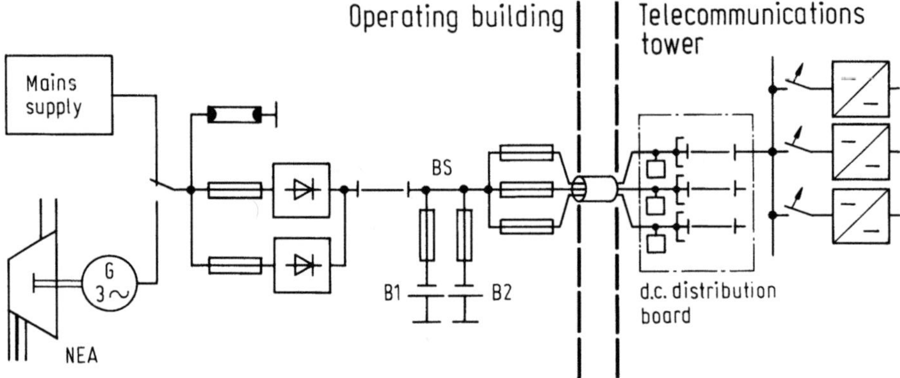


**Fig. 16.2.** Basic circuit of a power supply system for telecommunications towers (standby parallel mode)

the mains supply is restored the standby generator is shut down. The rectifiers supply the power to the loads as well as charging current for the battery.

The power is distributed by the battery switching panel BS through fuses. The battery switching panel is linked to the d.c. distribution board by a screened cable, which runs from the power installation building to the tower through a metal conduit. The distribution board, on the equipment floor of the telecommunications tower, carries lightning protection assemblies and also caters for the power distribution to the rack rows. Here the current is distributed to the individual items of transmission equipment from distributor units with automatic circuit breakers via d.c./d.c converters.

## 16.2 Power Supply for Ground Communication Stations for Telecommunications Satellites

Ground communication stations represent a special type of radio link installation. They are supplied with a.c. from an uninterruptible power supply system (UPS) (Fig. 16.3). In normal operation the inverters draw their power through rectifiers from the supply mains and, in the event of a supply failure, from a battery connected to the d.c. link. To increase the reliability and/or the power of the a.c. power supply the inverters are used in a half-load parallel arrangement or an $n+1$ operation system. In addition a revert-to-mains unit (static transfer switch STS) is provided, which on overload (a short-circuit, for example), or in the event of failure of both inverters, transfers the loads without a break to the mains (see Sect. 4.3.5 and Chapter 14).

To guarantee continuous operation of the earth station a power supply system consisting essentially of the following main parts is available (see Fig. 16.4).

– standby dual diesel-generator power plant,

**Fig. 16.3.** Ground communication station for telecommunications satellites. (Photo by courtesy of Mr. Alfhart Amberger and Siemens AG)

- a.c.-uninterruptible power supply (a.c. UPS) and
- a.c. distributions.

The main components of the diesel generator power plant are:

- two diesel-generator sets,
- two diesel-generator control modules,
- switch gear with load distribution,
- fuel supply system and
- engine lubricating oil supply system.

The standby diesel-generator power plant consists of two diesel-generator sets with its control modules which work in a sequence switching operating mode within intervals of 24 hours. This means that in the case of mains failure the diesel-generator set which is selected as a first priority set will start and take the load. If the mains outage period is longer than 24 hours the running diesel-generator set will stop and the second diesel-generator set will start and take the load until mains recovery. The mains monitoring control timing for start and stop commands are adjustable. If one of the diesel-generator sets is stopped, in case of failure the priority switching equipment is then overriden.

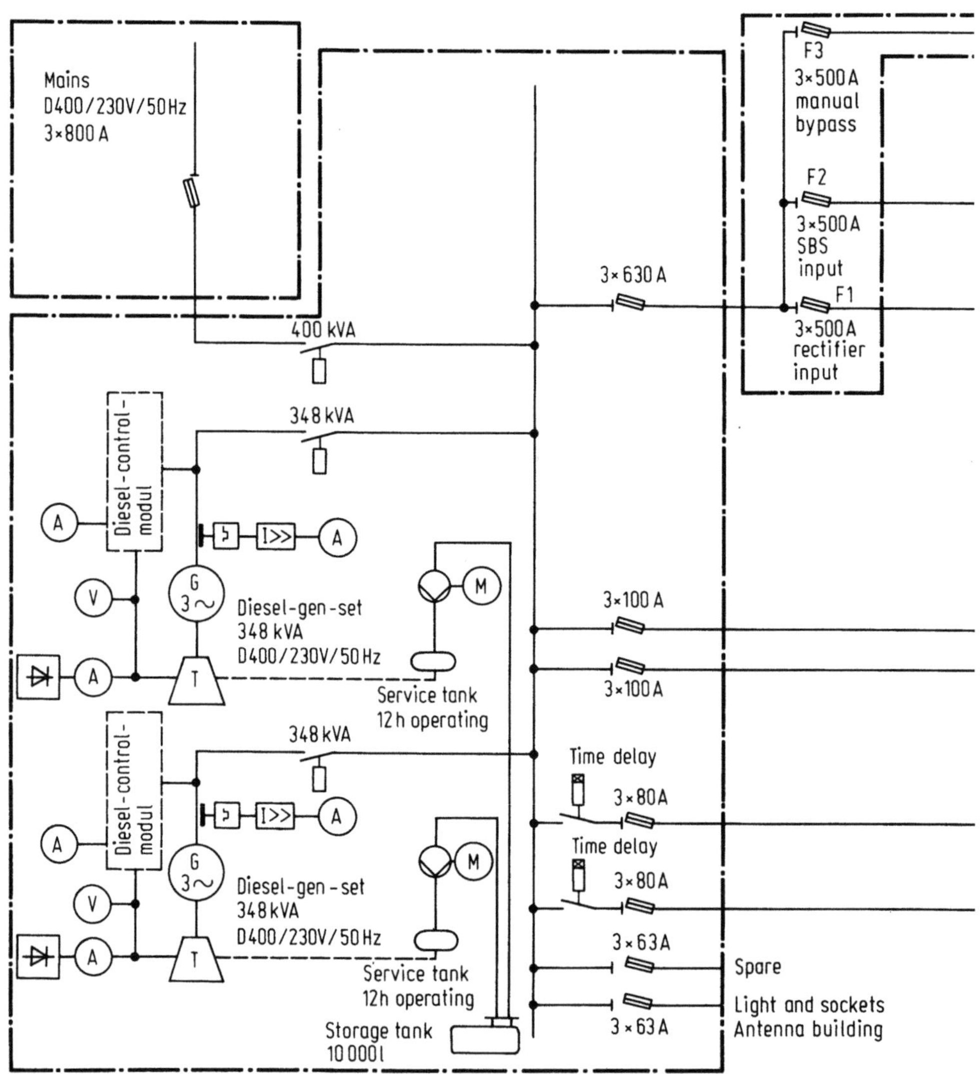


**Fig. 16.4.** Survey diagram of a power supply system for telecommunications satellites ground station. (*Source*: Siemens AG)

Fault monitoring circuits for supervision of the diesel-generator sets are provided.

The switch gear contains a sufficient number of load circuit breakers for connecting the generators, several pieces of diesel auxiliary equipment, the a.c.-UPS power supply, air conditioners and supply of generators and equipment within the building.

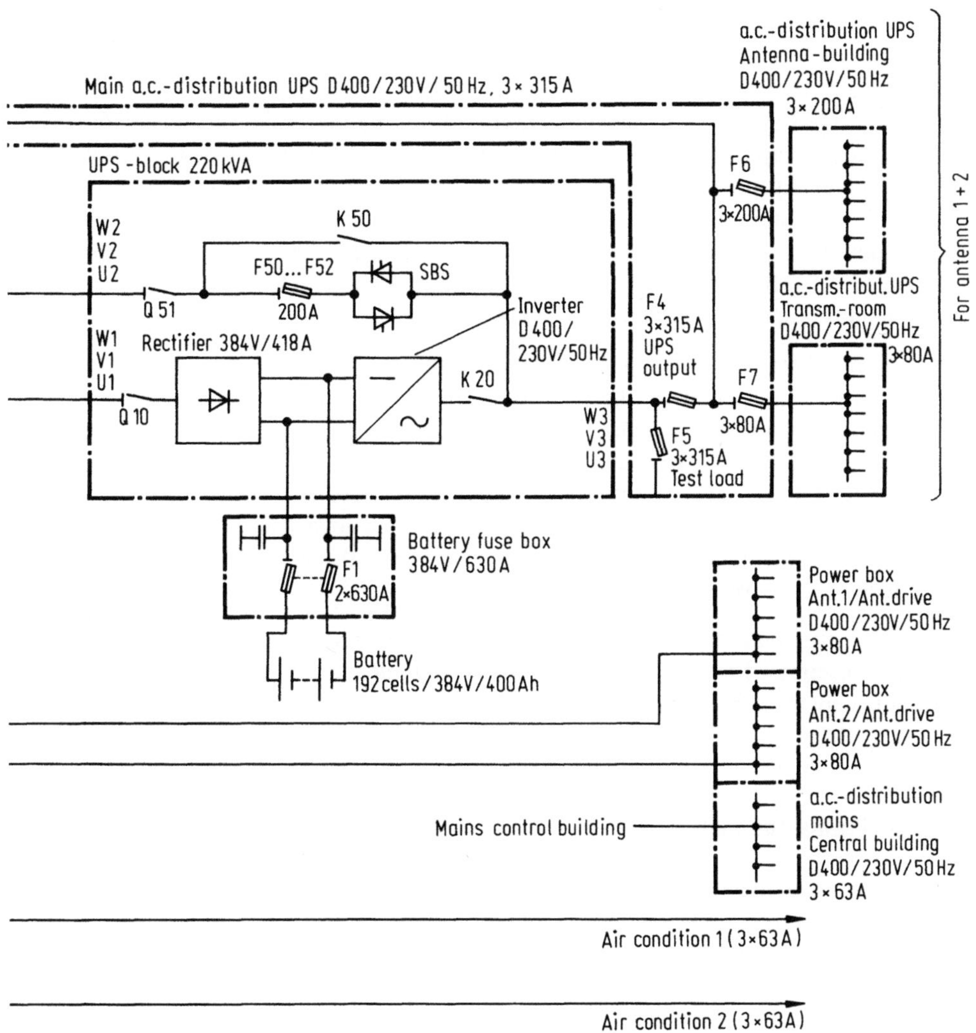


**Fig. 16.4.** *(Continued)*

Each diesel engine contains all the necessary components such as the day tank stand equipped with electric and manual operated fuel pumps and the necessary valves and fuel pipes (see Fig. 15.8). The day tank has a capacity for 12 hours' operation at full load. An outdoor storage tank, mounted above ground level with a capacity for at least 7 days' operation at full load, equipped with all the necessary armatures and fuel pipes for connection to the day tank is available.

For maintenance-free operation of each diesel-generator for say 168 hours the engines are equipped with enlarged oil pans and suitable oil filters. For oil change the drainage valves and outlets are easily accessible.

A.C.-UPS systems are installed where a fully-uninterruptible independent and secure power supply is required. One or more UPS-units are provided. They work in a redundancy parallel operating mode. The concept of parallel operating mode in the UPS system 42 (see Chap. 14) guarantees maximum independence for the individual units. It ensures that when faults occur the faulty units is detected and switched off. The remaining unit takes the load without interruption.
The main components of the UPS system are:

- the rectifier for rectifying the mains voltage,
- inverter which converts the d.c. voltage into a sinusoidal a.c. voltage,
- a static bypass switch (SBS) which can provide an immediate direct connection between the output and the mains feed,
- service bypass (manual override) which can be used to isolate the UPS equipment manually without interrupting the supplied load and
- a battery system which has a capacity to cover the connected UPS load for at least 15 minutes in case of mains fault.

The a.c. distributions are equipped with suitable magnetic air circuit breakers. Fuse load circuit breakers are provided at the inputs of the distributions. Voltage is indicated by signal lamps. All live parts of the distributions are covered.

## 16.3 Mains-Independent Power Supply Systems

As previously mentioned, radio link stations and repeater stations, if they are located where no public mains supply is available, must be provided with mains-independent power supply systems (Figs. 16.5 and 16.6).

In contrast to standby power supply systems, mains-independent power supplies in these circumstances signify small, continuously operating local 'power stations' (primary power sources).

### 16.3.1 Primary Power Sources

Primary power sources for use at operating locations are considered in the following paragraphs (see Table 16.1).

*Diesel generating set systems.* In power supply systems for stations with power sources with > 1000 W generally diesel generating sets are used.

*Continuously operating diesel generators.* In systems employing diesel generators, two or more diesel sets operate in rotation on a continuous basis in conjunction with rectifiers and parallel-connected batteries.

**Fig. 16.5.** Mains-independent-radio repeater station with (from right) equipment shelter with air conditions system, tower, shelter with dual diesel generating set 2 × 20 kVA and storage tank. (Photo by courtesy of Siemens AG)

**Fig. 16.6.** Mains-independent power supply system with dual diesel generator set in a shelter. (Photo by courtesy of Siemens AG)

**Table 16.1.** Technical features of primary power sources

| Power source | Power range | Fuel | Availability | Maintenance cost | Other features |
| --- | --- | --- | --- | --- | --- |
| Diesel generator | $\geq 2\,kW$ | Diesel oil | Adequate | High | A.C. supply available for air-conditioning plant and lightning; rectifier and generally a battery necessary; building or shelter necessary |
| Small steam turbine | 200 to 3000 W | Gas or diesel oil | High | Low with gas; somewhat high with diesel oil | D.C. output-a.c. only through an inverter; in a dual system only 2 hour maximum battery standby time required; no building required |
| Thermogenerator | 10 to 400 W | Gas, radio-isotopes (diesel oil on trial) | High | Low | D.C. output; no building required; battery not always necessary |
| Solar generator | $\leq 1000\,W$ | None required | Dependent on duration and intensity of radiation | Low | D.C. output; supporting structure required; large-capacity battery necessary |
| Wind generator | approximately 10 kW | None required | Dependent on local wind conditions | Low | D.C. or a.c. output; mast required; battery necessary |

Particular features of this mode of operation are:

- maintenance is only necessary at weekly intervals, or every two or three weeks, since each generator set (without special provision) can only continue in operation for about a week and
- an a.c. supply is available.

*Diesel generators with battery mode.* The diesel generators charge the batteries alternately in a relatively short time. The diesels run for only a few hours a day, while the loads are supplied continuously from the battery (see Fig. 4.5).
   Particular aspects of this mode of operation are:

- the maintenance intervals are extended to months. The battery capacity has to be significantly larger than in the systems mentioned previously,
- the diesel fuel consumption is relatively low and
- a.c. supplies are available if required.

*Steam turbines.* The low-power steam turbine operates with a closed steam circuit and can be fuelled with liquefied gas, natural gas, diesel oil or kerosene. The turbines are used either individually with batteries or in the parallel mode with or without batteries.

*Thermoelectric generators.* Thermoelectric generators comprise a large number of thermoelectric elements which convert heat into electrical energy; they thus contain no moving parts. The greater the temperature difference between the hot and cold sides, the higher is the efficiency. The output power of thermoelectric generators is strongly dependent upon the ambient temperature.
   A number of units can be connected in parallel to form a larger system and can be used with or without a battery.

*Solar generators.* The solar generator converts solar energy directly into electrical energy. It consists of a number of solar modules (see Fig. 16.15), each of which contains a large number of solar cells; these are protected from mechanical damage on the light-sensitive side by special high-transmission glass with a low-reflection surface. On the back of the module, a moisture barrier is provided by a plastic-laminated aluminium sheet, which at the same time assists heat dissipation. Optimum forms of solar modules and generators for particular requirements can be assembled by connecting cells in series and parallel.

*Solar cells.* A distinction is made in regard to the use of solar energy between direct and indirect utilization. Direct utilization is represented by photothermal solar technology and by photovoltaic (photoelectric) technology. The short description that follows is confined to the latter.
   Solar cells are large-area semiconductor diodes which convert energy from the sun (global irradiation $\triangleq$ direct and indirect light) directly into electrical energy.The basic material principally used for the manufacture of solar cells is

silicon. Different types of solar cell are made respectively from monocrystalline silicon (mainly used at present), from polycrystalline and from amorphous silicon. There are also solar cells made from cadmium sulphide/copper sulphide or gallium aluminium arsenide.

Further explanation will relate to the monocrystalline silicon solar cell.

Figure 16.7a shows the arrangement of layers in the solar cell. The incident sunlight generates charge-carrier pairs in the differently doped silicon semiconductor material, so that as much of the radiation as possible is captured. The proportion of the light that is reflected is kept low by an anti-reflection layer (a rough surface). The carrier pairs are separated by the space-charge zone; the electrons drift to the front, the holes to the back of the cell. This gives rise to the open-circuit voltage $U_L$. The short-circuit current of a circular solar cell of monocrystalline silicon with $100\,\text{mm} \times 100\,\text{mm}$ as an example, is over 3 A with an incident irradiation of $1\,\text{kW/m}^2$ (Fig. 16.7b). The photoelectric current is primarily dependent upon the irradiated area of the cell and the intensity of the incident light. The open-circuit voltage is about 0.5 V. The cells are provided with tinned contacts.

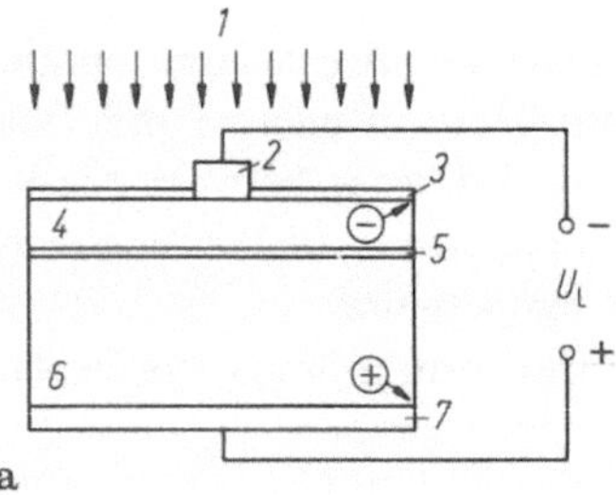

**Fig. 16.7a, b.** Layer construction of a solar cell (**a**), solar cell (**b**). *1* Solar global radiation, *2* metallization (front), *3* anti-reflection layer, *4* N-silicon ($\approx 0.3\ \mu\text{m}$), *5* PN junction, *6* P-silicon ($\approx 350\ \mu\text{m}$), *7* metallized back, $U_L$ photovoltage (open-circuit voltage). (Photo by courtesy of Siemens AG)

Figure 16.8 shows the typical current/voltage characteristic of a solar cell. Depending on the voltage or current required, solar cells are connected in series and parallel arrangements to form solar modules or generators.

The current/voltage characteristics of a solar generator are shown in Fig. 16.9; those in Fig.16.9a are based on a constant cell temperature ($T_C = 60\,°C$) and various irradiation levels, while in Fig.16.9b a constant irradiation $E$

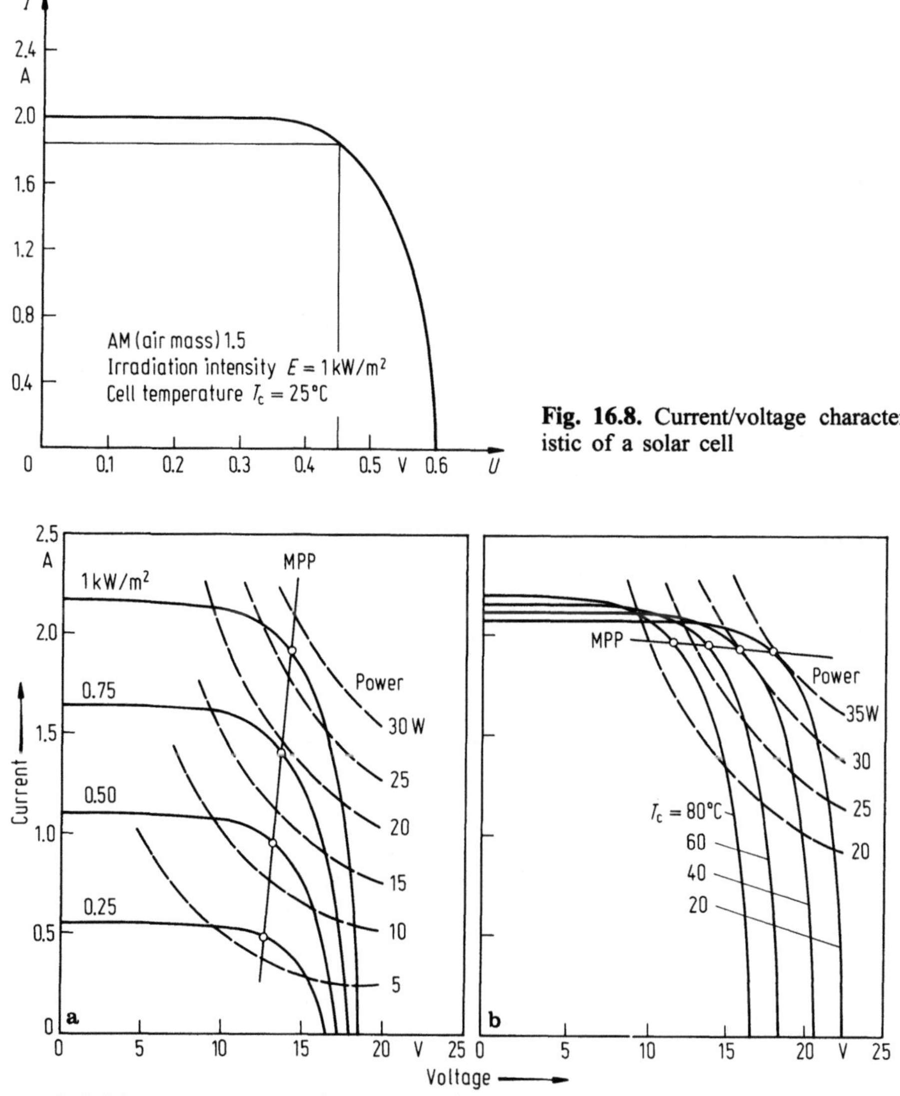


**Fig. 16.8.** Current/voltage characteristic of a solar cell

**Fig. 16.9.** Current/voltage characteristic of a solar generator

is assumed and the characteristics are drawn for various cell temperatures. Full utilization of the power capability of the generator is possible if it is operated at the point of maximum power (MPP) for every irradiation level and temperature. For this purpose an additional MPP controller is required in the control system of the d.c./d.c. converter associated with the solar generator.

A battery must be provided to cater for the hours of darkness and for sunless days.

In present technology, it is possible to use solar generators economically for stations with approximately up to 500 W power requirements in latitudes with average-to-good sunshine (Fig. 16.10).

*Wind-generator systems.* The wind generator consists of a generator coupled to a rotor or a turbine. There are several forms of construction:

- the *Savonius rotor* is noted for its robust construction and easy starting. It has a vertical shaft and its operation is not dependent upon the direction of the wind; the power obtainable is small,
- the *Darrieux turbine* is simple in construction, has a vertical shaft, operates dependently of wind direction and has a high efficiency. It is not self-starting and
- *horizontal-shaft turbines* are used in considerable numbers. Because of the experience gained from them, this pattern is today the most highly developed. The turbines are characterized by high efficiency.

Favourable locations for wind generators are mountain ridges, open plains and coastlines or funnel-shaped valleys (Fig. 16.11). If long periods of operational time are required it is necessary to observe the wind values at the predeterminated location if planning a wind driven generating system.

The *fuel cell* must be mentioned in this connection. It is not yet normally used these days in alternative power supply systems.

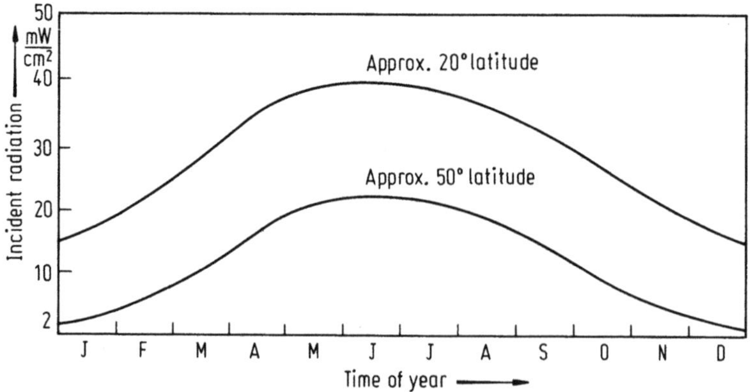


**Fig. 16.10.** Examples of average yearly variation of solar radiation intensity

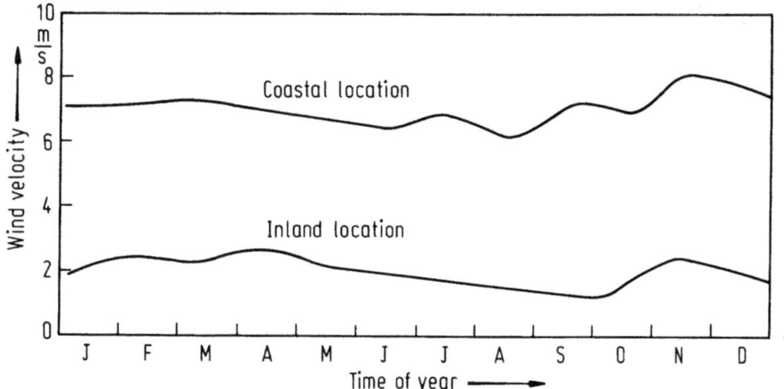


**Fig. 16.11.** Examples of average yearly variation of wind velocity

The idea of using fuel cells originates from the year 1930. Development has been pushed ahead in the last twenty years primarily from the point of view of its use in electric vehicles, in space and for small power stations. Fuel cells work, for example, with hydrogen and oxygen. The fuels can be fed from pressure storages, the quantity of fuel stored being matched to the desired operating time. Fuel cells can also function as reversible units by the addition of electrolyzers, i.e. as when charging a battery water can be decomposed into hydrogen and oxygen, which are then available again when there is a need for energy.

### 16.3.2 Solar Power Supply Systems

Photovoltaic power supply systems (Figs. 16.12 and 16.13) are designed to feed low power electrical consumers, such as transreceiving and relay stations in places where there is no (or inadequate) local power supply. These systems chiefly consist of the following elements:

- solar array to convert solar radiation into electricity,
- charging controllers to adapt the output of the solar array to the demands of telecom equipment
- storage battery to store surplus electricity generated during the day and to feed the telecom equipment at night or during short periods of bad weather.

*Solar Power Control Cubicle.* The solar power control cubicle as shown in Figs. 16.13 and 16.14 contains all the necessary installations for the management of energy from the solar power generator via its charging controller to and from the storage battery, and to the telecom equipment.

The charging controller, the storage battery and the telecom equipment are connected via fuses to a common battery busbar. Measuring instruments are provided to indicate the battery voltage, and the telecom equipment's current consumption. A special bidirectional ampere-hour counter indicates the charge

**Fig. 16.12.** Solar generator of a mains-independent power supply system. (Photo by courtesy of Siemens AG)

or discharge current to or from the battery and monitors the battery's state of charge.

A voltage monitoring unit monitors the voltage at the busbar and, should this either fall below 44.4 V or rise above 58.3 V a respective signal will be transmitted to the control room.

*Charging controller.* In order to adapt the solar modules' output to the demands of storage battery and telecom equipment, a charging controller is necessary. According to the battery's state of charge represented by the voltage at the busbar, the charging controller transmits energy from zero to the full range of energy offered into the system and so prevents any overcharging of the battery. The controllers contain integrated protections against surges from atmospheric overvoltages. Output terminals of more than one controller may be connected to the common battery busbar.

*Method of operation.* The voltage at the battery busbar is determined by the supply and demand of energy and the charge of the battery. If the solar modules offer more energy than is needed by the load, the difference will be fed into the battery. If the battery charge is low, the voltage at the busbar will be determined by the battery. The charging controller will not permit the cell voltage to rise above 2.33 V, and the charging current will then be determined by the battery.

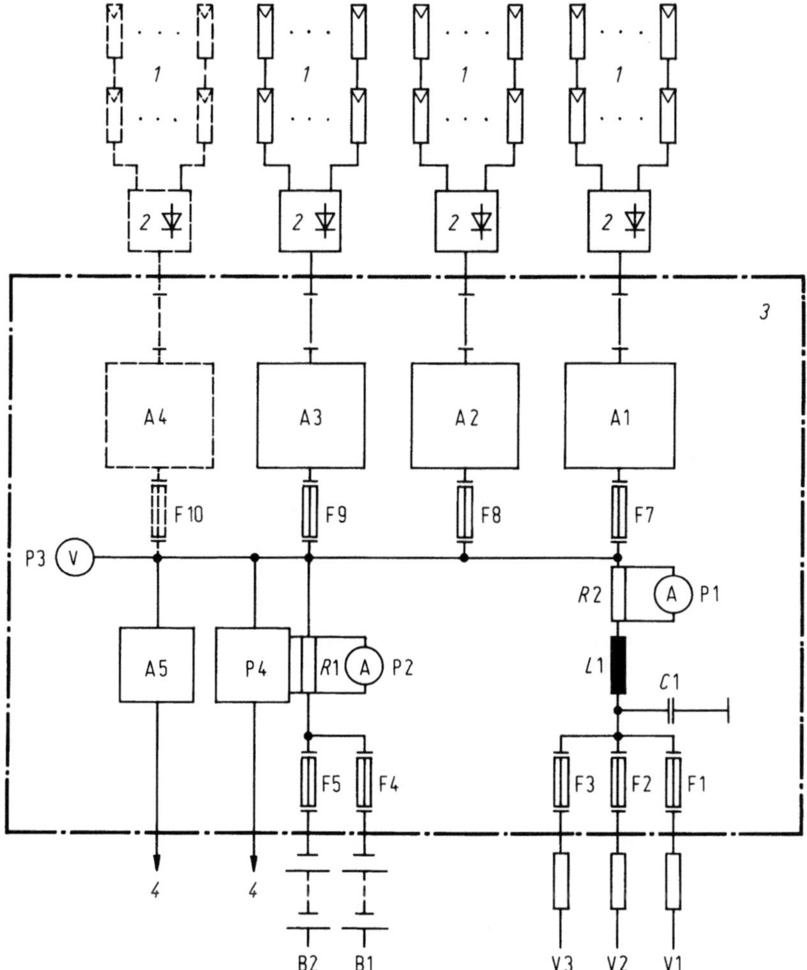

**Fig. 16.13.** Survey diagram of solar power supply. *1* Solar generator/solar modules (e.g. 4 in series), *2* array connecting box with decoupling diodes, *3* solar power control cubicle, *4* remote signalling, A1, A2, A3, A4 charging controller (voltage regulator), A5 d.c. voltage monitor, B1, B2 battery 1,2 (24 cells), P4 ampere-hour counter, V1, V2, V3 load (transmission system)

This ensures fast charging of a low battery and prevents overcharging as well. If the solar modules can not supply sufficient energy for the load, the difference will be drawn from the battery, which then will determine the voltage.

A voltage monitor will send out a signal in case the battery voltage should fall below 44.4 V.

*Ampere-hour counter.* A special bidirectional ampere-hour counter (programmable type) is installed to monitor the battery's state of charge. The display shows the

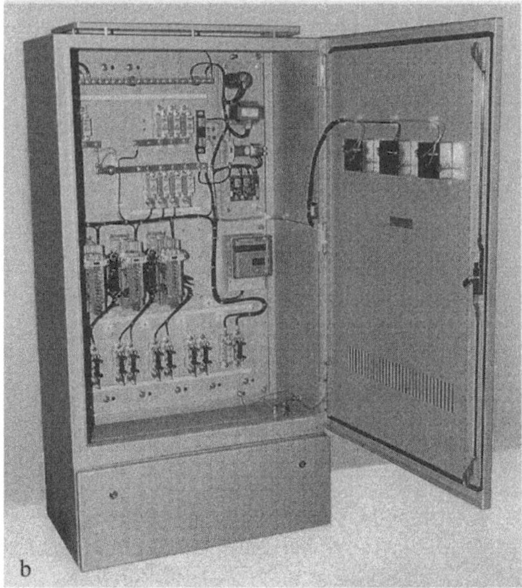

**Fig. 16.14a, b.** Solar power control cubicle with controlsystem for solarenergie, distribution, Ah-counter and three solar power regulators with a total power of 3600 W. (Photo by courtesy of Siemens AG)

actual current and capacity of the installed battery. The ampere-hour counter sums up the charge that has gone into the battery and subtracts the charge that has been taken out. The ampere-hour counter indicating 'zero' represents a fully charged battery. At a certain discharge condition, which is programmable, a relay will be activated and a signal can be issued. The lamp 'relay on' will light up to indicate this state. In order to compensate deviations from varying load/unload factors or other irregularities the ampere-hour counter will automatically reset to full charge when the battery voltage attains 53.5 V or 2.23 V per cell.

The button 'reset' allows the ampere-hour counter to be reset to full charge by hand.

### Fault signalling

*Voltage monitoring unit.* A voltage monitoring unit monitors the voltage at the busbar. Should this voltage either fall below 44.4 V or rise above 58.3 V respective signals are transmitted.

*Ampere-hour counter.* The ampere-hour counter will send out a signal to the control room if the battery has been discharged to an extent previously set at the counter. The lamp 'relay on' on the counter itself will also light up. Both signals will disappear if the battery has been charged again sufficiently by the solar power generator.

*Storage batteries.* Stations are equipped with valve-regulated (maintenance free) batteries of such type as 'Sonnenschein' dryfit A600. The main technical features of these batteries are: electrolyte immobilized in a gel, internal recombination, long life, high energy density and a high discharge circle rate as required for solar-power stations. Each station is equipped with two battery banks of 24 cells of 350 ampere-hour of a type such as 50 PzV350.

*Solar power generator.* Solar electric module features are:

- silent operation,
- sunlight as 'fuel',
- high power density,
- easy installation,
- rugged, durable construction,
- simple, reliable operation,
- easy to expand systems,
- low maintenance,
- no moving parts to wear out and
- no environmental pollutants.

These features are shown in such solar modules as the M 55 type (see Fig. 16.15). One solar module contains 36 monocrystalline silicon solar cells embedded in a soft transparent plastic (ethylene vinyl acetate) and is covered by a sheet of low iron tempered glass. This protects against corrosion and moisture and has superior light transmission, UV stability and electrical isolation.

The cells are textured and have an antireflection coating. Multiple redundant contacts provide a high degree of fault tolerance and circuit reliability. Cells within a module are electrically-matched for increased efficiency. The rugged anodized aluminum frame is designed for exceptional strength. Side rails with multiple mounting holes allow for easy installation. The tough, multi-layered polymer backsheet is used for environmental protection, resistance to abrasion, tears and punctures. Two junction covers with lids are designed for easy field wiring, safety and environmental protection. Wired-in bypass diodes reduce the potential loss of power from partial array shading. The charging voltage is achieved with as little as 5% full sunlight, resulting in more usable power available everyday.

One module has a typical load voltage of 17.4 A, an open circuit voltage of 21.7 V, a load current of 3.05 A, a short-circuit current of 3.4 A, and a rated power output of $53 W_p$[1]. Sets of four solar modules connected in

---

[1] $W_p$ Watt peak = Peak power under standard test conditions:
Air mass                    AM = 1.5
Irradiation intensity     $E = 1000$ W/m$^2$
Cell temperature         $T_c = 25\,^\circ$C

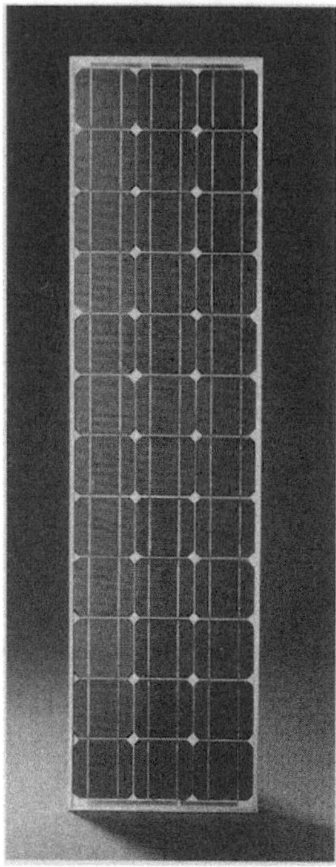

**Fig. 16.15.** Monocrystalline solar module M 55. (Photo by courtesy of Siemens AG)

series will match with the input data of the charging controllers. A connecting box is installed to connect the individual strings of modules to the cables via decoupling rectifiers. The terminal block which houses the string of modules contains easy to open terminal jumpers in order to disconnect individual strings. A reverse current into the solar modules is prevented by a diode in each controller. Up to $4 \times 4$ modules may be combined to form an array that is supported by a structure made of hot galvanized steel. One or more arrays may be used, depending on the energy demand of the installed telecom equipment.

Figure 16.16 shows the current-voltage characteristic of solar-module M 55.

### 16.3.3 Passive Cooling System

A natural-convection cooling system enables the temperature in the shelters to be restricted to a suitable level, even in hot temperatures (see Figs. 16.17 and 16.18).

The principle of natural-convection cooling depends upon the different specific densities of a liquid at different temperatures; the specific density of a liquid is reduced by heating and increased by cooling.

The cooling fluid in the system absorbs the heat generated by the radio link equipment in the internal heat exchanger (1); the heated fluid flows upwards to a tank (2), from which the cooler fluid descends to the heat exchanger. During the day, so long as the external temperature is higher than that of the tank, no

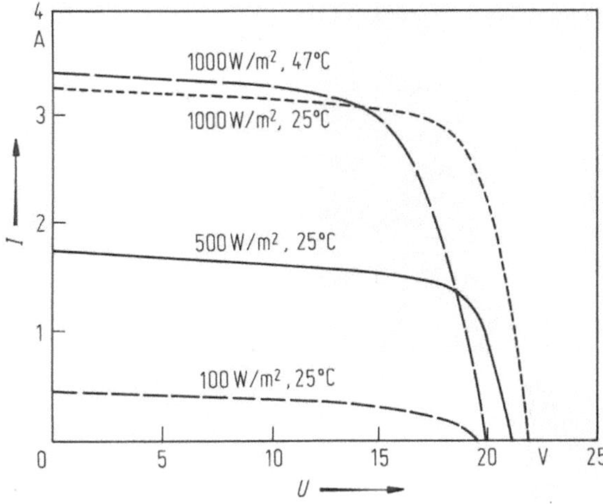



**Fig. 16.16.** Current-voltage characteristic of solar module M 55

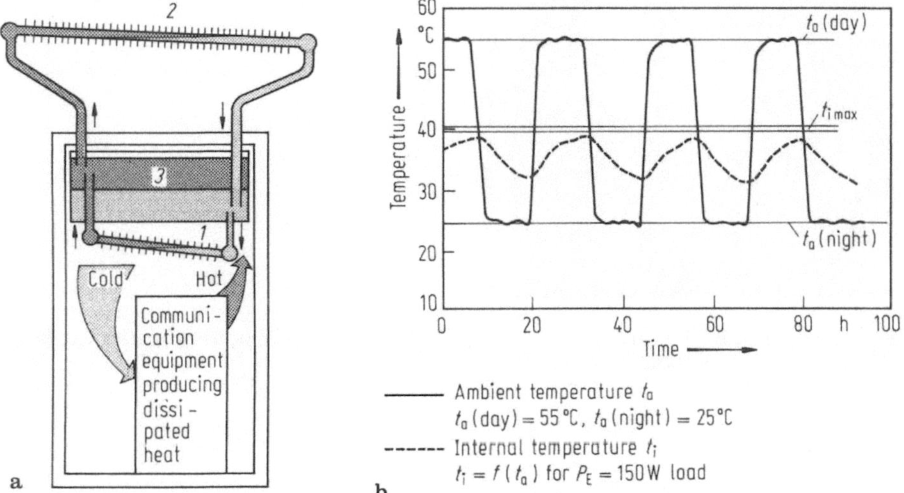



**Fig. 16.17a, b.** Operating principle of the passive cooling system. *1* Internal heat exchanger, *2* tank, *3* external heat exchanger

cooling fluid can flow to the external heat exchanger (3). Only when the external temperature falls below the temperature of the tank at night does the fluid ascend to the external heat exchanger, where it is cooled and then descends again. The cooling system thus operates entirely by virtue of the losses in the shelter, and requires no additional power.

**Fig. 16.18a, b.** Shelter with passive cooling system (**a**) for 150 W powerless after test in the climatic test chamber (**b**)

# 17 Grounding and Potential Equalization

*Grounding* (*earthing*) embraces all the means and methods whereby conducting parts are connected to earth through a grounding system.

*Potential equalization* means 'bonding' to eliminate potential differences between conducting parts. In telecommunications practice, potential equalization is mostly at earth potential.

The system of grounding and potential equalization must:

- prevent the occurrence of dangerous contact potentials and effect the disconnection of the faulty circuit[1] in the event of faults in any electrical equipment and installations,
- ensure the undisturbed operation of all central and decentralized functions of the telecommunications system,
- provide a sufficiently low ground resistance for all systems (e.g. signalling) that use earth as a return conductor,
- ensure the electromagnetic compatibility (EMC) of the system,
- be suitable for screening electronic equipment and
- assist in reducing the effect of lightning strokes, particularly in high telecommunications buildings.

A distinction is made between grounding for protective purposes and grounding for functional reasons (see Fig. 17.1).

## 17.1 Protective Grounding

In telecommunications practice the term 'protective grounding (earthing)' denotes the measures which in power engineering may be referred to as 'cut-off protection' against indirect contact (see Chap. 18).

As the term 'protective grounding' conveys, the grounding of the conductive housings of power equipment is intended to prevent the appearance of dangerous contact potentials. This is achieved, on the one hand, by potential equalization through the ground conductor and, on the other by disconnecting the faulty equipment. In the event of an insulation failure, a short-circuit current flows in the

---

[1] DIN 57 100/VDE 0100 and DIN 57 800/VDE 0800 in Germany are especially applicable for the protection of persons and equipment.

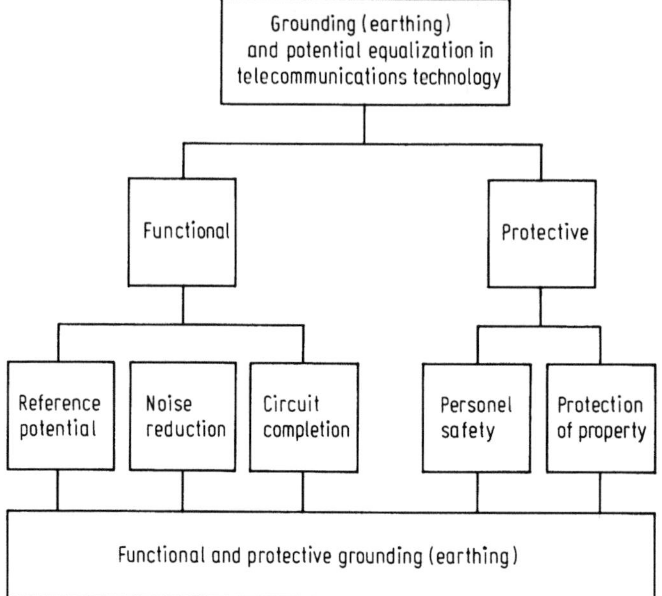


**Fig. 17.1.** Grounding (earthing) and potential equalization

connection to the ground conductor, which causes either a fuse or an automatic circuit breaker to operate.

Equipment with a protective ground connection is classified in, VDE 0106 as protective class I equipment. Such equipment may be used with supplies in the following categories:[2]

- TT supply system (protective grounding),
- TN S supply system (neutral grounding),
- TN C S supply system and
- IT supply system (protective line system).

In the TT supply system, in most cases, the neutral point of the three-phase supply is grounded directly. Equipment in protective class I is grounded directly or through a ground conductor. Fault current flows back to the power source via earth. Disconnection is effected either by a fuse or, preferably, by a fault current-operated protective circuit breaker.

In the TN supply system the neutral point is usually grounded directly. Equipment in protective class I is grounded directly through a connection with a cross-sectional area of at least 10 mm$^2$ or through a protective conductor to the PEN conductor (formerly neutral earth conductor). Fault currents flow back to the

---

[2] These symbols are used by international agreement to describe the grounding of supplies and loads (see Chap. 18).

power source mainly through the PEN conductor. To prevent the PEN conductor from assuming a dangerous contact potential in the event of its being broken, it should be grounded at as many points as possible, particularly at the entrance to the building. For the overall resistance of the supply system to ground, a value of $2\,\Omega$ is adequate. Disconnection depends upon a fuse. Fault current-operated protective circuit breakers are acceptable in a TN supply system as a means of improving the disconnection characteristics.

The IT supply system is operated without a ground connection. Protection depends entirely upon potential equalization. IT supplies are not widely adopted, and are only used in special circumstances, where disconnection is not acceptable (e.g. hospitals).

The voltage drop to ground permitted by the protection arrangements in a TT supply system should not exceed 50 V a.c.

These protective measures cannot be applied beyond a certain load power level, because the necessary ground resistances cannot generally be achieved economically. This limitation does not apply when fault current-operated protective switches are used. The ground resistance $R_E$ can in this case be significantly higher, depending upon the trip current ($\approx 10$ mA to 1 A):

$$R_E = \frac{\text{permissible contact potential (V)}}{\text{fault current (A)}} = \frac{U_B}{I_F} \qquad \text{(see Table 17.1)}.$$

Higher powers in the medium-voltage range are in any case supplied through a transformer. A local TN supply system is then provided on the low-voltage side. Potential equalization represents an effective extension of 'protective disconnection arrangements'; it is prescribed at every house lead-in as the 'main potential equalization', and connects the PEN or ground conductor PE via the potential equalization busbar, to the foundation ground, the metal structures and the lightning ground.

A 'supplementary potential equalization' is prescribed, in addition to the main potential equalization, if the conditions laid down for automatic disconnection as a protection against indirect contact cannot be met. Supplementary potential equalization requires the bonding of metal structures, reinforcing bars in reinforced-concrete structures, etc., to all conducting parts of fixed equipment that can be touched simultaneously. This equalization corresponds to the functional and protective grounding that is normally provided in any case in telecommunications systems (see Sect. 17.3).

Table 17.1 indicates the maximum earth resistance for protective grounding.


## 17.2 Functional Grounding

As is well known, signal circuits in telecommunications engineering are often arranged as single-wire circuits with a common ground return. This offers technical, but primarily economic, advantages. This common signal ground return uses, within the telecommunications system, the mesh network of reinforcing

**Table 17.1.** Maximum values of ground resistance in telecommunications systems. LU Line units, RS Repeater station, RT Radio transmission station, DCS Data conversion station, TX Transit exchange, PABX Private automatic branch exchange, LX Local exchange. <u>Note</u>: For digital trunks a ground resistance $\leq 10\ \Omega$ is required regardless of the size of the system

| Protective grounding (earthing) | Functional grounding (earthing) | | Functional and protective grounding[b] (earthing) | | | Lightning protection grounding (earthing) |
|---|---|---|---|---|---|---|
| Medium-voltage supply through transformer with TN system on low-voltage side $R_E \leq 2\ \Omega$ | Single wire ground signalling | | Single wire ground signalling | with lightning protection | without lightning protection | $\leq 5\ \Omega$ |
| | LX and PABX up to 500 LU | $\leq 10\ \Omega$ | LX and PABX up to 500 LU | $\leq 5\ \Omega^c$ | $\leq 10\ \Omega^c$ | |
| | LX and PABX from 500 to 1000 LU | $\leq 5\ \Omega$ | LX and PABX from 500 to 1000 LU | $\leq 5\ \Omega^c$ | $\leq 5\ \Omega^c$ | |
| | LX from 1000 to 2000 LU and PABX with more than 1000 LU | $\leq 2\ \Omega$ | LX from 1000 to 2000 LU and PABX with more than 1000 LU | $\leq 2\ \Omega$ | $\leq 2\ \Omega$ | |
| | LX with more than 2000 LU | $\leq 0.5\ \Omega^a$ | LX with more than 2000 LU | $\leq 0.5\ \Omega^a$ | $\leq 0.5\ \Omega^a$ | |
| | RS, RT | $\leq 2\ \Omega$ | RS, RT | $\leq 2\ \Omega$ | $\leq 2\ \Omega$ | |
| | TX, DCS | $\leq 0.5\ \Omega$ | TX, DCS | $\leq 0.5\ \Omega$ | $\leq 0.5\ \Omega$ | |
| Current-operated protection circuit | Loop signalling | | Loop signalling | with lightning protection | without lightning protection | |
| e.g. $R_E = \frac{50}{0.3}\ \Omega$ | Up to 1000 LU | $\leq 10\ \Omega$ | Up to 1000 LU | $\leq 5\ \Omega^c$ | $\leq 10\ \Omega^c$ | |
| $\approx 166\ \Omega$ | Up to 2000 LU | $\leq 5\ \Omega$ | Up to 2000 LU | $\leq 5\ \Omega^c$ | $\leq 5\ \Omega^c$ | |
| with $I_F = 0.3$ A | More than 2000 LU | $\leq 2\ \Omega$ | More than 2000 LU | $2\ \Omega$ | $\leq 2\ \Omega$ | |

[a]Provided that there is no three- or two-wire junction traffic with other systems through d.c. repeaters, the resistance in systems with more than 2000 subscribers may be as high as $2\ \Omega$.
[b]Measured at the ground ring main or busbar, allowing for all connected grounds.
[c]$2\ \Omega$ in the case of connection to a medium-voltage supply.

mesh, rack suites and the grounded 48 or 60 V distribution system. Because of the interconnected areas of all the grounded parts (hence the alternative term, area earth, as opposed to the radial ground), the result is a mesh network with low impedance, which gives rise to very low potential differences with both d.c. loading and a.c. loading up to the high-frequency region. It thus meets the requirements for use as a potential reference plane. The ground is necessary in order to stabilize the reference potential in relation to the surroundings, to protect against corrosion and to make use of the ground as a return conductor between geographically separated telecommunications installations. Common return paths, however, introduce the possibility of interference, especially from power circuits. Nevertheless, the use of ground returns in conjunction with single-wire circuits and signal circuits is widespread – e.g. between exchanges in local telephone networks. With the introduction of digital exchange systems (e.g. the EWSD system) the use of ground return signal paths is avoided by using PCM links between exchanges.

A further reason for grounding telecommunications systems is the reduction of interference. External interference can, for example, be reduced by the equalization of potentials or by diverting voltages to ground. This kind of interference reduction embraces also the suppression of overvoltages resulting from atmospheric effects or from the effects of power circuits. To enable the requirements of the functional ground to be met, the ground resistance must not exceed a certain maximum value (Table 17.1). If the necessary low values are not achieved, a loss of quality may result. This relates especially to noise pick-up and message-switching performance.

With increasing ground resistance, the introduction of noise deteriorates the signal-to-noise ratio.

The message-switching characteristics are adversely affected in systems with three-wire communications traffic, where the ground serves as a return conductor because of signal asymmetry. In this case current in the ground connection produces a voltage drop which is in opposition to the exchange voltage. This can lead to malfunctions.

An increase in the ground resistance by one step (Table 17.1, functional grounding) is acceptable if truly symmetrical signal processes are employed, with or without unbalanced ground relationships in the junction or exchange lines. For digital trunks a ground resistance $\leq$ 10 $\Omega$ is required regardless of the size of the system.


## 17.3 Functional and Protective Grounding

The ground resistance, from the point of view of the operation of a telecommunication system, must not exceed a certain value (see Table 17.1, functional grounding). This value is usually significantly lower than that required for protective grounding. The telecommunications equipment is grounded either through the power supply conductors or by means of a separate functional ground conductor

FE (functional earth). Because of the low permissible voltage drop, these conductors usually have a large cross-sectional area and can therefore be extended to a combined functional and protective ground (see Table 17.1). The functional and protective ground is used e.g. in the German Postal Administration Telekom Network (since January 1st, 1995: Telekom AG) for all exchange and transmission installations and often in installations outside Germany as well.

This method of grounding has the following characteristic features:

– the functional and protective grounding are combined in one conductor or conductor system,
– all conducting parts that require grounding for protective or functional reasons are connected to the functional and protective earth connector FPE (ground plane),
– through the connection of the grounded positive conductor to the racks, the rack structure in telecommunications equipment becomes part of the grounding

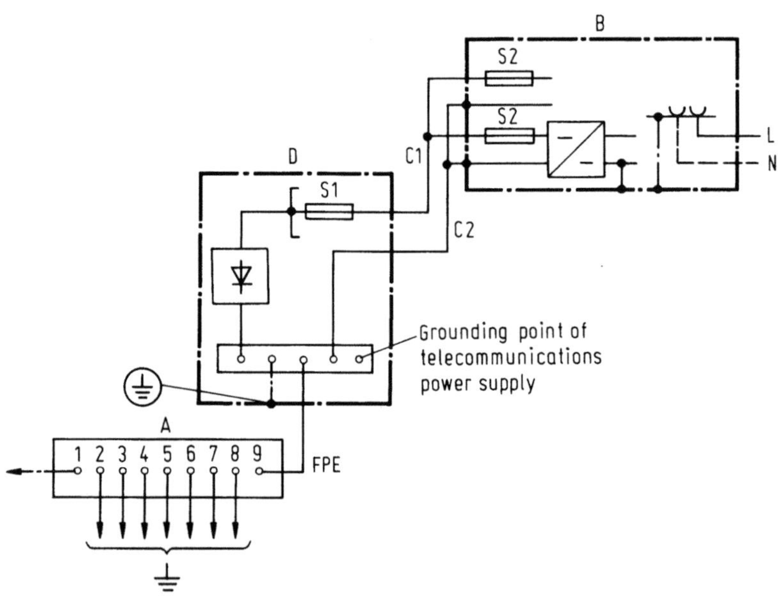


**Fig. 17.2.** Functional and protective grounding (earthing) of a telecommunications system (in accordance with VDE 0800 Part 2). A Grounding main conductor, B telecommunications equipment, C1, C2 telecommunications operating circuit: grounded conductor C2 connected to FPE and the reference conductors of the telecommunications equipment, D telecommunications power supply: the nominal direct and alternating voltages may exceed 120 V and 50 V respectively, L1, N power circuit, S fuse. Illustrated connections to grounding main conductor A: *1* Protective conductor (PE) of interior installation or potential equalization conductor of building, *2* telecommunications ground, *3* foundation ground, *4* conducting sheath of telecommunications cable, *5* reinforcement of building, *6* conducting water pipes in building, *7* heating system, *8* lightning-protection ground, *9* functional and protective earth conductor (FPE)

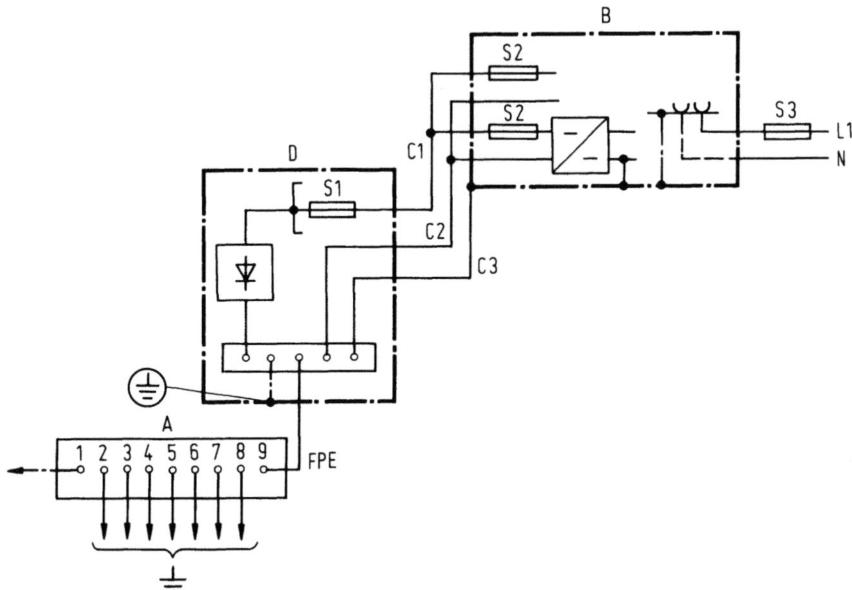


**Fig. 17.3.** Functional and protective grounding, with an additional ground conductor, of a telecommunications system (in accordance with VDE 0800 Part 2). C1, C2 Telecommunications operating circuit, grounded conductor C2 insulated and *not* connected to the reference conductors of the telecommunications equipment, C3 additional ground conductor, not carrying supply current, connected to FPE and the reference conductors of the telecommunications equipment. For A, B, D, L1, N, S, see Fig. 17.2.

system and assumes the protective grounding function for power system loads in protective class I. Then the ground conductor PE (protective earth) of the power system is not connected and

- the housings of the power supply equipment – e.g. rectifiers, mains switch panel and battery switching panel – are similarly bonded to the grounded positive conductor. The grounding connection is provided by the FPE conductor between the grounded conductor in the power supply system and the ground main conductor (ground bus). Also effected at the ground main conductor is the matching to the protective devices of the supply system (see Sects. 18.1.1 to 18.1.3). Thus, for example, the PE ground conductor of the house wiring and the potential equalization conductor of the building are equally connected at this point (see Figs. 17.2 and 17.3).

## 17.4 Selection and Design of Grounds and Grounding Systems

The choice and design of the grounding (earthing) system is determined by the requirements placed upon it and the local conditions. The components of a grounding system are:

- the *actual* ground (a ground electrode or groundly metalwork, pipes, etc.) which conducts current into the ground,
- the *grounding main conductor*, to which the ground and the parts of the installation that are to be grounded are connected,
- the *grounding conductors* (or potential equalization conductors), which connect the parts of the installation to the grounding main conductor.

### 17.4.1 Types of Grounds

In the case of a newly-erected building, the electricity supply undertaking will provide a foundation ground. This should be connected to the main potential equalization system and can be used for protective and functional grounding purposes. To the extent that the resistance of the foundation ground does not meet the requirements, additional grounds have to be provided.

These grounds may take the form of surface grounds (ground bands), vertical (strip) grounds or a combination of both.

Also these types of ground, pipe and cable networks, provided that they consist of conducting material and are laid in contact with the ground, function effectively as grounds because of the large area which they cover; it is then a question of socalled 'natural ground'. Included in this category are the metal sheathing of telecommunications cable and water pipelines, but not gas pipelines; these may be used only within buildings, and only for potential equalization. For small telecommunications systems the water main (or the heating pipes connected to it) is frequently an adequate functional ground.

In recent times the use of metal pipeline systems and cable sheathing has become progressively less common. Water pipes are often of plastic material, and lead-sheathed cable has mostly been replaced by communications cable with plastic covered aluminium sheathing. For these reasons, the German Postal Administration Telekom Network, for example, specifies a foundation earth for new telecommunications installations.

By *foundation ground* (earth) is meant a conductor which is embedded in the foundation level of a building, and therefore makes a large-area contact with the ground; it is incorporated as a closed ring in the external foundations and consists of galvanized steel strip. In general, foundation grounds provide sufficiently low ground resistances.

*Surface grounds* consist of galvanized steel strip or rod, and are generally installed at a depth of 0.5 to 1 m. The length of these grounds depends upon the necessary ground resistance; they can be arranged as radial, ring or mesh earths, as well as straight. Radial grounds should have equal divisions; more than three branches (with an angle not less than 60°) are not recommended because of mutual interference.

*Vertical grounds* can be installed at a depth of up to 30 m, depending on the subsoil; they consist of rod, tube or profiled bars and are driven into the ground as nearly as possible vertically; their length depends upon the necessary ground resistance. If several vertical grounds are necessary (in order to obtain a sufficiently low ground resistance) a separation of at least twice the length of an individual electrode is desirable.

*Plate grounds* were formerly often employed. They entail a greater expense than other types of ground and will not be described here.


### 17.4.2 Grounding Main Conductor

In the simplest case, the grounding (earthing) main conductor consists of a *ground terminal* on the equipment. In small or medium-sized systems a *ground busbar* of copper, brass or galvanized steel is used, whose length and cross-sectional area are chosen in accordance with the number of grounds and conductors to be connected to it.

The most effective form of grounding main conductor is represented by the grounding ring main conductor, which is installed in the basement or the ground floor of buildings containing extensive telecommunications systems. The metal sheaths of cables, conduits, water and heating pipes and the like are connected to the grounding ring main conductor by the shortest paths.

Because of its cross-sectional copper area of at least 50 mm$^2$ (usually 95 mm$^2$), the grounding ring main conductor has a very low resistance and effects a *potential equalization* in respect of all potentials acting upon it, external as well as internal.

The grounding main conductor may also be installed close to the telecommunications system in a storey of a high office building, in which case the connection to the ground line of the power system – assuming that one is provided for functional and protective grounding purposes – is made at the floor distribution board of the low-voltage supply.


### 17.4.3 Grounding and Potential-Equalization Conductors

Certain stipulations should be noted in regard to the installation of ground conductors, particularly their cross-sectional area (Table 17.2).

In most cases ground conductors are used for potential equalization for functions associated with communications operations in that they provide a practically uniform reference potential within the telecommunications system. This is particularly the case when one pole of the power supply line is also used as a ground conductor over long distances.

This kind of ground is illustrated in Fig. 17.2.

In transmission practice it is customary to ground the equipment by means of a separate conductor, which is kept free of power supply currents (Fig. 17.3).

By virtue of the exclusive use – in this case – of d.c./d.c. converters, 'galvanic separation' is achieved between the power supply and the communications equipment, which is especially advantageous in high buildings susceptible to lightning.

**Table 17.2.**[b] Minimum cross-sectional area of functional earth conductor FE (according to VDE 0800 Part 2A[a])

| Current rating of overcurrent protection device (A)[a] | Minimum cross-sectional area of PVC isolated copper conductor $(mm^2)$ | Current rating of overcurrent protection device (A)[a] | Minimum cross-sectional area of PVC isolated copper conductor $(mm^2)$ |
|---|---|---|---|
| Up to   35 | 2.5 | Up to   400 | 35 |
| Up to   50 | 4 | Up to   500 | 50 |
| Up to   80 | 6 | Up to   630 | 70 |
| Up to 125 | 10 | Up to   800 | 95 |
| Up to 200 | 16 | Up to 1000 | 120 |
| Up to 250 | 25 | | |

[a]The overcurrent protection device is not in the ground conductor path.
[b]Table 17.2 is used if, via the use of the FE or FPE conductor in the case of normal operation, (no fault) flow current supply.

In dimensioning the ground conductor C3, however, it is necessary to check whether, in the event of a ground short-circuit, the resistance of the fault path is such that the fuse can be relied upon to rupture in an acceptable time (5s). In systems such as that shown in Fig. 17.2 this check is not necessary, because the ground conductor C2 carries the power supply current, and therefore has the same cross-sectional area as the conductor C1.

From Table 17.2 the cross-sectional areas can be chosen according to the current rating of the operative overcurrent protection device. This will be explained with reference to Figs. 17.2, 17.3 and 17.4.

In the event of a ground short-circuit fault on conductor C1 in Fig. 17.2, the fault current would flow through the fuse S1, through conductor C1 to the fault location in the communications equipment B and back to the power supply through conductor C2. The conductor C2 must therefore be specified in conformity with the fuse S1. In the case of a ground short-circuit fault after fuse S2, the ground conductor concerned has, of course, only to be dimensioned to correspond to the fuse S2.

If a ground short-circuit fault occurs on the conductor C1, not the communications equipment B but to a part of the building (not shown) – a grounded rack, for example – the conductor shown as FPE between the grounding main conductor and the power supply equipment carries the ground fault current, and so must be dimensioned in conformity with fuse S1.

The cross-section of conductor C3 in Fig. 17.3 is specified in accordance with the larger of the two fuses S1 and S3.

Figure 17.4 shows the variation range of fuses in accordance of Table 17.2 and dependent from the lengths of the supply conductor, the permitted lengths of FE- or FPE-conductor.

The explanation of examples of grounding systems in telecommunications and power supply systems respectively FE- and FPE-conductors are presented in Figs. 17.5, 17.6 and Table 17.3.

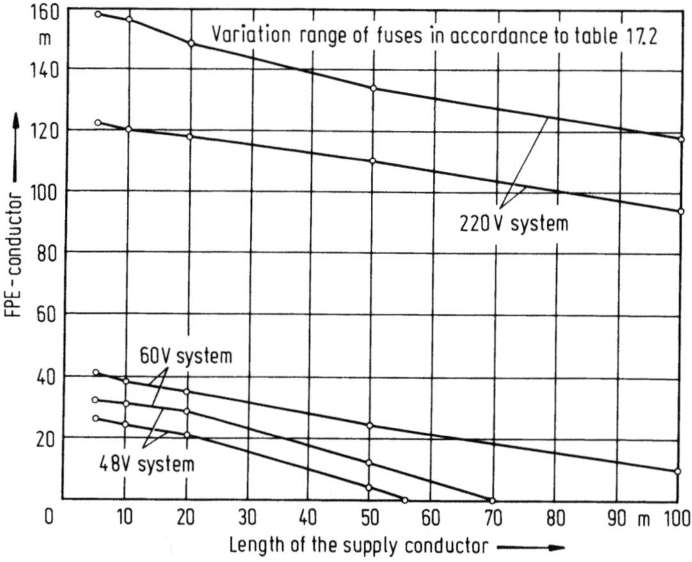



**Fig. 17.4.** Dependent from the lengths of the supply conductor *1*, the permitted lengths of FE- or FPE-conductor *2,3* variation range of fuses in accordance of Table 17.2. (*Source*: Siemens AG)

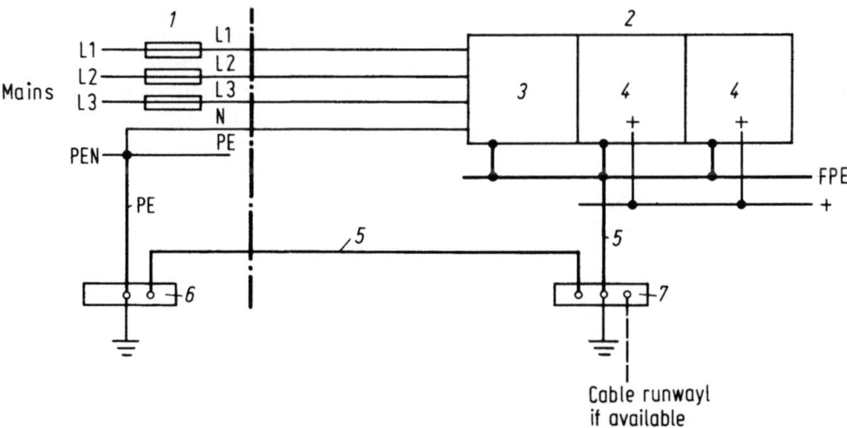



**Fig. 17.5.** FE- and FPE-conductors in power supply systems (*Source*: Siemens AG). *1* Domestic power area, *2* telecommunications with power supply system, *3* mains switch panel, *4* rectifier unit, *5* FPE-conductor, *6* grounding main conductor of the supply entry, *7* collective grounding main conductor

In the case of short circuit to ground the FE- or FPE-conductor must be able to take over the hole short circuit current. This is the reason why if there is a short circuit the cross-sectional area of the FPE-conductor must be calculated in direction of the next larger fuse of the current distribution circuit.

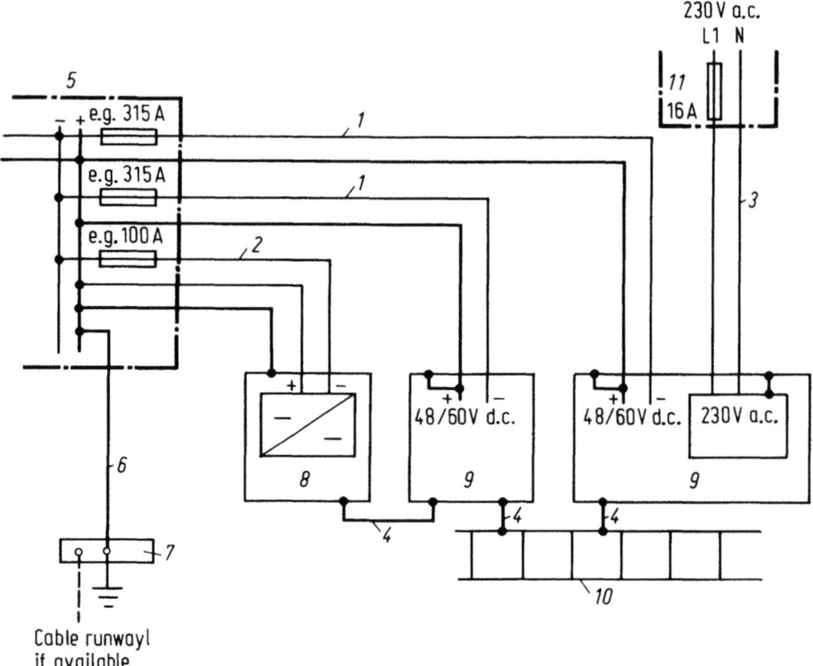

**Fig. 17.6.** FE- and FPE-conductors to and in telecommunications systems (*Source*: Siemens AG). *1* Example of FPE-conductor, the grounded positive conductor is additional used for protective reasons, typical used in d.c. supply circuit in telecommunications systems like digital switching systems EWSD, *2* example of separated FPE-conductor which is kept free of power supply currents, typical used in d.c. supply circuit of transmissions systems like PCM, *3* a.c. supply circuit (e.g. socket-outlets and lighting circuits), *4* potential equalization conductor, *5* battery- or d.c. - distribution panel (48 V/60 V d.c.), *6* FPE-conductor of the power supply system, *7* collective grounding main conductor, *8* equipment of transmission systems, *9* equipment of switching systems, *10* constructions parts of the telecommunications system, e.g. cable grid, *11* mains distribution switchboard NVT

**Table 17.3.** Coordination of cross-sectional area of protective earth conductor PE to cross-sectional area of outer conductor (according to VDE 0100 Part 540/11.91)

| Cross-sectional area of outer conductor of the system $S$ $(mm^2)$ | Minimum cross-sectional area of protective earth conductor PE $S_{PE}$ $(mm^2)$ |
|---|---|
| $S =< 16$ | $S$ |
| $16 < S =< 35$ | 16 |
| $S > 35$ | $S/2$ |

Table 17.3 is used, if via the use of FE- or FPE-conductor in the case of normal operation (no fault) flow current supply

## 17.5 Design Pointers for Grounding Systems

In designing a grounding system it is necessary to determine whether the ground that is available for the connection of the main grounding conductor, such as a foundation ground, lightning ground or cable sheath, has a sufficiently low resistance. This entails measurement. Inasmuch as these considerations arise before the building construction begins, an estimate has to be made, for which the specific ground resistance must be known.

### 17.5.1 Specific Ground Resistance

The specific ground resistance can vary greatly according to the depth at which the grounding electrode is buried; it also depends upon the composition of the soil, its degree of dampness and the temperature. Figure 17.7 shows values of specific ground resistance $\rho_E$ that can be used to estimate the resistance of a ground electrode. In each case the lower values relate to damp ground conditions.

### 17.5.2 Measurement of Specific Ground Resistance

The specific ground resistance $\rho_E$ is quoted in ohm-metres. Special ground-measuring equipment is used to determine it; this usually employs the voltage-balance method. A measuring current is caused to flow in the ground by means of a ground electrode and an auxiliary ground electrode. The voltage drop in the ground resistance is compared with that across an adjustable resistor. The measurement is made using a.c. (e.g. from a hand generator) to avoid errors due to polarization which could occur with d.c. Probes are required to measure the voltage drops (Fig. 17.8).

The value of the resistance between the probes $S_1$ and $S_2$ can be measured on the balancing resistor 3.

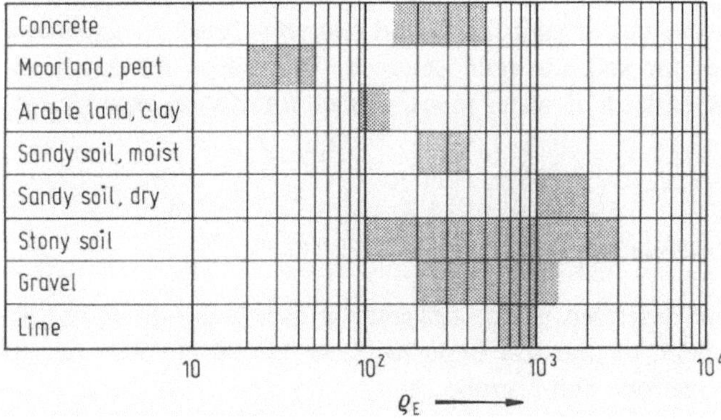



**Fig. 17.7.** Specific ground resistances $\rho_E$ in various ground conditions

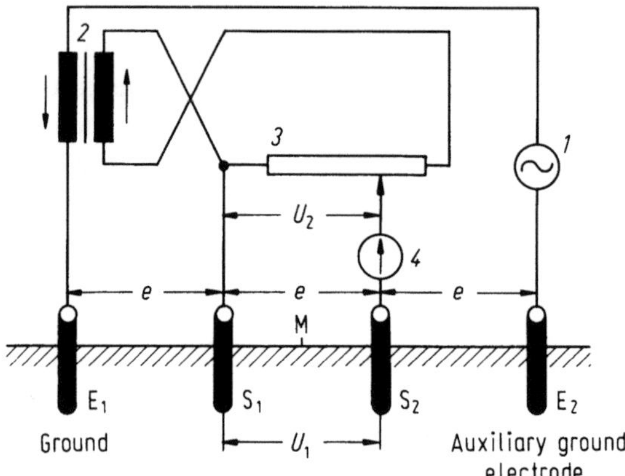


**Fig. 17.8.** Arrangement for the measurement of soil specific resistivity. *1* a.c. source (generator), *2* transformer, *3* balancing resistor, *4* null detector, $E_1$, $E_2$ current electrodes, $S_1$, $S_2$ voltage probes

The *mean specific ground resistance* $\rho_E$ is given by:

$$\rho_E = 2\pi e R$$

where

$R$ = measured resistance ($\Omega$)

$e$ = probe distance (m)

$\rho_E$ = mean specific ground resistance ($\Omega$m) up to a depth equal to the probe distance $e$.

Four ground spikes are driven into the ground at equal distances in a straight line with its mid-point at M. Measurements are made initially with a distance $e$ of 1 m. For further measurements the distance of the spikes is increased in steps of 1 m with the same mid-point M. If the value of ground resistance $\rho_E$ so measured is constant with varying probe distance, the soil is homogeneous. If $\rho_E$ increases, the resistance of the deep soil strata is higher, and vice versa.

A rapidly falling value of resistance indicates that the water table has been reached.

In large towns lower values of specific ground resistance than are accounted for by the constitution of the soil are often obtained. The reason for this is the large number of conductors, such as water pipes, cables, foundation grounds etc., laid in the ground.

### 17.5.3 Measurement of Grounding Resistance

The grounding equipment described is also suitable for measuring grounding resistances. This measurement, of practical importance, is carried out with the aid of an auxiliary ground electrode and a probe.

As in the measurement of specific ground resistance (see Fig. 17.8) a measuring current is passed between the ground electrode and the auxiliary electrode, and the voltage drop is determined by means of the probe. To obtain reliable results, the probe must be applied outside the areas of high potential gradient surrounding the ground and auxiliary ground electrodes, in which large variations of potential occur as a result of the proximity of the electrodes or non-uniformity of the soil. It is therefore necessary first to establish the size and shape of these areas by exploratory measurements.

Beyond the surrounding areas of high potential gradient the potential is approximately constant (neutral zone, reference potential; see Fig. 17.9). The probe is then moved in steps of about 5 m, beginning at the ground electrode, towards the auxiliary electrode, and the measurements are plotted on a curve. The ground resistance is obtained by drawing a straight line parallel to the horizontal axis through the point of inflection. The lower part of the curve then gives the resistance of the ground electrode.

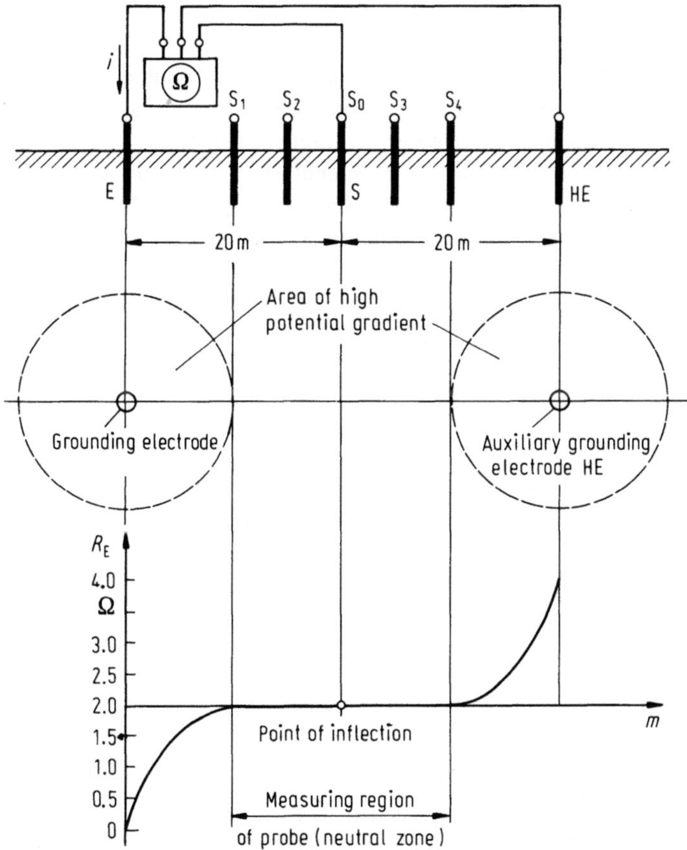


**Fig. 17.9.** Variation of resistance obtained from ground resistance measurement. E Ground electrode, S probe, HE auxiliary ground electrode, $R_E$ ground resistance at particular location, $S_1$ to $S_4$ locations of probe in measurement sequence

### 17.5.4 Calculation of Ground Resistance

Rule-of-thumb formulae for calculating ground resistances are given in Table 17.4.

The formulae are valid so long as the specific ground resistance is constant along the length of the ground electrode. In the case of vertical electrodes this is often not so. For the purpose of a preliminary calculation, with a known specific ground resistance $\rho_E$, the vertical ground electrode may in this case be considered as divided into sections 1 m long. The resistance of one section is then equal to the specific ground resistance $\rho_E$ at the relative position.

The total resistance of a vertical ground is obtained by adding the conductances of the individual sections.

Thus the ground resistance is given by:

$$\frac{1}{R_E} = \frac{1}{\rho_{E1}} + \frac{1}{\rho_{E2}} + \cdots + \frac{1}{\rho_{En}}$$

The relationship holds for d.c. and low-frequency a.c.

With impulse (lightning) currents, the impedance of an extended ground electrode ($> 30$ m) can be considerably increased. From this point of view a number of short electrodes connected together are more satisfactory.

The resistance of a *combined ground* can be calculated sufficiently accurately for practical purposes by first determining, in the case of a vertical electrode and a surface electrode coupled together, their separate resistances, but taking into consideration only a part of the conductance of the surface electrode. To this end the length of the surface electrode is reduced to half the length of the vertical electrode. The total resistance $R_G$ is then calculated as:

$$R_G = \frac{R_1 R_2}{R_1 + R_2}$$

The resistance of a *foundation ground* is given approximately by the formula for the ring or mesh electrode (see Table 17.4), according to the nature of the foundations, if the value of $A$ is taken as the area enclosed by the foundation

**Table 17.4.** Ground resistances obtained with various kinds of ground electrode.

$R_E$ ground resistance ($\Omega$),
$l$  length of ground electrode (m),
$D$ diameter of ring electrode or equivalent circular area (m) (equivalent diameter)
    $= 1.13\sqrt{A}$,
$A$  area enclosed by ring or mesh electrode (m$^2$)

| Type of ground electrode | Formula | Type of ground electrode | Formula |
| --- | --- | --- | --- |
| Vertical electrode | $R_E \approx \frac{\rho_E}{l}$ | Ring electrode | $R_E \approx \frac{2\rho_E}{3D}$ |
| Surface electrode | $R_E \approx \frac{2\rho_E}{l}$ | Mesh electrode | $R_E \approx \frac{\rho_E}{2D}$ |

ground. If the resistance of the foundation ground is not sufficiently low, it must be supplemented by a vertical or surface ground electrode.

The question of which kind of ground electrode to use can only be decided from a knowledge of the subsoil. Given a relatively homogeneous soil, the cost of grounding with vertical or surface grounds is about the same.

In most cases, however, vertical grounds are more economical, because the deeper soil strata are generally of higher conductivity. For the same ground resistance, a vertical ground needs to be only about half the length of a surface ground.

Surface grounds are useful when the subsoil is stony or rocky, when the specific ground resistance increases with increasing depth, or when it is necessary to extend the area of the grounding system because of an expansion of technical equipment and buildings.

In mountainous country, surface grounds frequently offer the only possibility of installing a grounding system.


## 17.6 Protection against Overvoltage and Interference

An aspect of grounding previously mentioned in passing is its role as an important part of the measures adopted by way of protection against overvoltage and interference.

The introduction of electronic devices and systems increased the demands made upon overvoltage protection.

Since that time it has been necessary to adopt a comprehensive, coordinated protection system against overvoltage; it embraces the overall and specific protection of the conductor system, constructional features in the design of apparatus and the incorporation of grounded parts of the building, such as reinforcing steel, structural steelwork, lightning protection systems, etc., in the total grounding scheme.

If this system of grounded parts is sufficiently fine-meshed, it can also be used as a screen to enhance EMC (electromagnetic compatibility). By the term EMC is meant here the limitation of emission of electromagnetic interference and immunity to external fields (e.g. radio and broadcasting).

Through the interference path, the *source of interference* is coupled to the *affected apparatus* either 'galvanically', inductively, capacitively or by radiation.

There are three possible ways of reducing interference when necessary:

- reducing the interference at source – e.g. by suitable suppression of relays or contactor coils,
- reducing the degree of coupling; this can be done, for example, by screening, modification of wiring layout, etc.,
- increasing the immunity to interference of the system; this may, for example, be achieved by circuit techniques, such as the incorporation of voltage-limiting components.

At this point the grounding and potential equalization is being considered primarily as part of the means of reducing interference. It is, however, worth considering briefly the sources of interference and the methods of increasing immunity to interference.

### 17.6.1 Overvoltage Sources

*External conductor networks.* Overhead lines are by their nature the most frequently affected by atmospheric discharges.

More numerous than overvoltages due to close discharges (a direct stroke on the conductor or flashover in nearby apparatus) are those caused by remote strokes. In the case of mains cables these are strokes in the medium-voltage or high-voltage system, or they can be produced in any conductors by lightning discharges between two clouds. The overvoltage spreads from its place of origin in both directions with the speed of light. The magnitude of the transmitted voltage is determined by the insulation level of the conductor system. In the event of a lightning stroke on a low-voltage conductor, a flashover to ground will occur on the nearby insulators. In low-voltage networks and overhead telecommunications cables, overvoltages higher than 10 kV occur but rarely.

Telecommunications cables laid in the ground are affected by power systems as well as lightning strokes.

A lightning stroke raises the potential of the ground at the striking point relative to a remote ground. If a cable route passes through the area of high potential gradient at the location of the stroke and the cable sheath, in the course of the cable run, is in contact with the remote ground, a transient current flows in the sheath. Through the inductive effect of the lightning current and the capacitance of the conductors, voltages are also induced in the cores of the cable. Depending upon the magnitude of the potential difference between the cable sheath and the cores, the result may be breakdowns or even destruction of the cable.

Particularly frequent is interference with telecommunications cables from the indirect effect of lightning – i.e. through the effect of lightning discharges parallel to the cable route.

In relation to interference from power systems, electric railways should be particularly mentioned; in these, a portion of the operating current returns to the transformer or rectifier substation through the ground. As a result, the sheaths of telecommunications cables laid in the vicinity of the railway installation can assume a proportion of these currents, which give rise to interference and overvoltage of longer duration.

Short-duration interference (of 0.1 to 1 s duration) with high-voltage transients is produced mainly by short-circuits in power systems.

The overvoltages in the cable cores are also dependent upon the construction of the cable.

*Conductor networks within buildings.* Because of the smaller area covered by conductor networks within buildings, and the screening effect of modern concrete structures, consideration is principally directed at this point, leaving aside direct lightning effects, to interference caused by the switching of currents.

Every conductor possesses an inductance which depends upon the length of the conductor, its geometrical arrangement and the magnetic properties of the surrounding materials. Every change in current produces a voltage in the inductance of the conductor which opposes the applied voltage. The magnitude of the voltage depends upon the rate-of-change of current and the inductance. When a short-circuit occurs, large current variations are caused by the rupture of the fuse, and high overvoltages are consequently produced. Overvoltages of up to 230 V have been measured in 60 V telecommunications systems where the arrangement of the conductors is not satisfactory. Even in well-laid-out systems with low inductance, considerable overvoltages still occur; e.g. the following typical levels were measured in a 60 V system (before the fuse):

– rupture of a 10 A rack fuse: 150 V, duration 0.4 ms,
– rupture of a 63 A rack row fuse: 130 V, duration 0.7 ms.

Significantly higher overvoltages occur as a result of short-circuits in 220/380 V systems. The following average overvoltages were measured in tests on low-voltage systems, according to the fuse rating and the length of cable between the distribution transformer and the short circuit:

– 10 A fuse: 2 to 7 × nominal voltage,
– 35 or 100 A fuses: 1.5 to 4 × nominal voltage.

The overvoltages measured with the 10 A fuses were surprisingly high: the reason lies in the rapid current interruption in these fuses.

Overvoltages do not occur only on d.c. or a.c. power supply conductors; they are also produced on signal conductors and ground conductors through inductive effects or 'galvanic' coupling.

Through the grounding of one pole of the d.c. power supply (e.g. +60 V) for functional reasons, this is connected at many points to structural parts of the telecommunications installation, and hence to grounded structural parts of the building. If the building is struck by lightning, components of the lightning current may flow in the grounded power supply conductors and, as a result of the voltage drop across the impedance of these conductors, give rise to potential differences between parts of the telecommunications system.

At the same time, the lightning current may induce voltages in ungrounded signal conductors. The magnitude of these voltages depends upon the extent of the conductor loop. Widely ramified conductor systems are therefore especially affected.

### 17.6.2 Interference in Telecommunications Systems

Telecommunications installations must be protected against interference voltages deriving from the conductor network. Limiting values for acceptable interference levels were given in VDE recommendations at a very early stage (from 1920).

The permissible limits in regard to risk from power installations are contained in VDE 0228 Part 1.

If the interference levels exceed the limits, or if atmospheric overvoltages have to be allowed for, protective measures are necessary (in accordance with VDE 0845); these may be:

- in the case of conductor networks, a suitable choice of cable routes to avoid proximity,
- the use of cables or conductors of lightning-protected construction,
- the use of isolating transformers,
- the use of overvoltage suppressors (surge diverters).

With the change to switching systems with central electronic control and electronic digital systems, the question of EMC has become serious.

In the peripheral circuits such systems operate with signal levels of $\pm 12$ V, and in the processors with $\pm 5$ V, at speeds in the high-frequency range e.g. 5 MHz. Transient overvoltages on ground and signal lines thus lie, in terms of voltage and frequency, in the operating range of the signal sequences of the system, and can therefore lead to errors in signal processing.

Measures must accordingly be adopted to make the system immune to interference.

These are principally design features, such as the disposition of conductors, potential equalization, screening and the incorporation of overvoltage protection devices.

The fundamental problem area in digital systems is not so much in the control of the interference on the live power supply conductors and ungrounded signal leads – this can be achieved by means of suppression circuits and screening – as in the effect on the grounding system. It must be ensured, for example, that transient currents due to the discharge of overvoltage suppressors (surge diverters) do not induce voltages in the grounding system of such a magnitude as to cause signal-processing errors. Precautions against interference are described below.

*Limitation of overvoltages by overvoltage protection devices.* To protect against externally derived overvoltages which could affect the telecommunications system, voltage-limiting components must be provided. The classical overvoltage protection device is the *spark gap*, originally an air spark gap – in modern practice a *gas-discharge surge diverter* (arrester, overvoltage suppressor). This is a suppressor in which two electrodes are mounted with a small separation in a capsule filled with an inert gas. When an overvoltage occurs, the voltage across the suppressor first rises to the triggering level, without a significant flow of current. Ionization of the gas in the discharge path leads to a breakdown between the electrodes, and consequently a highly conducting connection ($< 0.1$ $\Omega$), such that the voltage across the suppressor is brought down to the arc burning voltage of 10 to 15 V. At the conclusion of the discharge process, the suppressor reverts to the non-conducting condition. A disadvantage of gas-discharge surge diverters is that the triggering voltage depends upon the rate-of-rise of the voltage.

When gas-discharge surge diverters are applied to power conductors, triggering may lead to a follow-up current through the surge diverter to ground. Such diverters may not therefore be used on a supply with a short-circuit current greater than 0.5 A or an operating voltage greater than 20 V without an appropriate means of current interruption; this purpose may be served either by fuses or by (series-connected) varistors.

To suppress overvoltages below the triggering level of the gas-discharge surge diverter, it may be combined with other voltage-limiting devices. For this purpose a number of components are available, e.g.:

- metal oxide varistors,
- surge-suppressor diodes (transient-absorption zener diodes),
- zener diodes.

*Metal oxide varistors* are voltage-dependent resistors with symmetrical characteristics.

A block of sintered zinc oxide, with an admixture of other metal oxides, is mounted between two contact plates. The voltage dependency is due to the variable contact resistance between the sintered oxide crystals. The resistance ($> 1$ M$\Omega$) collapses very rapidly ($< 25$ ns) to values in the region of 1 $\Omega$. Metal oxide varistors are used in combination with gas-discharge surge diverters for primary protection, while specific protection is afforded in the circuits by means of suppressor diodes or zener diodes.

The characteristics of overvoltage protection components are compared in Table 17.5.

As an example of 'graded protection' with various components, Fig. 17.10 shows the protection scheme against lightning interference for an exposed telemetry cable. The gas-discharge surge diverter, as primary protection, absorbs the main energy associated with the overvoltage, while the following combination, consisting of a varistor and a suppressor diode, clips the voltage wavefront up to the point where the gas-discharge surge diverter triggers.

In this connection, the *diverter valve* should be mentioned. This consists of a combination of a spark gap and a varistor and is used, for example, to protect mains supplies against the effects of lightning strokes.

### 17.6.3 Design of the Distribution Network for EWSD Systems

In conventional switching systems one common line (main load line) serves to distribute current to all the equipments in a room. From this line, lines of smaller cross-section, branch off to the rack rows (branch load lines). Because of the reduction in cross-section, fuses are inserted into the branches. In this distribution system, if the installation has to be extended, work has to be carried out on the live main load or branch load lines. Undervoltages or overvoltages caused by fuse blowing as a result of short-circuits or ground faults, which cannot be avoided in this kind of work, affect the whole distribution system. Under

**Table 17.5.** Comparison of the principal characteristics of overvoltage protection components

| | Inert gas-filled discharge surge diverter | Metal oxide varistor | Suppressor diode | Zener diode |
|---|---|---|---|---|
| Range of protection level (V) | 65 to 12 000 | 20 to 2000 | 6 to 400 | 2.4 to 200 |
| Surge current (A) | 60 000[a] | 25 000[a] | 1000 | 200 |
| Energy adsorption capacity (J) | 60 | 1800 | 1 | 0.05 |
| Range of continuous loading (W) | 800 | < 1.5 | > 1 | 0.5 to 50 |
| Response time (ns, ps) | $du/dt$-dependent, approx. 500 ns | < 25 ns | < 10 ps[b] | 10 to 1000 ps |
| Capacitance (pF) | 1 to 7 | 40 to 15 000 | 300 to 12 000 | 8 to 1500 |
| Leakage current (nA) | < 15 | < 200 000 | < 5000 | < 100 |
| Application | Primary protection for following close-protection devices | Primary and close protection in mains circuits | Close protection | Close protection |
| Advantages | High surge current, low capacitance | High surge current, very good extinction characteristics, no follow-on current | Very short response time, very good extinction characteristics, no follow-on current | Short response time, relatively high loading capability |
| Disadvantages | Low-impedance supplies require assisted extinction. Moderate response time falling with increasing $du/dt$ | High capacitance, affected by ageing | High capacitance. Relatively low surge current capability, dependent on voltage | Relatively low current loading capability, low protection level |

[a]With 8/20 $\mu$s overvoltage impulse.
[b]With low-inductance connections.

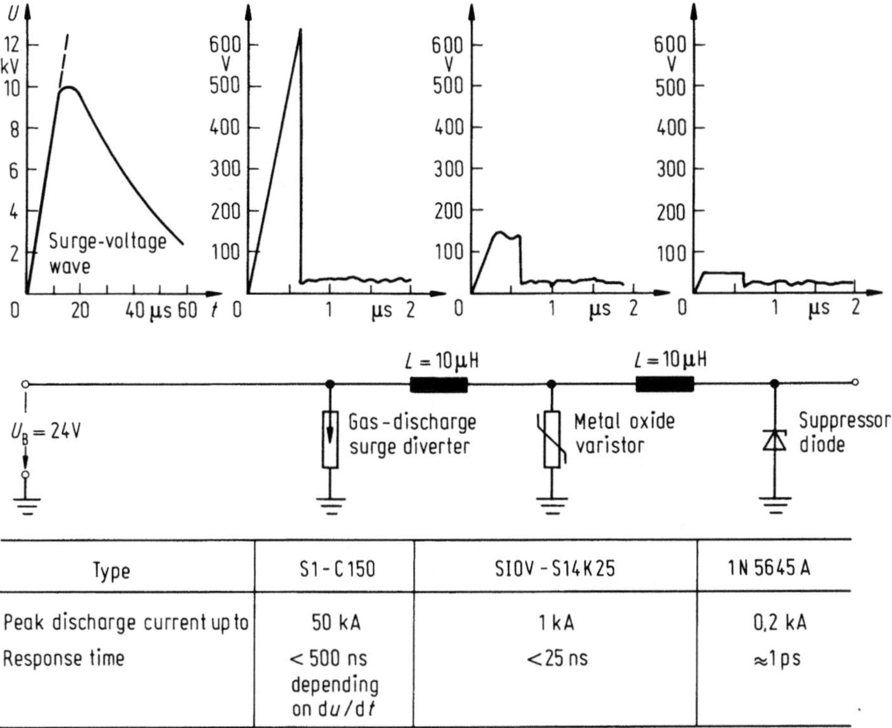


| Type | S1 - C 150 | SIOV - S14K25 | 1N 5645 A |
|---|---|---|---|
| Peak discharge current up to | 50 kA | 1 kA | 0,2 kA |
| Response time | < 500 ns depending on $du/dt$ | < 25 ns | ≈ 1 ps |

**Fig. 17.10.** Example of graded protection with voltage-limiting components

unfavourable conditions an operational breakdown of the electronic system is then possible. To limit the incidence of operational failures of the system due to short-circuits in the distribution network, therefore, the power distribution for the EWSD system has to be arranged in a fundamentally different way (Fig. 17.11). Each cubicle row has its own feeder. A choice of two standard cross-sections (95 and 150 mm$^2$) is available. To facilitate the laying of cables for the purpose of extensions, feeder cables with fine-strand cores are used. To keep the d.c. voltage drop within acceptable limits, up to four cables can be connected in parallel.

To minimize the overvoltages which can be produced by the blowing of a 10 A fuse, the distribution network must be so arranged as to introduce the lowest possible impedance. This is achieved in the case of the positive line by means of large surface-area return conductors, and in general through close proximity of the positive and negative conductors. In addition, the total combination of the 2.2 mF capacitors incorporated in each of the system cubicles effects a useful degree of suppression. The separate power circuit for each cubicle row also affords good decoupling. The voltage dip caused by the blowing of a fuse affects principally the associated current-carrying negative line.

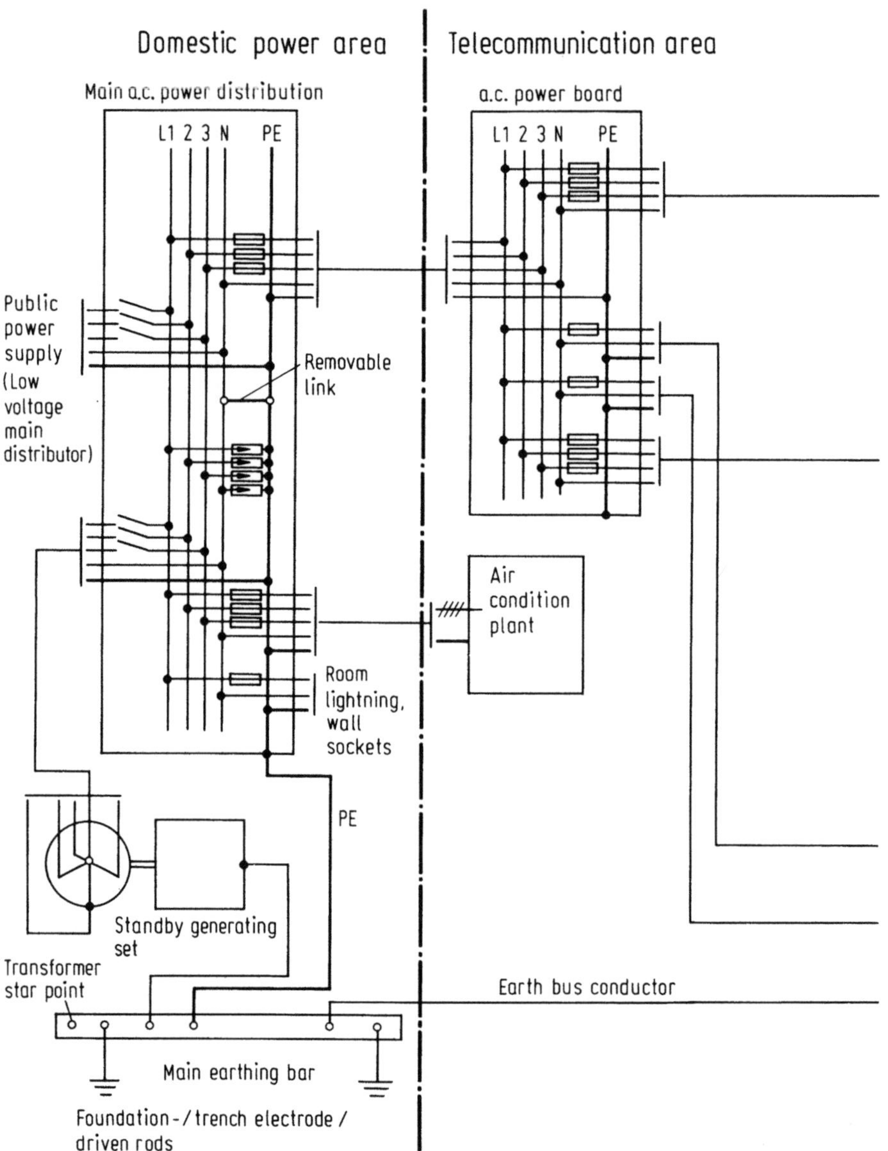


**Fig. 17.11a.** Grounding and potential equalization in a telecommunications building for EWSD systems (*Source*: Siemens AG). MDF Main distribution frame, O&M operation and maintenance centre, PE protective earth conductor, FE functional earth conductor, FPE functional and protective earth conductor

For Fig.17.11b see page 362

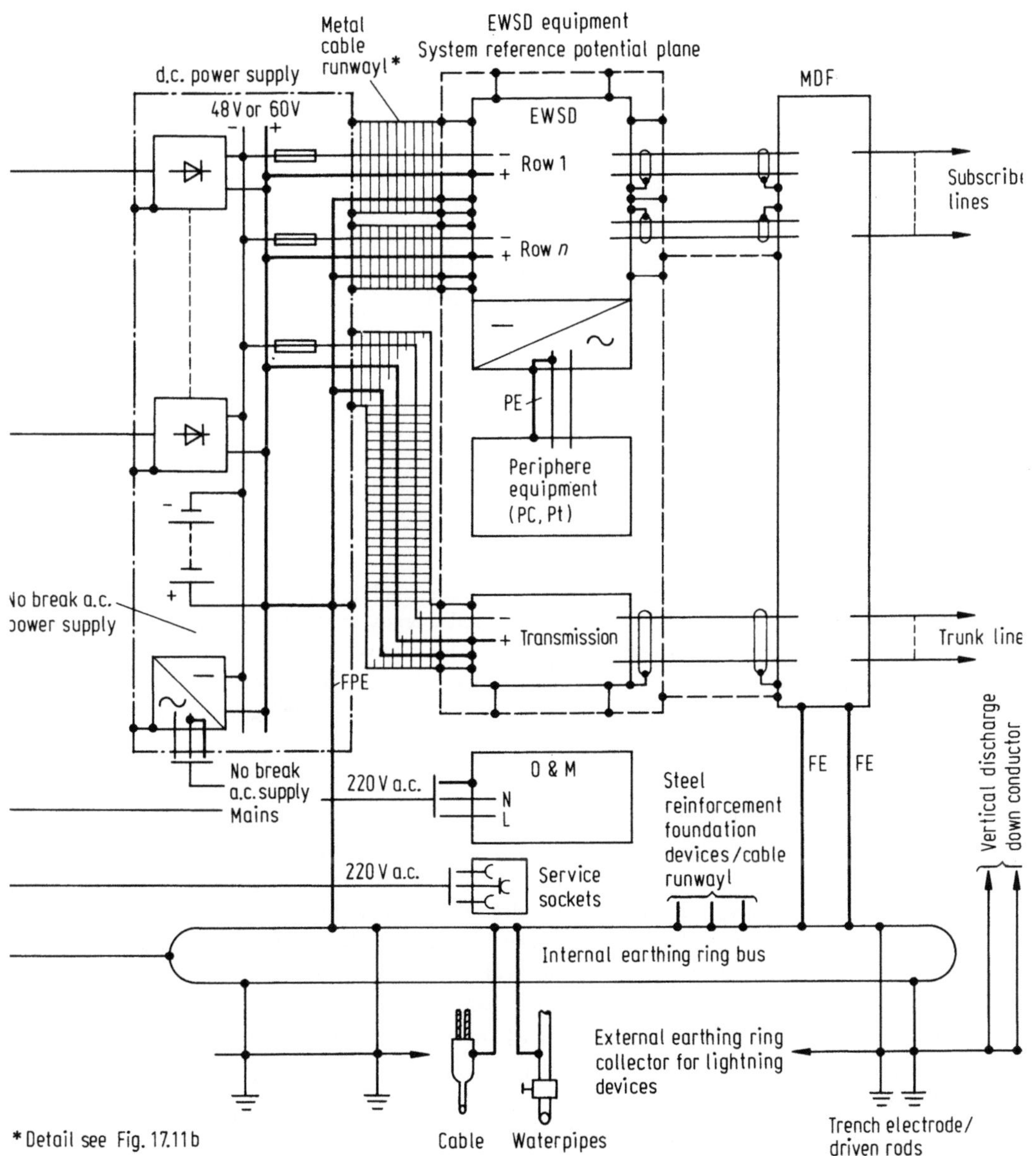


### 17.6.4 Design of Grounding, Potential Equalization and Lightning Protection for EWSD Systems

**Internal system provisions**

To ensure undisturbed operation of all the central and decentralized functions of the system, and for the purposes of screening, a potential reference plane is necessary, which also permits no appreciable potential differences within the

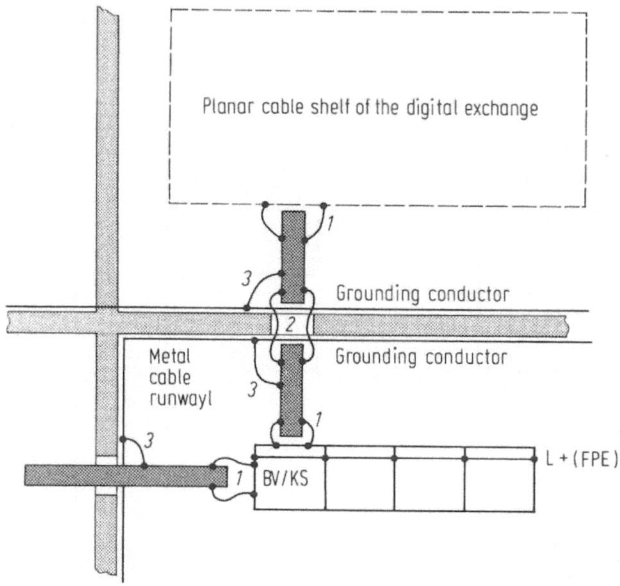


**Fig. 17.11b.** Grounding the cable runwayls. BV Battery and distribution cabinet, KS Compact power supply, *1* metal cable runwayls screwed directly to the planar cable shelf and to the power supply system or connected by $2 \times 16$ mm$^2$ gnye cables, *2* do not route cable runwayls through fire barriers; use $2 \times 16$ mm$^2$ gnye cable for such connections, *3* connect cable runwayls to the grounding conductor at least $1 \times$ with 16 mm$^2$ gnye cable

plane as a result of pulse currents (fuse-blowing, atmospheric overvoltages). In former systems the design of the potential reference plane was characterized by cross-connection of the grounded positive lines with the rack structure and the floor reinforcing mesh. In the case of digital systems, because of the compact construction that is possible and the consequent high heat loss per unit area, a false floor is sometimes provided in the exchange room for air conditioning; since this is consequently available also for the cabling, the floor mesh in its structural form would not be effective in respect of the cabling. The electrical function of the mesh as significant part of the potential reference plane cannot, however, be dispensed with.

To fulfil the electrical function of the floor mesh, a large-area conducting network of aluminium strips is laid in the double floor.

Each system cubicle is connected to this grounding screen. A grounding network of typical dimensions can limit the potential difference produced by a peak current of 1 kA (straight across the area of the exchange room) to less than 10 V. This is the highest value, according to experience, that can result from interference in earth cables from atmospheric discharges.

Figure 17.12 shows the relationships that apply on the occurrence of an overvoltage due to an atmospheric discharge. The current that flows as a result of the triggering of a gas-discharge surge diverter is divided according to the impedances of the conducting paths represented by the main distribution frame (MDF)-system-power supply equipment – ground and the main distribution frame ground. The voltage drop due to the peak current in $Z_2$ (cable tray between the main distribution frame and the system) is applied to the subscribers' or junction lines, while the impedance $Z_1$ (the grounding line to the main distribution frame) determines the overall magnitude of the overvoltage. This means that these connections must be made with the minimum impedance. The main distribution frame must therefore be provided with multiple grounds. The cable trays between the main distribution frame and the system or between parts of the system must be bonded with low-resistance area joints. If the cables have to pass through walls, and it is not possible to carry the longitudinal tray members through, wide copper strips must be provided for electrical connection. Where they enter the exchange room, the trays should be connected to the grounding network over a broad area.

With the interconnections described a substantial uniformity of potential in the d.c. area will be achieved. This is a prerequisite for the two-way connection of the screens of distribution cables in order to utilize their attenuating properties.

To make further use of the attenuation of the grounding network, the connecting cables must be laid directly on the metal sheets; in addition, the cable runs must lie within the area of the network. To the extent that in large installations individual cables are laid above the cubicles (e.g. bus connections) a cable with a braided screen must be used for this purpose.

**Provisions in the building**
The erection of lightning-protection systems is covered by VDE regulations DIN 57185/VDE 0185 Parts 1 and 2; these have replaced the previous general lightning-protection regulations (published by the lightning conductor construction committee, ABB).

Lightning-protection systems installed in accordance with the former ABB recommendations ensured protections only for persons and against fire.

The design of lightning protection described below takes into consideration both the 'normal' building protection and the additional measures necessary for the protection of electronic systems.

*Lightning conductor installation.* On flat roofs, or on the ground under ridged roofs, lightning conductors should be erected at a regular spacing of about 0.6 m. A conducting roof curb, in conjunction with the closely interconnected network of lightning conductors, affords the best distribution of the lightning current to the diverter discharge cage.

*Diverters.* In steel-framed concrete buildings the individual stanchions of the outer walls are included in the earthing system and used as diverters. To this end it is necessary to weld the vertical reinforcing bars or girders. Where this is not possible from considerations of strength, 10 mm galvanized steel rod should be laid with the reinforcement of the reinforced-concrete pillars and bonded. The distance between the diverters must be less than the distance between floors.

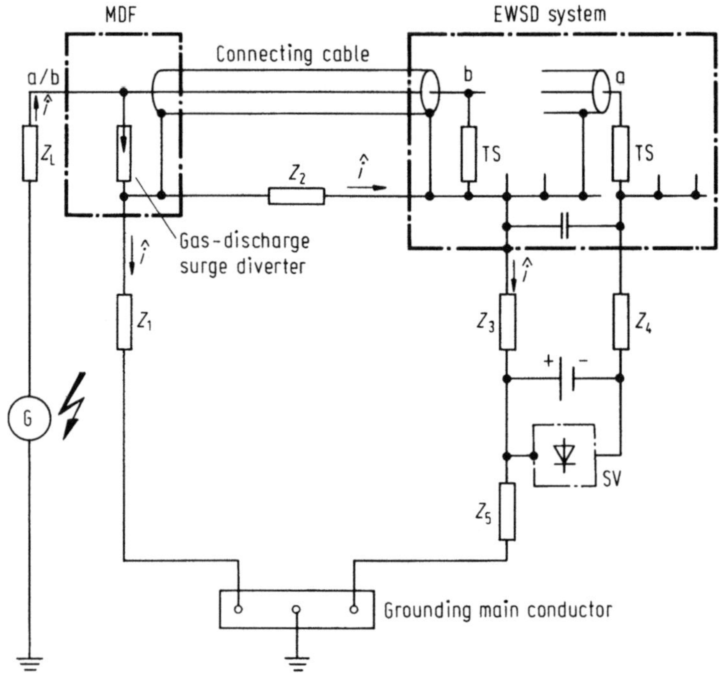


**Fig. 17.12.** Essential representation of conditions on overvoltage. $Z_L$ Line impedance, $Z_1$ impedance of ground conductor between main distribution frame and grounding main conductor, $Z_2$ impedance of cable tray between main distribution frame and system, $Z_3$ impedance of positive conductor between system and power supply installation, $Z_4$ impedance of negative conductor between system and power supply installation, $Z_5$ impedance of FPE conductor between power supply installation and grounding main conductor, MDF main distribution frame, TS subscriber circuit, SV power supply installation

In framed buildings with wider spacing, additional diverters should be introduced into the brick bays of the walls. For horizontal potential equalization the reinforcing bars of the floors and ceilings must be bonded to the vertical diverters. The same applies to the reinforcing bars in the foundations; these should be carefully welded or tied with binding wire to the individual mats.

*Foundation ground.* A foundation ground should be laid in the footings of the outer walls. The reinforcing bars of the reinforced-concrete stanchions must be bonded to the foundation ground. For connection to the external lightning conductors terminal lugs should be brought out at the spacing of the diverters. Terminal lugs should also be provided for the internal grounding ring main conductor.

*Grounding ring main conductor.* For the lowest floor of the building a grounding ring main conductor through all the rooms is recommended. The ring main takes the form of an insulated copper conductor of 95 mm$^2$ cross-section.

To this conductor, for the purposes of potential equalization, are connected metal piping systems (except gas pipes), ventilating ducts, cable trays and the conducting sheaths of the external cables.

To avoid introducing potential differences in these external conductor systems into the building, where the risk from lightning is high they should, as far as possible, be brought into the building at one point, where the sheaths can be connected directly to one another and to the grounding ring main.

*Screening.* Insofar as precast reinforced-concrete parts or conducting cladding panels are used for outer walls and included in the grounding system, a screening attenuation of about 20 dB is to be expected in the broadcast radio-frequency range. This attenuation is sufficient, on the one hand, to meet the requirements of the interference limit class A or B and, on the other hand, to render the EWSD apparatus, for example, immune to an external field of more than 3 V/m.

Brick-built outer walls or bays between concrete stanchions, at least in the vicinity of telecommunications apparatus, necessitate additional screening. This can be achieved with suitable metal cladding, or else the separation between the diverters must be reduced to about 1.2 m.

The diverters should be joined at roof level through the lightning rods and at ground level through a ring main.

In special cases an equipment room can be lined with copper foil as a retrospective measure.

In larger installations, accommodated in separate parts of the building, the connecting cable must be installed with adequate protection. The cables should be laid either in steel conduit, in ducts with conducting linings or in reinforced-concrete cable ducts. In each case, at the point of entry into the building, the screen (conduit, reinforcement) must be connected with a large-area contact to the grounding system of the building.

The ground lines of the exchange and transmission apparatus and the power supply equipment should be connected to the grounding ring main in the basement by the shortest route. Since with the desired dense network of diverters the close proximity between the diverters and the system cannot be avoided, the potential reference plane of a system larger than 20 m in height should be connected to the diverters.

With the measures described, substantial potential differences within a building are avoided. Cables entering the building from outside, however, are connected to a remote ground. If the building is struck by lightning, its potential can be raised by the voltage drop in the grounding system.

For telecommunications buildings, which are particularly susceptible to direct lightning strokes because of their position and height, an especially low ground resistance is desirable; in addition, the cable cores should be protected by diverters.

Since mains supply cables also represent a remote ground in the event of a lightning stroke, the ungrounded conductors must be protected by diverter valves.

### 17.6.5 Special Requirements for Lightning Protection in Telecommunications Towers

Usually the power supply installation for the transmission system is in the operating building beside the base of the tower, and the very long power supply conductors introduce the risk of induced overvoltages.

A lightning stroke on a telecommunications tower produces a large voltage drop in the steel reinforcement. The overvoltage thus produced in the cable may damage the input circuit of the d.c./d.c. converter.

To avoid this, the following protective measures must be adopted:

– the d.c. cables between the operating building and the base of the tower are run in a metal tube. This tube, which is 'galvanically' connected to the reinforcement of the tower and the operations building and also, via a specially formed cable duct in the operating building, to the battery switching panel, ensures that the tower and the building, including the power supply system, are always at approximately the same potential with respect to ground (Fig. 17.13);

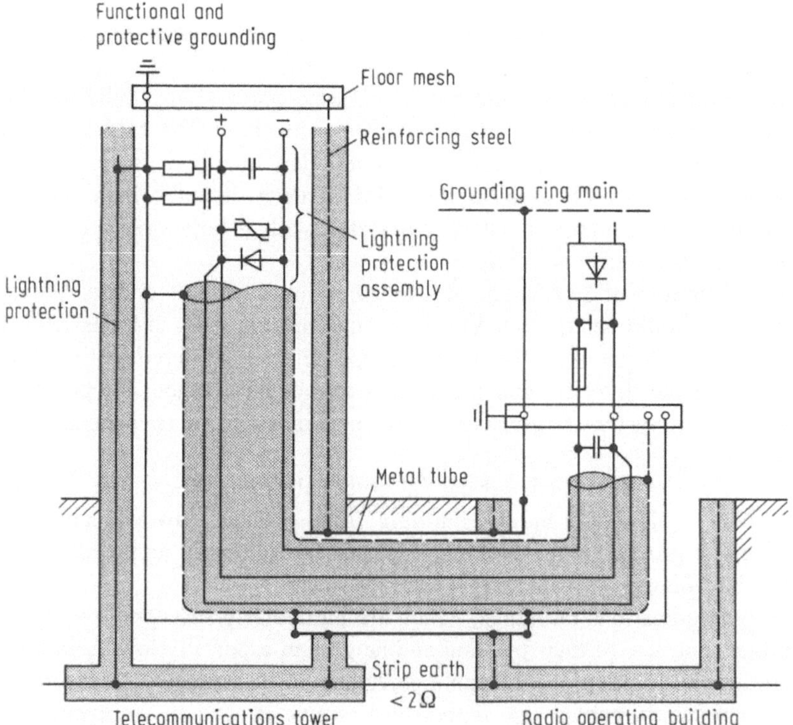


**Fig. 17.13.** Diagrammatic representation of grounding system in a telecommunications tower and an operating building

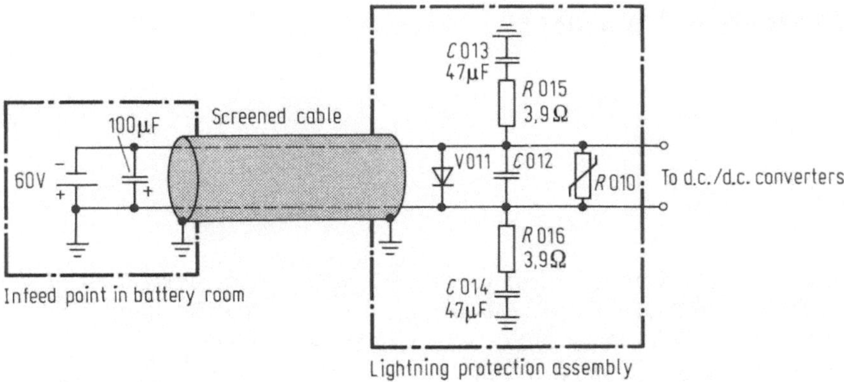



**Fig. 17.14.** Lightning-protection assembly for d.c./d.c. converters with and without 'galvanical' separation

– the d.c. power supply conductors between the battery switching panel in the operating building and the operating floor in the telecommunications tower are in the form of screened cables. The screen of the cable is connected by the shortest route to the structural metalwork of the switching panel, to the steel tube in the cable duct and the base of the tower, and in the operating floor to the floor mesh.

By these expedients the ground resistance is reduced considerably.

Overvoltages which exceed the permissible level in spite of the measures described above are limited on the operating floor by means of a lightning-protection assembly.

In the case of d.c/d.c. converters with 'galvanic' separation, adequate protection is obtained if the two lines are balanced with respect to ground; interference voltages between the positive and negative poles are then sufficiently small. This balancing is achieved by the capacitors $C013$ and $C014$ (Fig. 17.14).

For the protection of converters without 'galvanic' separation, the positive pole should be connected to the functional and protective ground within the exchange wiring. Additional protective measures therefore have to be applied to prevent an unacceptable rise of voltage between the negative pole and the functional and protective ground; these take the form of a capacitor ($C012$) and a number of varistors ($R010$) connected between the positive and negative lines. In this arrangement the capacitor limits the high-frequency component of the interference voltage while the varistors limit the low-frequency component which represents greater energy.

It is possible, as a result of the polarity of the interference voltage, for the polarity at the input to the converter to be reversed. The possibility that the input transistors may be destroyed before the varistors respond cannot be ruled out; a diode ($V011$) is therefore provided as additional reverse-polarity protection.

# 18 Protective Measures

To avoid accidents from electrical causes, the *rules for the prevention of accidents* of the professional association[1], the *VDE regulations* (e.g. DIN 57 100/VDE 0100) and all other *safety rules* should be observed.
The VDE regulations are divided into:

- equipment regulations, for the manufacture of equipment,
- erection regulations, for installation, etc., on site,
- operating regulations, for operation and maintenance of equipment or installations.

Decisive factors affecting the magnitude of the consequences of an electrical fault are the current level, the duration of the effects of the current, the current path and the nature of the current. Mains alternating currents with a magnitude of more than 50 mA are particularly dangerous if they persist for more than 0.1 s and the current path includes the heart.

The regulations pertinent to protection against dangerous body currents should be strictly observed.

The corresponding measures are divided into:

- protection against direct contact. These measures should prevent contact with live parts,
- protection against indirect contact. These measures should protect against danger arising from contact with conducting parts, referred to as 'bodies', which may become live under fault conditions,
- protective low voltage or functional low voltage. This method rests upon the operation of circuits at voltages of not more than 50 V a.c. or 120 V d.c.

Working on live parts or systems represents an especial hazard. Such work is therefore not permitted. Exceptions are possible if the installation cannot be made dead and only qualified personnel are working on it (VDE 0105).

**Protection against direct contact**
These regulations should prevent the inadvertent touching of live parts, specifically through the 'insulation' and 'covering' of such parts.

The requirements for insulation (test voltage, leakage currents, etc.) are laid down in the relative equipment specifications. Covers must comply with creepage

---

[1] For example, in Germany: VBG 4 (electrical equipment and systems).

distance and air-gap requirements and must not anywhere permit the ingress of a hand.

In rooms through which unqualified people may pass, covers must be such that they can only be removed with the aid of tools.

Less onerous conditions are permitted in electrical operating locations. Thus, for example, a cover is adequate which protects against inadvertent contact and can be removed without tools.

In locked electrical operating locations, protection against contact may be completely dispensed with. However, such relaxations are only permissible when these special operating locations comply with all the requirements stipulated by the VDE regulations.

## Protection against indirect contact

In electrical installations, parts which in normal operation carry no voltage with respect to ground (e.g. housings) may become dangerous contact potentials in the event of faults. A means of protection for indirect contact is therefore necessary (formerly known as 'protection against excessive contact potential').

Protective measures against indirect contact are required in installations and equipment with voltages in excess of 50 V a.c. or 120 V d.c. with respect to ground.

On the other hand, no protective measures are required in relation to indirect contact in installations and equipment:

- with voltages below 50 V a.c. or 120 V d.c. with respect to ground (protective low voltage or functional low voltage),
- with voltages up to 250 V with respect to ground in public supply mains equipment for electrical energy and power measurement,
- in domestic installations with insulating floors, where no coincidental contact with grounded water, gas or heating equipment is possible,
- with alternating voltages below 1000 V and direct voltages below 1500 V for steel or reinforced-concrete poles, metal conduits and metal cases with insulating coatings, metal conduits for the protection of multiple conductors or multi-core cables, and sheathing or armouring of conductors and cables, so long as the cables are not laid in the ground.

Protective measures for indirect contact may be classified as follows:

*Protection by disconnection or indication*

- Disconnection by overcurrent protection device:[2]
  TN S supply system (neutral earthing),

---

[2] Supply systems in which protective measures involving protective conductors are used are designated by two basic letters. The first letter indicates the relationship to ground of the power source: T direct ground connection: I insulation. The second letter indicates the relationship to ground of the loads: T direct ground connection: N direct connection to star or neutral point. The meaning of the additional abbreviations is: S separate; C common.

TN C S supply system,
TT supply system (protective earthing),
IT supply system (protective-line system),
– disconnection by protective switch:
  fault voltage–operated ($FU$) protective circuit breaker (not described),
  fault current–operated ($FI$) protective circuit breaker.

One of these protective measures is a prerequisite for the use of equipments in protective class I.[3] All such equipments are provided with a ground connection.

*Protection without disconnection or indication*

– Protective insulation (protective class II[3]),
– equipments with protectively insulated parts,
– protective isolation.

Equipment in protective class II can be used on any supply system.


**Protective low voltage and functional low voltage**
It is stipulated that it must not be possible for the voltage on the low-voltage system to exceed 50 V a.c. or 120 V d.c. This necessitates the use of specified power sources and the observance of regulations covering the arrangement of circuits (e.g. isolation from higher-voltage circuits; see Section 18.2.3).

In general, protection against direct contact can be dispensed with if the nominal voltage does not exceed 25 V a.c. or 60 V d.c.


**Safety measures instructions**
Before starting work very important is to observe properly the correct sequence of the five safety measures:

---

1. *disconnect from power,*
2. *protect against reconnection of power,*
3. *check whether power is off,*
4. *grounding and short–circuiting and*
5. *cover, or prevent access to, adjacent live parts.*

---

In all technical documentations like troubles shooting manuals, description and operating instructions remarks to the safety regulations can be found e.g.:

---

[3] Protective classes I and II are defined in several VDE specifications – among others. VDE 0106 and VDE 0804.

*Example 1:*

    ⚠ CAUTION   ***Hazardous voltages are present in this electrical equipment during operation. Failure to properly maintain thet equipment can result in death, severe personal injury or substantial property damage.***

*– maintenance shall be performed only by qualified personnel,*
*– always de-energize and ground the equipment before maintenance, and*
*– use only authorized spare parts in the repair of the equipment.*

*Example 2:*

***Elevated voltages are inevitably present at specific points in this electrical equipment. Some of the parts can also have elevated operating temperatures.***

***Non-observance of these conditions and the safety instructions can result in personal injury or in property damage.***

***Therefore only trained and qualified personnel may install and maintain the system.***

*Example 3:*
*Attention!*

***Before putting the devices into service, the correct position of the modules, plug-in connections, screw-connections and terminal connections has to be carefully checked.***

**Instructions for measuring on printed circuit boards (PCB)**

Measuring and testing operations have to be effected very carefully, as distorting the voltages or causing a short-circuit may lead to the destruction of the components. Therefore, any work on PCBs has to be carried out in a de-energized condition as far as possible. Furthermore, it is important to know that even for a switched-off device the filter capacitors can still be energized. Thus, before starting work the existing automatic circuit breakers must be cut off.

Before starting the measurement, the measuring leads have to be connected first in order to avoid short circuits during their application. Then the device can be powered up.

All voltage data wave forms relate to a measuring effected between the indicated test point and the reference voltage of the switch.

If the measuring is performed with an oscilloscope, attention has to be paid to the fact that no ground potential is carried into the circuit via the measuring leads. Therefore, a totally insulated oscilloscope (no earth on ground) must be used, or the oscilloscope has to be connected via an isolated transformer to the mains.

When exchanging modules an adjustment to the device in question is necessary.

The modules are equipped with circuits for example using the CMOS-technique, which can be destroyed by electrostatic discharge; for this reason, corresponding protective measures (ESD rules) have to be observed.

## 18.1 Protection by Disconnection or Indication

Bodies[4] which do not form part of the functional circuit may, as a result of insulation failure, assume a potential relative to ground or to other conducting parts, referred to as the contact potential $U_B$ (Fig. 18.1); this can drive a dangerous current through the human body.

Common to all protection methods relying on disconnection is a *protective conductor* PE[5] (protective class I). The protective conductor connects the body to ground (see, for example, Fig. 18.5) or with the PEN conductor.[6]

In the event of a fault, the fault current is carried from the body by the protective conductor. The disconnection must be so affected that the supply voltage is removed as quickly as possible.

All bodies must be connected to the protective conductor. The terminal for the connection of the protective conductor is distinguished by the symbol ⏚ (DIN 40 011). To prevent rupture of the protective conductor, it must have a certain minimum cross-sectional area (depending upon its length and the rating of the fuse which precedes it). In the case of flexible lines it should be contained within the common sheath and coded green/yellow. Structural parts may be used as protective conductors if they fulfil the requirements as to cross-sectional area, provided that the protective connection is not broken by the removal of structural parts and the connection remains highly conductive with the passage of time.

---

[4] Bodies, in the terminology of the VDE specifications, are touchable conducting parts of an installation which are not live in normal operation (e.g. equipment housings)

[5] The designation SL was formerly used for this

[6] The PEN conductor combines the functions of the neutral conductor N (formerly designated Mp) and the protective conductor PE. The description 'Nulleiter' ('zero or neutral conductor') (SL/Mp) was formerly applied to the PEN conductor

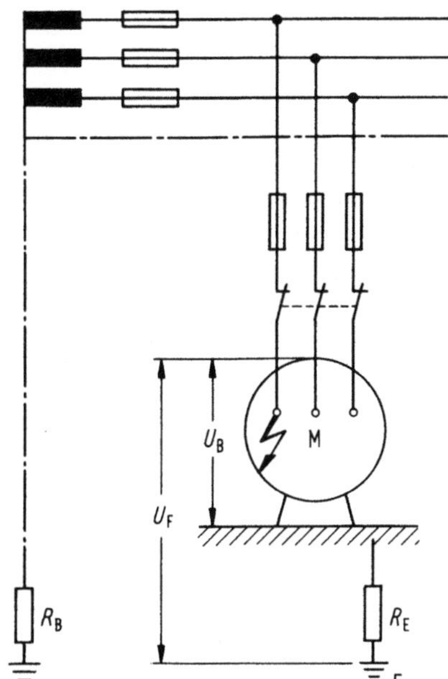



**Fig. 18.1.** Fault voltage $U_F$ and contact potential $U_B$ with a non-insulated floor. E Reference ground, $R_B$ operating ground resistance, $R_E$ sum of ground resistances

The fault voltage $U_F$ is the voltage that appears between a body and the reference ground, or between bodies, in the event of a fault. By contact potential $U_B$ is meant that part of the fault voltage that can be embraced by the human body.

In telecommunications equipment, the insulation between power circuits and accessible conducting parts must be specified for a test voltage of 1.5 kV (in accordance with VDE 0804 for equipment in protective class I). For functional reasons one pole of the d.c. supply line – usually the positive pole – is directly grounded. From the point of view of the functional efficiency of the system, this line should not introduce more than a certain voltage drop. Depending on the current level and the length of the line in question, this can lead to large cross-sections.

Since the conductors thus grounded are connected together in each rack, through the rack structure, for example, the result is a very low-resistance (grounded) conductor network. Hence this 'area ground', apart from its operational function, is also applied in general to protective purposes, in that the enclosures of equipments in protective class I are bonded to the rack structure. The whole system is designated as functional and protective grounding (see Chap. 17). This represents a generic term for all protection measures based on disconnection, and should not be confused with the special protective measures associated with the TT supply system (protective grounding) in accordance with DIN 57 100/VDE 0100 (see Sect. 18.1.2).

By virtue of the functional and protective grounding, telecommunications systems can be designed *in standard form,* irrespective of the protective measure of the public mains supply. The coordination with the various protective systems is effected centrally at the grounding main conductor.

In the following Sects. 18.1.1 to 18.1.4, the power system protection measures according to DIN 57 100/VDE 0100 are compared with the corresponding implementation of the functional and protective grounding scheme in accordance with VDE 0800 Part 2.

### 18.1.1 Protective Measures in the TN Supply System

The TN system corresponds to the type of supply in which neutral grounding was formerly applied; it refers to the most widely used protection method. In this scheme all bodies are connected, either directly or through a special protective conductor PE, to the PEN conductor (formerly the neutral conductor). TN S supply system see Fig. 18.2 and TN C S supply system Fig. 18.3. The connection to the protective line ensures that all accessible parts are at the same potential in the event of a fault to frame in any equipment. The short-circuit current to which it gives rise causes the preceding fuse to operate and thereby to disconnect the circuit from the supply.

The following time-delays are appropriate for this purpose:

– 0.2 s for socket-outlet circuits of up to 35 A current rating,
– 5 s for all other circuits.

To ensure that the potential of the PEN or protective conductor deviates as little as possible from ground potential, this conductor should be grounded at numerous points to the public distribution network, particularly at the point of entry into the building.

When a protective measure is applied to a TN supply system *with a* separate ground conductor PE in accordance with DIN 57 100/VDE 0100, a connection should be made between the PEN conductor terminal at the connection point of the incoming low-voltage mains supply and the grounding main conductor of the telecommunications system (Fig. 18.3). The cross-sectional area of this

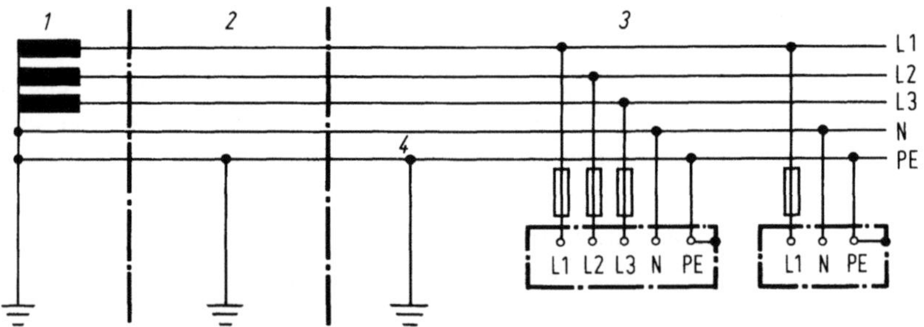


**Fig. 18.2.** TN S supply system (formerly neutral grounding), Regulation VDE 100 part 300 or international regulation IEC 364 part 312.2, in the hole supply separate neutral- and protective earth conductor. (*Source*: Siemens AG). *1* Power source (distribution transformer, generator), *2* distribution supply network, *3* load system, *4* in the hole distribution supply network at every point of house entry connection box must the protective conductor be grounded, L1, L2, L3 outer (voltage supply) conductors, N neutral conductor (formerly Mp-conductor), PE protective earth conductor

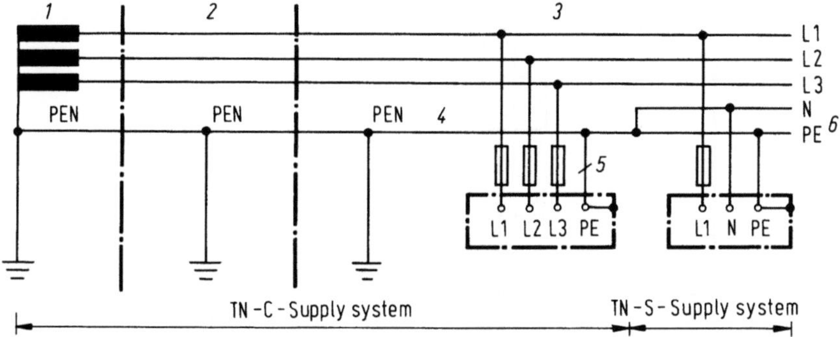



**Fig. 18.3.** TN C S supply system* Combination of common neutral- and protective conductor in a part of the supply (*Source*: Siemens AG). *1* Power source (distribution transformer, generator), *2* distribution supply network, *3* load system, *4* in the hole distribution supply network at every point of house entry connection box must the protective conductor be grounded, *5* only permitted if connection cross-sectional area $\geq$ 10 mm², *6* a connection between N- and PE-conductor is after the distribution in N- and PE-conductor not permitted, L1, L2, L3 outer (voltage supply) conductors, N neutral conductor (formerly Mp-conductor), PE protective earth conductor, PEN combinated special neutral and protective conductor (formerly zero conductor)

* VDE regulation VDE 100 part 300 or international regulation IEC 364 part 312.2

connection must be the same as that of the PEN conductor of the incoming feeder, but in any case at least 16 mm² (copper). Further connections between the PEN conductor and the telecommunications grounding system (the functional earth) at other points in the building are not permitted, because they would introduce the risk of interference in the telecommunications system. The separate earth conductor PE is connected to the PEN conductor only at the point of entry into the building; it is thus also connected to the grounding main conductor of the telecommunications system through the required connection.

Figure 18.4 shows the connection of the power supply system to different kinds of mains systems.

In the case of connection to a medium-voltage supply through a transformer substation in the region of the telecommunications system, the low-voltage grounding system of the substation (installed in accordance with DIN 57 100/ VDE 0100) should be connected to the grounding main conductor of the telecommunications system (Fig. 18.4c).

Only a *single connection* is permissible between the directly grounded star point of the low-voltage side of the transformer substation and the telecommunications grounding system. The cross-sectional area of this connection must be such that the overcurrent protection device connected in the feeder to the telecommunications power supply or to the power system loads in the telecommunications racks operates in the event of a fault to frame or to ground in the feeder. To this end the cross-section should be specified in accordance with VDE 0800, but it must in any case be at least 16 mm² (copper).

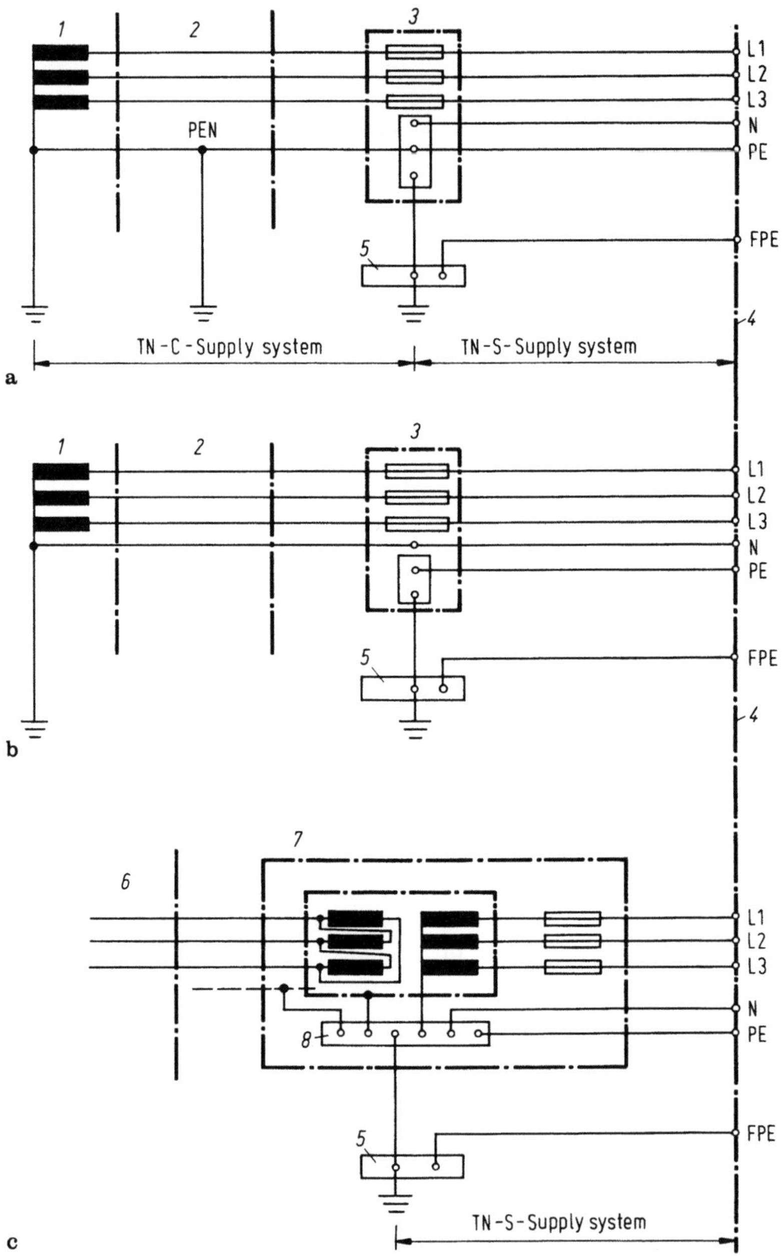


**Fig. 18.4a, b.** Connection of the power supply system to different kind of supply systems. (*Source*: Siemens AG). **a** Mains infeed: low voltage system with TN-supply system. **b** Mains infeed: low voltage system with TT-supply system. *1* Power source (distribution transformer, generator), *2* distribution supply network, *3* house connection box, *4* intersection to telecommunications system, *5* grounding main conductor, FPE functional and protective earth conductor, L1, L2, L3 outer (voltage supply) conductors, N neutral conductor (formerly Mp-conductor), PE protective earth conductor. **c** Mains infeed: medium voltage system $\geq$ 1 kV with TN S supply system. (*Source*: Siemens AG). *6* Mains infeed: medium voltage system $\geq$ 1 kV, *7* transformer substation (above 1 kV) in the area of the telecommunications system, *8* potential equalization of transformer substation

### 18.1.2 Protective Measures in the TT Supply System

The protective measures in the TT supply system (protective grounding) entail the direct connection of bodies to ground or grounded parts, so that, as with the TN supply system (see Sect. 18.1.1), the supply is disconnected by overcurrent protective devices in the event of an insulation failure (a direct breakdown to body). In the example shown in Fig. 18.5 the faulty circuit must be isolated by the fuse F1. In a TT supply system one line of the power supply is directly grounded, which means that the fault current flows through the ground. The contact potential $U_B$ may not exceed a level of 50 V a.c. To this end the ground resistance $R_E$ (in ohms) must be calculated according to the following relationship:

$$R_E \leqq \frac{U_B}{I_A}$$

Nowadays the protective device is generally the $FI$ protective circuit (see Sect. 18.1.4). If overcurrent protective devices are used, disconnection is required additionally in the neutral line. If this is not possible, a 'supplementary potential equalization' is used; its requirements are fulfilled in telecommunications systems by the functional and protective earth (see Sect. 17.3).

An example of the protective measures in a TT *supply system* in telecommunications installations is shown in Fig. 18.6. Telecommunications installations

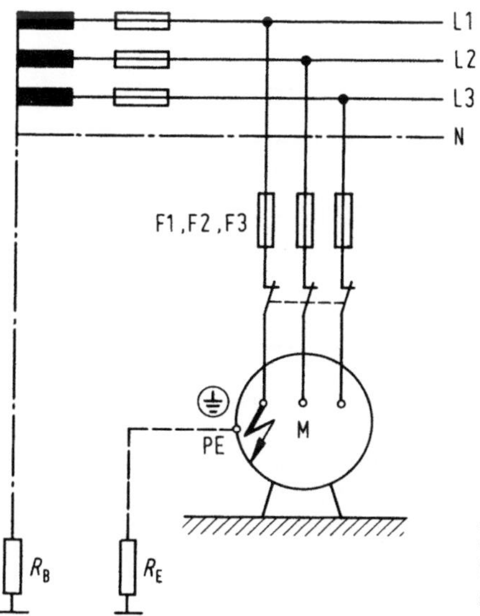


Fig. 18.5. Principle of protective measures in the TT supply system. $R_B$ Operating ground resistance, PE protective line, $R_E$ ground resistance

incorporating functional and protective grounding can be connected without any special precautions to TT supply systems that comply with DIN 57 100/VDE 0100.

If there are ground conductors (PE) for other power loads in the vicinity of the telecommunications installation (in the same room), they also may be connected directly to the grounding main conductor instead of the functional and protective ground. The cross-sectional area for this purpose should be in accordance with VDE 0800 Part 1.

The neutral conductor N associated with a TT supply system must *not* be connected to the functional ground of the telecommunications system (see DIN 57 100/VDE 0100).

### 18.1.3 Protective Measures in the IT Supply System

In the IT supply (protective-line) system, to prevent the occurrence of excessive contact potentials, all bodies are connected together and to the accessible conducting parts of the building, pipe systems, etc., and to grounding electrodes in the case of grounded supplies (no diagram).

A detailed explanation of the protective measures in the IT supply system is omitted here, since the arrangement is only used in special supply systems – in hospitals, for example, in which it is not acceptable to interrupt the low-voltage supply immediately on the occurrence of a fault.

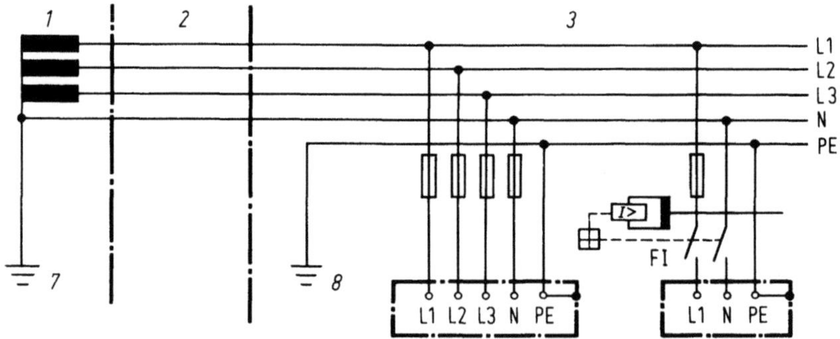


**Fig. 18.6.** TT supply system* (formerly protective grounding). (*Source*: Siemens AG). *1* Power source (distribution transformer, generator), *2* distribution supply network, *3* load system, *7* grounding of the power source (functional grounding), *8* grounding system of the load, *FI* fault current-operated protective circuit-breaker, L1, L2, L3 outer (voltage supply) conductors, N neutral conductor (formerly Mp-conductor), PE protective earth conductor

* Regulation VDE 100 part 300 or international regulation IEC 364 part 312.2

### 18.1.4 Fault Current-Operated Protective Circuit

In the fault current-operated (F$I$) protective circuit, a fault current-operated protective circuit breaker (switch) opens automatically on the occurrence of a fault current in excess of its rated current (the circuit being interrupted within a period of not more than 0.2 s).

The basic mode of operation of the fault current-operated protective circuit breaker SS may be explained with reference to Fig. 18.7.

All the a.c. lines to the equipment to be protected are passed through a summing current transformer. In the absence of a fault (not illustrated) current flows from the main conductor L1 through the fuse, fault current-operated protective circuit breaker and the load, and returns completely to the neutral conductor.

The current-transformer core is not magnetized, since the magnetizing effects of the currents cancel. No voltage, therefore, is induced in the secondary winding of the transformer and the protective circuit breaker is not tripped.

If an insulation failure occurs in the protected equipment, the fault current $I_F$ flows through the earth to the star point of the supply transformer (Fig. 18.7). The current that flows to the load is larger than the return current by the amount of the fault current. This difference produces a magnetic field in the core of the current transformer, which energizes the relay of the protective circuit breaker.

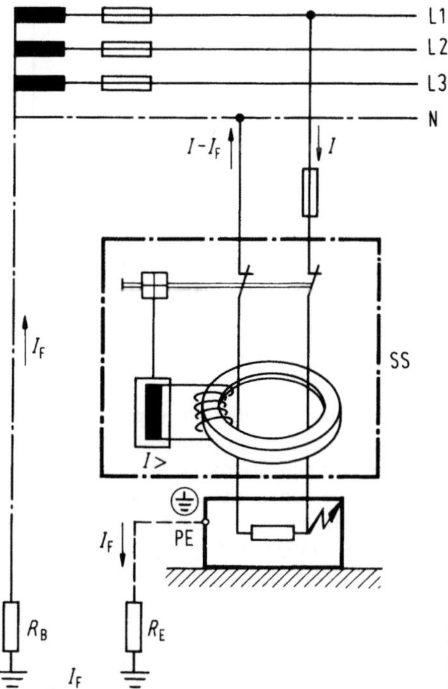


Fig. 18.7. Principle of fault current-operated protective circuit (shown with fault current flowing)

If the fault current reaches the rated value, the relay operates and the protective circuit breaker trips. The circuit is then broken by the opening of the contacts.

Modern fault current-operated protective circuit breakers trip very rapidly ($\leq$ 0.03s) in the event of a fault. With high-sensitivity types, operating on, for example, 30 mA fault current, it only requires an 'grounded' person to touch a live part to cause the protective circuit breaker to trip.

Figure 18.8 shows an example of a fault current-operated protective circuit breaker in a telecommunications system.

When a fault current-operated protective device is used, the neutral conductor N of the low-voltage mains supply must not be connected to the grounding system of the telecommunications installation behind the protective circuit breaker, since it would in that case by-pass the circuit breaker and render it ineffective. The

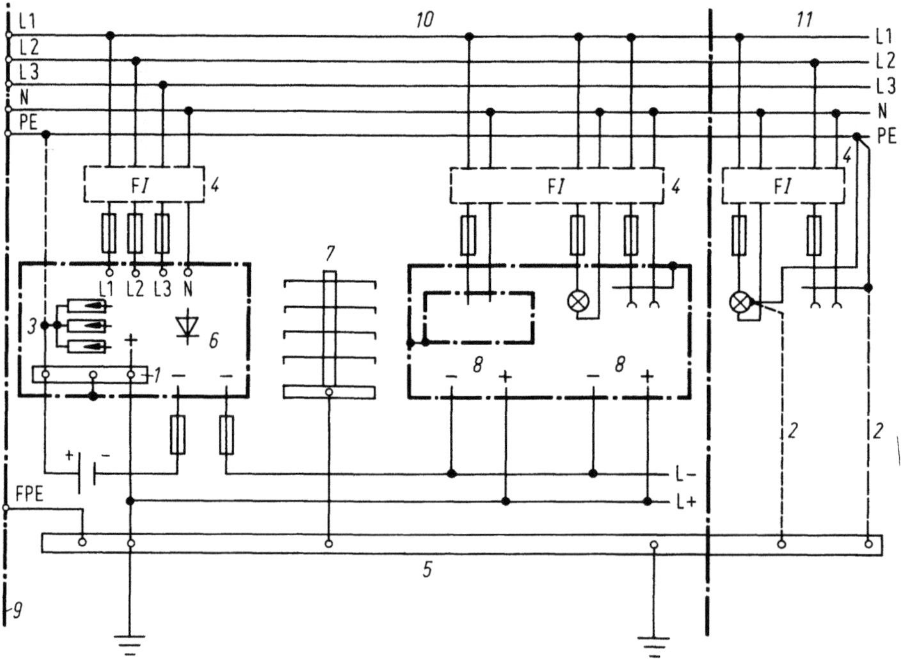

**Fig. 18.8.** Survey-grounding telecommunications systems at different supply systems. (*Source*: Siemens AG). *1* Grounding bus bar of the power supply system, *2* the functional conductors of other power loads near the telecommunications system are possible to connect to the protective earth conductor PE or to the functional and protective earth conductor FPE, *3* overvoltage protectors, *4* if TT supply system is used, the load must be connected via fault current-operated protective circuit-breakers F*I*, *5* grounding main conductor, *6* power supply system, *7* main distribution frame MDF, *8* telecommunications equipment, *9* intersection to different supply systems, *10* telecommunications systems, *11* other power loads near the telecommunications system, FPE functional and protective earth conductor, L− negative bus bar, L+ positive bus bar, L1, L2, L3 outer (voltage supply) conductors, N neutral conductor (formerly Mp-conductor), PE protective earth conductor

earth conductor PE for other power system loads in the vicinity of the telecommunications system should be connected to its grounding main conductor. The cross-sectional area must be in accordance with VDE 0800 Part 1.

The fault current-operated protective circuit can also be used to obtain improved tripping in TN supply systems.

## 18.2 Protection without Disconnection or Indication

Protective measures without any disconnection device relate to equipments which have no protective conductor connection; they are 'intrinsically safe'. Protection in this case depends neither upon an effective potential equalization nor upon disconnection of the deranged circuit.

### 18.2.1 Protective Insulation

The regulations covering 'protective insulation' (protective class II) indicate the measures to be adopted in equipment and in the construction of electrical installations in order to ensure protection against excessive contact potential through insulation.

In protectively insulated equipment, protection depends not only upon primary (basic) insulation; additional, or at least increased, insulation is required (Fig. 18.9). No connection is provided for a protective conductor.

Such an equipment must have a housing of insulating material which encloses all metal parts, apart from those such as nameplates, screws, rivets, etc., which must, however, be isolated by extra insulation.

If for functional reasons a metal housing or accessible metal parts are unavoidable (e.g. a chuck on a hand-drill), these parts must, without exception, have 'double insulation' (basic plus additional insulation) from live parts. The insulation must not become ineffective in the event of the breaking and springing apart of wiring.

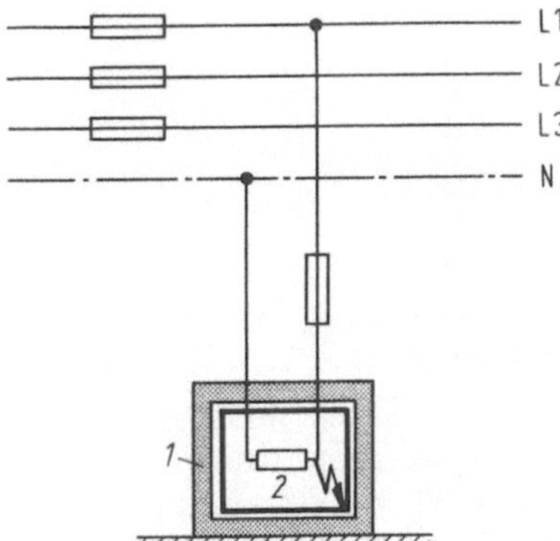

**Fig. 18.9.** Principle of protective measures by protective insulation. *1* Protective insulation, *2* load

The leakage current must not exceed 0.5 mA in telecommunications equipment (VDE 0804) or 0.25 mA in domestic equipment (VDE 0700 Part 1).

Protectively insulated equipment must be marked with the appropriate symbol ▣ (in accordance with DIN 40 014).

The statements above refer to telecommunications and domestic equipment. Protective insulation can also be applied to the construction of systems with equipments which are accommodated within an installation from the outset. In this case several further requirements apply:

- parts that are conducting, but not live, within the protectively insulated housing must not be connected to a protective conductor inserted or looped into it,
- the protectively insulated housing must not at any point be pierced by conducting parts in such a way that a voltage can be introduced,
- unused access openings should be closed in such a way that a tool is necessary to open them.

The requirements listed above do not, however, preclude the possibility of looping a protective conductor into protectively insulated equipment so long as the associated live conductors are carried with it (e.g. in protectively insulated distribution boxes including protective conductor bars).

### 18.2.2 Equipment with Protectively Insulated Parts

Equipment with protectively insulated parts, in accordance with VDE 0804, are also to be found in communications engineering (Fig. 18.10).

This equipment is similar in construction to that with protective insulation (protective class II) – see Sect. 18.2.1. Unlike protectively insulated equipment

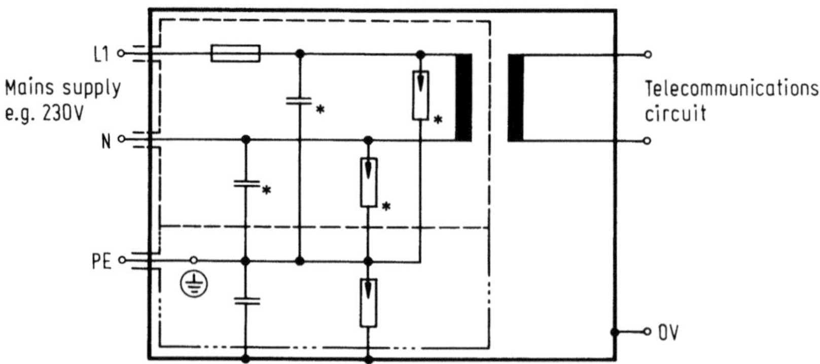


**Fig. 18.10.** Power supply equipment with protectively insulated parts. *Insulation test voltage:* – – – – At least 2.5 kV between mains and telecommunications circuits or between mains and frame (double insulation), * 1.5 kV between mains and PE, – ··· – ··· – 0.5 kV between PE and frame, ——— accessible metal parts (frame)

it has, a protective conductor connection or *protective conductor*, and must *not*, therefore, be classified as protectively insulated equipment.

The protective conductor in this case is for the purpose of EMI suppression or screening, or is required for following equipment in protective class I.

In this equipment the following conditions must be met:

- telecommunications circuits and any accessible conducting parts (frame) connected to them must be insulated from the protective conductor and accessible conducting parts connected to it for a test voltage of 500 V,
- the protective conductor must be coded green/yellow and its connecting terminal marked with the symbol ⏚ in accordance with DIN 40 011,
- equipment with protectively insulated parts must *not* be marked with the symbol ▢ of DIN 40 014.

The same requirements apply to protectively insulated parts of equipment as to the whole (see Sect. 18.2.1).

Equipment with protectively insulated parts may have accessible conducting parts and an accessible conducting enclosure, which are not part of the protection arrangement against excessive contact potential (by indirect contact). The following should be observed for such parts:

- if the parts referred to possess a connection terminal for functional grounding or potential equalization, it must be marked with the frame symbol rh in accordance with DIN 40 016,
- instead of the functional ground, a functional and protective ground (in accordance with VDE 0800 Part 2) may be connected to this terminal,
- conductors used to interconnect accessible conducting parts of the frame must *not* be coded green/yellow.

### 18.2.3 Protective Low Voltage and Functional Low Voltage

Protective low voltage or functional low voltage refers to protective measures in which the circuit operates with a rated voltage of up to 50 V a.c. or 120 V d.c.; hence an excessive contact potential cannot arise as a result of an insulation failure. The protective low voltage may not be obtained by means of series resistors, autotransformers or voltage dividers, but only through safety transformers (protective isolating transformers) complying with VDE 0551 (Fig. 18.11) or from motor generators with separate windings, diesel generators and electrochemical power sources (e.g. secondary batteries). With these power sources are included certain electronic equipment in which the voltage at the output terminals does not exceed the permissible level, even under fault conditions. The disposition of the circuits must be such that the low-voltage circuits (e.g. 24 V) are reliably separated from higher-voltage circuits. If a point in the circuit is grounded for functional reasons, the system is classified as *functional low voltage* (e.g. the supply to a communications system at rated voltages of 48 or 60 V).

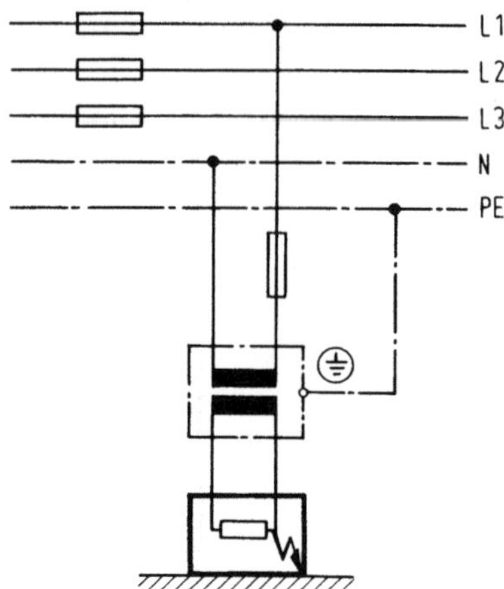

**Fig. 18.11.** Principle of protective measures by protective low voltage and protective isolation

### 18.2.4 Protective Separation

By protective separation is understood the 'galvanic' isolation of an operating equipment, also as shown in Fig. 18.11, but with a rated voltage of up to 1000 V, by means of an isolating transformer.

In the event of a fault to the body, protective isolation prevents the occurrence of a contact potential derived from the supply mains between earth and the faulty operating equipment.

Protective isolation is only effective so long as no ground fault occurs on the secondary side, e.g. through damage to cables; it is applied in telecommunications engineering mainly in mobile systems, such as military equipment or broadcasting vans. The advantage lies in the fact that the equipment can be operated without dependence upon the kind of protective measures embodied in the supply mains, making it unnecessary to provide an earthing system with defined characteristics at the point of installation.

# Bibliography

Benda, D.; Lipinski, K.: Oszilloskope für Praktiker. Berlin: VDE Verlag 1994

Berger, H.: Automating with the SIMATIC S5-115U. Erlangen: Siemens Publicis MCD Verlag 1992

Berger, H: Automating with the SIMATIC S5-135U. Erlangen: Siemens Publicis MCD Verlag 1993

Berndt, D.; Euler, K. J.; Siller, B.: Geschichte der Elektrotechnik; Band 13: Gespeicherte Energie. Berlin: VDE Verlag 1994

Bezner, H.: Dictionary of Power Engineering and Automation; Part 1: German/English; Part 2: English/German. Erlangen: Siemens Publicis MCD Verlag 1993

Dudenhausen, H.; Kloth, S.: Elektromagnetische Verträglichkeit; Eine anwendungsorientierte Einführung zur Gestaltung störungsarmer elektronischer Systeme. Renningen: Expert Verlag 1994

Ferretti, V.: Dictionary of Electronics, Computing and Telecommunications; German/English. Heidelberg: Springer-Verlag 1993

Ferretti, V.: Dictionary of Electronics, Computing and Telecommunications; English/German. Heidelberg: Springer-Verlag 1993

Freyer, U.: Professionelle Stromversorgung; Prinzipien zu maßgeschneiderten Problemlösungen. München: Franzis-Verlag 1989

Gißler, J.; Schmid, M.: Vom Prozeß zur Regelung; Analyse, Entwurf, Realisierung in der Praxis. Erlangen: Siemens Publicis MCD Verlag 1990

Goetzberger, A.; Voß, B.; Knobloch, J.: Sonnenenergie, Photovoltaik; Physik und Technologie der Solarzelle. Stuttgart: Verlag B. G. Teubner 1994

Häberlin, H.: Photovoltaik; Strom aus Sonnenlicht für Inselanlagen und Verbundnetz. Aarau: AT Verlag 1991

Hasse, P.; Wiesinger, J.: Handbuch für Blitzschutz und Erdung. Berlin: VDE Verlag 1993

Heier, S.: Windkraftanlagen im Netzbetrieb. Stuttgart: Verlag B. G. Teubner 1994

Heinhold, L.: Power Cables and their Application; Part 1: Materials, Construction, Criteria for Selection, Project Planning, Laying out Installation, Accessories, Measuring and Testing. Erlangen: Siemens Publicis MCD Verlag 1990

Heinhold, L.; Stubbe; Reimer: Power Cables and their Application; Part 2: Tables Including, Project Planning Data for the Cables and Accessories, Details for the Determination of the Cross-Sectional Area. Erlangen: Siemens Publicis MCD Verlag 1993

Herhahn, A.; Winkler, A.: Elektroinstallation nach DIN VDE 0100; mit Sicherheitslexikon. Würzburg: Vogel Verlag 1992

Heumann, K.; Stumpe, A.: Thyristoren; Eigenschaften und Anwendung. Stuttgart: Verlag B. G. Teubner 1974

Heumann, K.: Grundlagen der Leistungselektronik. Stuttgart: Verlag B. G. Teubner 1991

Hirschmann, W.; Hauenstein, A.: Schaltnetzteile; Konzepte, Bauelemente, Anwendungen. Erlangen: Siemens Publicis MCD Verlag 1990

Hofheinz, W.: Protective Measures with Insulating Monitoring. Berlin: VDE Verlag 1993

Hoffmann, A.; Stocker, K.: Thyristor-Handbuch. Erlangen: Siemens Publicis MCD Verlag 1976

Hotopp, R.; Oehms, K.-J.: VDE Schriftenreihe Bd. 9; Schutzmaßnahmen gegen gefährliche Körperströme nach DIN 57100/VDE 0100 Teil 410 und Teil 540. Berlin: VDE Verlag 1983

Jäger, R.: Leistungselektronik; Grundlagen und Anwendungen. Berlin: VDE Verlag 1993

Jungnickel, H.: Stromversorgungspraxis. Berlin: Verlag Technik 1991

Just, W.: Blindstrom-Kompensation in der Betriebspraxis; Ausführung, Wirtschaftlichkeit, Regelung, Umweltschutz, Überschwingungen. Berlin: VDE Verlag 1991

Kiefer, G.: VDE 0100 und die Praxis. Berlin: VDE Verlag 1994

Kiehne, H. A.: Batteries; Fundamentals and Theory, Running Techniques, Outlook. Renningen: Expert Verlag 1989

Kiehne, H. A.: Portables Batteries; Fundamentals and Theory, Running Techniques, Outlook. Renningen: Expert Verlag 1988

Kilgenstein, O.: Schaltnetzteile in der Praxis. Würzburg: Vogel Verlag 1992 (this book is also available in english at: Chichester/England: John Wiley & Sons Ltd., title in english: Switching-mode power supplies ...)

Kloss, A.: Auf den Spuren der Leistungselektronik; Erfinder und Erfindungen der Stromrichter-technik. Berlin: VDE Verlag 1990

Kloss, A.: Oberschwingungen; Beeinflussungssysteme der Leistungselektronik. Berlin: VDE Verlag 1989

Lappe, R.: Handbuch Leistungselektronik; Grundlagen, Stromversorgung, Antriebe. Berlin: Verlag Technik 1994

Lappe, R.; et al.: Leistungselektronik. Berlin: Verlag Technik 1991

Lappe, R.; et al.: Leistungselektronik-Meßtechnik; Berlin: Verlag Technik 1993

Ledjeff, K.: Brennstoffzellen; Entwicklung, Technologie, Anwendung. Heidelberg: Hüthig Verlag 1995

Marshman, Ch.: The Guide to EMC Directive 89/336/EEC. Berlin: VDE Verlag 1993

Meyer, H. J.: Stromversorgungen für die Praxis. Würzburg: Vogel Verlag 1989

Panzer, P.: Praxis des Überrspannungs- und Störspannungsschutzes; elektrische Geräte und Anlagen. Würzburg: Vogel Verlag 1986

Schmid, J.: Photovoltaik – Strom aus der Sonne; Technologie, Wirtschaftlichkeit und Marktent-wicklung. Heidelberg: Hüthig Verlag 1994

Seip, G.: Electrical Installations Handbook; Part 1: Power-Supply and Distribution Systems; Part 2: Power Cables, Protective Devices, Meters Power Factor Correction, Standby Power-Supply Systems, Lighting, Space Heating, Lift Installations; Part 3: Large Buildings and Out-door Areas, Special Installations, Installation Specifications and Safety Measures. Erlangen: Siemens Publicis MCD Verlag 1987

Siemens, no author: Switching, Protection and Distribution in Low-Voltage Networks; Handbook with selection criteria and planning guidelines for switchgear, switchboards and distribution systems. Erlangen: Siemens Publicis MCD Verlag 1994

Siemens, no author: Programmable Controllers; Basic Concepts. Erlangen: Siemens Publicis MCD Verlag 1989

Wagemann, G.; Eschrich, H.: Grundlagen der photovoltaischen Energiewandlung; Solarstrahlung, Halbleitereigenschaften und Solarzellenkonzepte. Stuttgart: Verlag B. G. Teubner 1994

# Subject Index

# Springer-Verlag and the Environment